AF352912

Advances in Low Carbon Technologies and Transition

Advances in Low Carbon Technologies and Transition

Editors

Shigemi Kagawa
Hidemichi Fujii

MDPI • Basel • Beijing • Wuhan • Barcelona • Belgrade • Manchester • Tokyo • Cluj • Tianjin

Editors
Shigemi Kagawa
Faculty of Economics,
Kyushu University, Fukuoka,
Fukuoka Prefecture
Japan

Hidemichi Fujii
Faculty of Economics,
Kyushu University, Fukuoka,
Fukuoka Prefecture
Japan

Editorial Office
MDPI
St. Alban-Anlage 66
4052 Basel, Switzerland

This is a reprint of articles from the Special Issue published online in the open access journal *Energies* (ISSN 1996-1073) (available at: https://www.mdpi.com/journal/energies/special_issues/Low_Carbon_Technologies_Transition).

For citation purposes, cite each article independently as indicated on the article page online and as indicated below:

LastName, A.A.; LastName, B.B.; LastName, C.C. Article Title. *Journal Name* **Year**, *Article Number*, Page Range.

ISBN 978-3-03943-557-9 (Hbk)
ISBN 978-3-03943-558-6 (PDF)

Contents

About the Editors

Shigemi Kagawa obtained his BA in Civil Engineering, MA in Information Sciences, and Ph.D. from Tohoku University, Japan, in 2001. Since 2020, he is a distinguished professor at Kyushu University. He is a board member of *Economic Systems Research* (Journal of the International Input-Output Association) and the *Annals of Regional Science*. He was awarded Leontief memorial prize and Sir Richard Stone Prize for his work on environmental input-output analysis from the International Input-Output Association.).

Hidemichi Fujii is an associate professor at the Faculty of Economics at Kyushu University. He has been awarded several national research grants on topics such as environmental innovation and energy economics. His work has appeared in academic journals including *Global Environmental Change*, *Energy Economics*, *Business Strategy and the Environment*, *Technological Forecasting & Social Change*, and *Energies*.

Preface to "Advances in Low Carbon Technologies and Transition"

Environmental and energy scientists are always asked to solve real-world environmental problems by suggesting concrete policies based on new findings. Many social and economic factors can significantly affect the environment. We think that technology and product are very important in solving environmental problems. In this context, the following questions are crucial: how can new technological advancements and new product developments contribute to improving energy efficiency at technology and product levels and reducing production- and consumption-based environmental emissions and what policies can be effective in promoting low-carbon technologies and products identified by the analyses? We believe that this book provides a novel contribution to these research questions. We gratefully acknowledge all authors who contributed to the Special Issue "Advances in Low Carbon Technologies and Transition" and this edited book.

Shigemi Kagawa, Hidemichi Fujii
Editors

Article

A Lifecycle Analysis of the Corporate Average Fuel Economy Standards in Japan

Mitsuki Kaneko

Graduate School of Economics, Kyushu University, 744 Motooka, Nishi-ku, Fukuoka 819-0395, Japan; kaneko.mitsuki.974@s.kyushu-u.ac.jp

Received: 9 January 2019; Accepted: 10 February 2019; Published: 20 February 2019

Abstract: This study estimated the corporate average fuel economy (CAFE) and CAFE targets of Japan's domestic automobile manufacturers and evaluated whether manufactures have achieved these estimated CAFE targets. Furthermore, an analysis framework was proposed for estimating what impact the introduction of the CAFE standards in Japan will have on motor vehicle-derived lifecycle CO_2 emissions. As a result, the following was found: (1) Automobile manufacturers can maximize their sales under the constraints of the CAFE standards, but vehicle sales plans based on sales maximization will lower their CAFE standard scores. (2) Economically optimal automobile manufacturer behavior—striving to achieve CAFE standards while maximizing sales—will increase the manufacturers' overall carbon footprint and actually worsen the environment.

Keywords: lifecycle analysis; CAFE standards; fuel economy; automobile manufacture; carbon footprint

---

1. Introduction

The Paris Agreement, adopted in December 2015, attempts to tackle the growing problem of global warming by setting carbon dioxide (CO_2) emission reduction targets for each country in order to meet the goal of limiting the rise in the average global temperature to below 2 °C relative to the pre-industrial revolution level [1]. To achieve this goal, limiting emissions from the transportation sector, which accounts for 29% of the CO_2 emissions of Organization for Economic Co-operation and Development (OECD) countries, is of paramount importance [2]. In Japan, the transportation sector accounts for 20% of total CO_2 emissions, and 90% of these emissions are generated by the motor vehicle sector [3]. Accordingly, reducing tailpipe CO_2 emissions derived from motor vehicles is essential, especially by means of improving motor vehicle fuel economy.

In the United States, the Corporate Average Fuel Economy (CAFE) standard has been in effect since 1975 [4]. This standard aims at improving the fuel economy of motor vehicles to ensure that the fuel economy of a relevant company does not drop below a fuel economy standard value (CAFE standard), a target which is a motor vehicle sales weighted average [4]. In Japan, on the other hand, the fuel economy values of the most efficient vehicle models in specific vehicle weight categories (i.e., the best performing vehicles) are adopted as targets to drive improvements in the fuel economy of each vehicle model [5]. This could be called the 'Top Runner Approach'. Japan plans to adopt CAFE standards in 2020, to use in addition to its current 'Top Runner Approach', with the dual objectives of reducing transportation sector CO_2 emissions and promoting more flexible motor vehicle sales by companies [5].

There are several problems with CAFE standards, however. The first is that even if the aggregated CAFE of a relevant company exceeds the CAFE target, the fuel economy of some of the company's vehicle models may still fall below the fuel economy standard value by vehicle weight category, and vehicle models with poor fuel economy will likely end up on the motor vehicle market. Increasing sales of hybrid vehicles is likely another factor that may drive up the CAFEs of companies. There is also a problem that hybrid vehicles (i.e., electric-petroleum hybrids) impose a heavier environmental burden in manufacturing than conventional gasoline vehicles because they require additional parts and materials (e.g., [6]).

Thus, CAFE standards may not well work toward reducing gasoline consumption and environmental burden through the fuel economy improvements over all the vehicle weight categories and vehicle types. Regarding this problem of CAFE standards, previous studies studied the optimal design of the CAFE standard [7–12] analyzed the welfare effects of tightening the CAFE standard in the U.S. [13–16] compared reductions in fuel consumptions through increasing gasoline taxes and tightening the CAFE standards.

It is important to note that since the demand-side energy policy of a higher gasoline tax has already been imposed in many countries, the supply-side energy policy of improving the CAFE is needed to reduce the environmental burdens associated with the automobile. Studies estimated direct CO_2 emissions associated with fuel combustions of the transport sector, e.g., [17,18], whereas an importance of the life cycle analysis has been increased [19]. To the best of our knowledge, there are very few studies evaluating how companies meeting the CAFE standards affects lifecycle CO_2 emissions through the automobile lifecycle.

It is essential to consider the lifecycle CO_2 emissions of motor vehicles rather than just fuel economy. In this study, CAFEs and CAFE targets of Japan's domestic automobile manufacturers were estimated and it was assessed how well the manufacturers met their targets. The impact that the introduction of the CAFE standards in Japan will have on motor vehicle-derived lifecycle CO_2 emissions was also analyzed.

Specifically, the 2015 sales performance figures of seven of Japan's automobile manufacturers (Toyota Motor Corporation, Nissan Motor Co., Ltd., Honda Motor Co., Ltd., Mitsubishi Motors Corporation, Mazda Motor Corporation, Suzuki Co., Ltd., and Subaru Co., Ltd.) and the published fuel economy values of the sold vehicle models were investigated, in order to estimate the CAFE of each company, as well as their CAFE target, and to assess how well the manufacturers met their targets in 2015.

The car sales of a specific company affect not only the CAFE based on the weighted-average fuel economies of the car sales but also the lifecycle CO_2 of motor vehicles sold by the company. To estimate the lifecycle CO_2 of motor vehicles, it is important to estimate the lifecycle CO_2 emission intensity of a specific vehicle model sold by the company expressed in ton-CO_2 per vehicle. This is because several studies treated a wide variety of vehicles as a specific homogeneous product and analyzed a life cycle assessment of the specific vehicle (e.g., aggregated conventional gasoline vehicle) with a comparison of the environmental burdens of conventional vehicles with vehicles equipping other engines, electric vehicles, hybrid vehicles, plug-in hybrid vehicles, and hydrogen fuel cell vehicles [20–23].

Using the pooled observations of cars sold by the above seven manufactures in 2015, a statistically specified relationship was created between car prices and car weights as a regression equation. When the car price of an 'average vehicle' described in the Japanese commodity-by-commodity input-output table (Ministry of Internal Affairs and Communications, Japan, 2010) was inserted into the specified relationship between car prices and car weights, a car weight of the average vehicle could be obtained. Using the ratio between the embodied CO_2 emission intensity of the 'average vehicle' provided by the Embodied Energy and Emission Intensity Data for Japan using Input-Output Tables [24] and the car weight of the 'average vehicle' estimated in this study, the embodied CO_2 emission intensity of the specific vehicle model of the company was proportional to the weight of the car. Using the proposed methodology, a new database of disaggregated lifecycle emissions of motor vehicles sold by the Japanese auto manufactures was compiled.

Estimating the disaggregated lifecycle inventory database of motor vehicles, the impact that the introduction of the CAFE standards in Japan is likely to have on motor vehicle-derived lifecycle CO_2 emissions was evaluated, to assess the validity of the CAFE standards from an environmental perspective.

Companies would maximize profits from car sales under the CAFE standards. This study therefore formulated an optimization problem of maximizing profit, as the objective function, under constraints with respect to both car sales and the CAFE standards and examined how the optimized car sales of each company differed from the actual car sales and what effect achieving the CAFE standards would have toward reducing lifecycle CO_2 emissions under the optimized car sales.

The remainder of this paper is organized as follows: Section 2 describes the methodology, Section 3 explains the data used in this study, Section 4 provides the results and discussion, and Section 5 concludes this paper.

2. Methodology

2.1. CAFEs and CAFE Targets for Automobile Manufacturers

The CAFE of each automobile manufacturer, C (km/L), was estimated based on the number of new vehicle sales and the published fuel economy values using the following equation:

$$C = \frac{X}{\sum_{i=1}^{N} \frac{x_i}{z_i}} \tag{1}$$

where X is the number of new vehicle sales of a particular automobile manufacturer j, z_i (km/L) is the fuel economy value of vehicle model i of the automobile manufacturer, and x_i is the number of vehicle model i that the automobile manufacturer sold for the year. Furthermore, N is the number of different vehicle models sold by the relevant automobile manufacturer. As the CAFE obtained from Equation (1) increases, the fuel economy of the 'average vehicle' sold by the relevant automobile manufacturer improves. The CAFE target, $\tilde{C}$(km/L), is calculated as follows [4]:

$$\tilde{C} = \frac{X}{\sum_{k=1}^{M} \frac{x_k}{\tilde{z}_k}} \tag{2}$$

where $\tilde{z}_k$ (km/L) is the target fuel economy value for a predefined passenger vehicle weight category k, x_k is the total number of sales of vehicle models belonging to vehicle weight category k by a particular automobile manufacturer, and M is the number of vehicle weight categories.

2.2. Sales Maximization

In this section, the optimal number of unit sales for each vehicle model for automobile manufacturers to maximize sales while satisfying the CAFE standards given by Equation (2) is estimated. This is done so by solving the linear programming problem illustrated in Equations (3) through (6) below.

$$\text{Max.} \sum_{i=1}^{N} p_i x_i \tag{3}$$

such that

$$\frac{\sum_{i=1}^{N} x_i}{\sum_{i=1}^{N} \frac{x_i}{(1+\varepsilon)z_i}} \geq \frac{\sum_{i=1}^{N} x_i}{\sum_{k=1}^{M} \sum_{i_k=1}^{N_j} \frac{x_i}{\tilde{z}_k}} \tag{4}$$

$$x_i \leq \alpha x_i^* \tag{5}$$

$$\sum_{i=1}^{N} x_i \leq \beta \sum_{i=1}^{N} x_i^* \tag{6}$$

where p_i is the price for each vehicle model, x_i^* is the actual number of units sold, α is a parameter for determining the upper limit of vehicle models i, β is a parameter determining the upper limit of total units sold, and ε represents the rate of fuel economy improvement. Equation (4) is a constraint for the linear programming problems in which the relevant company must meet the CAFE standards. In this study, four scenarios are considered: Scenario I, fuel economy for the vehicle models is the baseline value ($\varepsilon = 1.0$); and Scenarios II, III, and IV, in which fuel economy for the vehicle models is uniformly improved from the baseline fuel economy by 10%, 15%, and 20%, respectively ($\varepsilon = 1.1$, $\varepsilon = 1.15$, and $\varepsilon = 1.2$). Next, Equation (5) is the constraint for sales patterns in which the relevant company's current number of units sold for vehicle model i grows by a factor α, which is set as $\alpha = 2$ for this study. Finally, Equation (6) is the constraint for the total number of units sold, which is set as $\beta = 1$ for this study. This study solves the sales maximization problem within the four fuel economy improvement scenarios given above (Scenarios I–IV) to estimate the optimal sales pattern for the vehicle models of the relevant automobile manufacturers.

2.3. Lifecycle CO$_2$ Emissions of the Automobile Manufacturers

For gasoline-engine and hybrid vehicle models i, the average lifecycle emission intensity per vehicle is found as f_m by taking the weighted average by number of units sold for the lifecycle emission intensity ($f_{m,i}^g$ and $f_{m,i}^h$) derived from the manufacturing, transportation, and sales origin for a single vehicle. Here, one can estimate the lifecycle CO$_2$ emissions (t-CO$_2$) derived from the automobiles as sold by the relevant companies in Japan for 2015 as follows:

$$Q = \sum_{i \in N_g} x_i f_{m,i}^g + \sum_{i \in N_h} x_i f_{m,i}^h + \sum_{i=1}^{N} x_i f_{g,i} + \sum_{i=1}^{N} f_{h,i} \tag{7}$$

where N_g is the set of gasoline-engine vehicles models, N_h is the set of hybrid vehicle models, $f_{g,i}$ is the CO$_2$ emission intensity during travel for vehicle model i and $f_{h,i}$ is the CO$_2$ emission intensity during disposal of vehicle model i.

For a relevant automobile manufacturer, the weighted average fuel economy for a passenger vehicle i is defined as as e_i (km/L) and the lifetime travel distance of passenger vehicles as d (km). Thus, g_i (L), the lifetime gasoline consumption of a passenger vehicle i, is obtained as follows:

$$g_i = \frac{d}{e_i} \tag{8}$$

The CO$_2$ emissions due to gasoline consumption during travel per vehicle can then be estimated by multiplying the CO$_2$ emission intensity generated per liter of gasoline burned r_g by the quantity of gasoline consumed g_i from Equation (6):

$$f_g^{direct} = g_i r_g = \frac{d r_g}{e_i} \tag{9}$$

In addition, the CO$_2$ emissions associated with refining the gasoline necessary for travel per vehicle can be estimated by multiplying the CO$_2$ emission intensity generated per liter of gasoline refined r_c by the quantity of gasoline consumed g_i from Equation (8):

$$f_{g,i}^{indirect} = g_i r_c = \frac{d r_c}{e_i} \tag{10}$$

Thus, the embodied CO_2 emission intensity during travel per vehicle $f_{g,i}$ in Equation (7) is the sum of $f_{g,i}^{direct}$, the direct emissions generated by gasoline consumption during travel, and $f_{g,i}^{indirect}$, the indirect emissions generated in refining the gasoline:

$$f_{g,i} = f_{g,i}^{direct} + f_{g,i}^{indirect} \tag{11}$$

3. Data

In this study, the vehicle models of each company sold in 2015 were as follows: Toyota, 42; Nissan, 21; Honda, 17; Mitsubishi, 10; Mazda, 9; Subaru, 9; and Suzuki, 8. The number of vehicles of each model sold by each company, which is necessary for calculating the CAFE and CAFE target, can be obtained from data on the number of vehicles sold by brand [25]. For the fuel economy of each vehicle model the fuel economy figures for each vehicle model in JC08 mode cycle was used, as published in the Automobile Fuel Economy List [26]. The vehicle weight categories for the CAFE standards due to be introduced in MY2020 are shown in Table A1.

The CO_2 emission intensities per passenger vehicle in manufacturing, during travel, and in disposal were estimated using the Embodied Energy and Emission Intensity Data for Japan Using Input-Output Tables [24]. The passenger vehicle lifetime travel distance d was assumed to be 100,000 km and therefore estimated the emission intensity during travel r_g to be 0.00231 t-CO_2 and r_c to be 0.00063 t-CO_2. In addition, in accordance with a previous study [6], the emission intensity in disposal $f_{h,i}$ was set to be 0.0574 t-CO_2.

In order to estimate the life cycle CO_2 emission intensity of vehicles, the life cycle CO_2 emission intensity derived from both manufacturing and driving must be estimated for each vehicle model. While by no means a simple task, in this study, the lifecycle CO_2 emission intensity for each vehicle model was estimated by specifying a relationship for model sales prices and new vehicle weight. First, the sales price information was obtained for 82 gasoline-engine vehicle models sold by the automakers (Toyota, Nissan, Honda, Mitsubishi, Mazda, Subaru, and Suzuki) in 2015 from Autoc One [27], an informational site that releases comprehensive vehicle sales information. Vehicle weight information was also obtained for the same 82 models from the MLIT automotive information site [26]. From the sales price and vehicle weight data for the 82 models, a regression analysis was run and the following results were obtained.

$$\begin{aligned}
p_i^g = \quad & 0.35 w_i^g \quad - \quad 222 \\
& (7.46) \qquad\quad (-3.09)
\end{aligned} \tag{12}$$

$$\text{Adjusted} R^2 : 0.38$$

where w_i^g (kg) is the vehicle weight for vehicle model i and p_i^g (10,000 s of Japanese yen) is the sales price for vehicle model i. The numbers in parentheses below the parameters are the t-values, and each of the estimated parameters is statistically significant at the 1% level in a two-sided test. The relationship given in Equation (12) shows us that an increase of 100 kg in vehicle weight corresponds to an increase of 350,000 yen in sales price.

From the 2005 Input-Output Tables, the average vehicle sales price in 2005 was 2.2 million yen. Given this, the relationship specified in Equation (12) can be used to estimate the average vehicle weight as $w^g = (220 + 222)/0.35 = 1264$ kg. Meanwhile, from the Embodied Energy and Emission Intensity Databook (3EID) ([24]) as based on the 2005 Input-Output Table as released by the National Institute for Environmental Studies, the average lifecycle emission intensity for vehicle production is 1.93 t-CO_2 per 1 million yen, and the lifecycle emission intensity for transportation and sales services incidental to sales price for one vehicle unit is 1.2 t-CO_2 per 1 million yen. Accordingly, one can estimate a lifecycle CO_2 emission intensity of $1.93 \times 2.2 = 4.2$ t-CO_2 as derived from manufacturing one average vehicle in 2005 with a sales price of 2.2 million yen and vehicle weight of 1264 kg. Next, the lifecycle CO_2 emission intensity was estimated, as derived from manufacturing a relevant vehicle model by taking the ratio of the vehicle weight of that model to the average vehicle weight (1264 kg) and multiplying by the unit intensity derived from manufacturing. To estimate the lifecycle CO_2 emission intensity incidental to transportation and sales services for one unit of a relevant vehicle model, the lifecycle emission intensity for transportation and sales services was taken as 1.2 t-CO_2 per 1 million yen and multiplied this quantity by the sales price of the relevant vehicle model. The lifecycle CO_2 emission intensity $f^g_{m,i}$ for a single gasoline vehicle model i was then solved for by adding up the lifecycle CO_2 emission intensities derived from manufacturing and from transportation and sales for the relevant model. It is important to note that although we can estimate the lifecycle CO_2 emissions by multiplying the average lifecycle emission intensity for vehicle production (1.93 t-CO_2 per 1 million yen) by each vehicle price, and that the estimated emissions are not consistent with the vehicle weight important for the CAFEs.

Similarly, a separate regression analysis for 42 hybrid vehicle models was run and the following relationship for sales price and vehicle weight was obtained:

$$p^h_i = \underset{(8.12)}{0.41w^h_i} - \underset{(-3.54)}{282}$$

$$\text{Adjusted} R^2 : 0.62 \tag{13}$$

where w^h_i (kg) is the vehicle weight for hybrid vehicle model i and p^h_i (10,000s of yen) is the sales price for hybrid vehicle model i. Again, the numbers in parentheses below the parameters are the t-values, and each of the estimated parameters is statistically significant at the 1% level in a two-sided test. The relationship given in Equation (13) shows us that an increase of 100 kg in vehicle weight for hybrid vehicles corresponds to an increase of 410,000 yen in sales price. The lifecycle CO_2 emission intensity $f^h_{m,i}$ derived from manufacturing and from transportation and sales for a single hybrid vehicle model i was solved for with the same methods described above for calculating unit intensity for a gasoline vehicle model. The detailed lifecycle CO_2 emission intensity data by vehicle model as estimated in this study are described in Table S3 of the Supporting Information.

4. Results

4.1. Life-Cycle CO_2 Emission Intensities of Vehicle Models

Table 1 shows the data showing mean, standard deviation, maximum value and minimum value of the life cycle CO_2 emission intensities of vehicle models of seven automobile manufactures in Japan estimated by Equations (12) and (13). According to Table 1, the maximum value of the life cycle CO_2 intensity in seven firms is 60.74 t-CO_2/car Toyota CENTURY (gasoline vehicle) and the minimum value is 14.8 t-CO_2/car Toyota AQUA (Hybrid vehicle). Thus, there is a large difference in life cycle CO_2 intensities within a firm as well as between firms. The mean of the intensities of each firm is caused by the number attributes (e.g., body weight, fuel economy, etc.) of cars sold by the firm and it means that firms with a higher standard deviation of the intensities like Toyota have more varieties of cars.

Table 1. Lifecycle CO_2 intensities and life cycle CO_2 emissions in 2015.

Company Name	The Number of Vehicle Models	Estimated Life Cycle Emission Intensity					Baseline (2015) Life Cycle CO_2 Emissions (Million t-CO_2)
		Mean (t-CO_2/car)	Weighted Mean of the Number of Sold Vehicles	Standard Deviation (t-CO_2/car)	Maximum Value (t-CO_2/car)	Minimum Value (t-CO_2/car)	
Toyota	42	28.5	23.7	47.3	60.7	14.8	28.4
Nissan	21	32.7	24.9	39.3	54.4	18.3	8.0
Honda	17	22.8	21.2	13.8	32.6	15.9	8.0
Mitsubishi	10	28.0	27.7	32.6	45.3	18.0	0.7
Mazda	9	28.0	24.9	17.7	37.7	20.2	3.9
Suzuki	8	31.3	25.7	20.5	43.1	23.8	1.9
Subaru	9	28.1	27.7	9.0	35.7	20.9	3.4

The last column of Table 1 shows the life cycle CO_2 emissions of each firm in 2015 that is the benchmark emissions in this analysis. Importantly, Toyota has the largest number of vehicle models sold (see first column of Table 1) and it has the largest life cycle CO_2 emissions, amounting to 28.4 million t-CO_2 in 2015. This is because the life cycle CO_2 emissions depend on the number of sold cars as well as the number of sold vehicle models. The total of CO_2 emissions of Japan in 2015 was 1325 million t-CO_2 [28] and the sum of the life cycle CO_2 emissions of seven automobile manufactures in 2015 was 54.2 million t-CO_2 that accounts for 4% of the total CO_2 emissions of Japan. Therefore, it is essential to management the life cycle CO_2 emissions in automobile industry.

4.2. CAFEs and CAFE Targets of Seven Automobile Manufacturers in Japan

Table 2 shows the CAFEs and CAFE targets of Japan's seven major automobile manufacturers (Toyota, Nissan, Honda, Mitsubishi, Mazda, Subaru, and Suzuki), as estimated using Equations (1) and (2).

Table 2. CAFEs and CAFE targets of seven automobile manufacturers (unit: km/L).

Company Name	CAFE Target	CAFE
Toyota	17.6	19.0
Nissan	18.0	17.9
Honda	19.1	21.6
Mitsubishi	16.4	13.3
Mazda	20.6	18.2
Suzuki	23.2	21.2
Subaru	17.4	15.1

Table 2 shows that the CAFEs of Toyota and Honda exceeded their CAFE targets, while those of Nissan, Mitsubishi, Mazda, Subaru, and Suzuki fell below their CAFE targets. When the CAFE standards are introduced in 2020, Nissan, Mitsubishi, Mazda, Subaru, and Suzuki which cannot currently meet their targets, will need to step up their efforts to improve fuel economy. The relationships between fuel economy by vehicle model, vehicle weight, and the number of vehicle sales by model for the two automobile manufacturers that met their CAFE targets, Toyota and Honda, are plotted in Figures S1 and S2 of the Supporting Information, respectively. Figure S1 shows that Toyota sells a large number of vehicle models that have exceptionally good fuel economies. The fact that Toyota sells a much larger number of hybrid vehicles than the other six automobile manufacturers appears to be a factor in Toyota's success in meeting the CAFE standards. On the whole, Honda sells fewer vehicle models with poor fuel economies than does Toyota, and for that reason, it too managed to meet the CAFE standards (Figure S2). Thus, differences in sales patterns and fuel economy technology between companies account for the gaps in their ability to achieve their targets.

4.3. Sales Maximization Under the CAFE Standards

Before delving into the results for sales maximization, let us first review the state of Japan's seven major automobile manufacturers as of 2015. According to the Japan Automobile Dealers Association, approximately 2.7 million passenger vehicles (standard-sized vehicles (white plate vehicles) and Kei passenger cars (yellow plate vehicles)) were sold in 2015. It should be noted that Kei passenger car has an engine of 660 cc or smaller, whereas standard-sized vehicles has a larger internal-combustion engine than 660 cc. Sales shares by company were led by Toyota at 46% (1.25 million vehicles), followed by Honda at 14% (380,000), Nissan at 11% (290,000), Mazda at 7% (180,000), Subaru at 5%, (120,000), Suzuki at 3% (70,000), and Mitsubishi at 1% (30,000). Japan's automotive-related industries combined for a market scale of 64 trillion yen [29].

While these 2015 sales figures do not account for CAFE standards, as given in the previous section, fuel economy and sales patterns for each vehicle model are two necessary elements for achieving the CAFE standards. Thus, the sales for each scenario will now be given with regards to the CAFE standards by performing sales maximization as specified in Equation (3).

Figure 1 shows the rate of change in current sales for Scenarios I through IV compared to 2015 sales. Sales tend to increase with the rate of fuel economy improvements but are still decreasing for some companies; this likely depends on the sales patterns of the different companies. The slumping sales of certain companies can be explained by the poor fuel economy of each vehicle model and limited vehicle models that can be sold to satisfy the CAFE standard constraint. In contrast, sales for Nissan and Suzuki, two manufacturers who have not met their CAFE targets, increased in Scenario I, illustrating the vital importance of sales patterns (Figure 1). In Scenario IV (fuel economy improved 20%), total sales across all seven manufacturers increased by 13.7 trillion yen, with each manufacturer increasing as follows: 10 trillion yen at Toyota, 2 trillion at Nissan, 700 billion yen at Mazda, 600 billion yen at Subaru, 200 billion yen at Honda, 100 billion yen at Mitsubishi, and 100 billion yen at Suzuki. Overall, the automotive market would increase 20% (Figure 1). For the optimal sales patterns for each company, please refer to the Supporting Information.

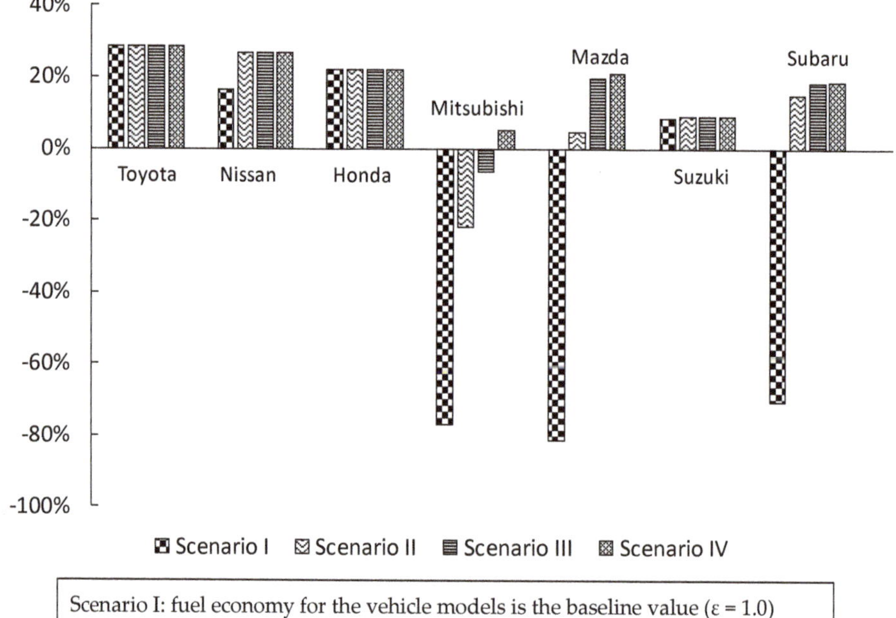

Figure 1. Percentage changes in car sales under optimal Scenarios I–IV relative to the actual sales.

Currently, five of the seven manufacturers—Nissan, Mitsubishi, Mazda, Suzuki, and Subaru—have not achieved their CAFE targets (Table 1). As shown in Table 3, however, all seven can implement sales plans for maximizing sales and still achieve the CAFE standards in all of the fuel economy scenarios. Even though the sales optimization has the CAFE standards imposed as an inequality constraint, note that the CAFEs, which are based on the endogenously determined optimal vehicle model sales figures, are the same as the CAFE target. One important point is that Toyota's CAFE target based on its actual units sold for 2015 is 17.0, whereas its CAFE target based on optimized units sold would have been 15.6. This illustrates that sales activity aimed at sales maximization will bring down the CAFE target and consequently lead to a lack of discipline.

Table 3. CAFEs and CAFE targets of seven automobile manufacturers for the actual and optimal cases.

	Actual Case			Optimal Case under Scenario I		
Company Name	CAFE Target	CAFE	Achievement Status (Yes/No)	CAFE Target	CAFE	Achievement Status (Yes/No)
Toyota	17.6	19.0	Yes	15.6	15.6	Yes
Nissan	18.0	17.9	No	16.6	16.6	Yes
Honda	19.1	21.6	Yes	17.3	20.0	Yes
Mitsubishi	16.4	13.3	No	22.6	22.6	Yes
Mazda	20.6	18.2	No	20.1	20.1	Yes
Suzuki	23.2	21.2	No	23.2	23.2	Yes
Subaru	17.4	15.1	No	16.9	16.9	Yes
Mean	18.9	18.0		18.9	19.3	
S.D	2.3	3.0		3.1	3.0	

If the above CAFE standards are instated, each company can fashion their sales activity to maximize sales by shifting their sales patterns. In the next section, the environmental loads brought about by the sales activity of each company if this happens are analyzed.

4.4. Lifecycle CO_2 Emissions Under the Optimized Sales Pattern

The original purpose of the CAFE system was to restrict CO_2 and air pollutant emissions by making fuel economy standards more flexible. Thus, a simple analysis of CAFE standard achievement rates would be insufficient; one needs to analyze how the CAFE standards relate to the lifecycle CO_2 emissions associated with vehicles. Therefore, this section analyzes the lifecycle CO_2 emissions derived from vehicles with the CAFE standards introduced.

As estimated with Equation (7), the lifecycle CO_2 emissions associated with vehicles manufactured by their relevant automobile manufacturer (the carbon footprint of that automobile manufacturer) in 2015 were as follows: 20 million tons for Toyota, 8 million tons for Honda, 7 million tons for Nissan, 3.6 million tons for Mazda, 3 million tons for Subaru, 1.4 million tons for Suzuki, and 730,000 tons for Mitsubishi. These constitute a footprint of approximately 40 million tons for all seven manufacturers. Thus, the Japanese automotive industry's carbon footprint accounts for roughly 30% of CO_2 emissions attributed to Japan's transportation sector [3].

Next, Figure 2 shows the rate of change in carbon footprint for each company from their baseline carbon footprints, based on the optimal units sold for each company in fuel economy improvement Scenarios I through IV if they maximize their sales while meeting the CAFE standards. From Figure 2, one can see that as the fuel economy improvement rates increase and gasoline consumption decreases, a company's carbon footprint will also tend to decrease.

In addition, from Figures 1 and 2, although optimal vehicle sales patterns under the CAFE standard constraint would help to increase sales, they would also increase carbon footprints and thus be bad for the environment (see the Toyota and Nissan values in Figures 1 and 2). Based on the estimated optimal sales patterns for each company in Scenario IV, where fuel economy for the vehicle models sold is improved 20%, the overall carbon footprint for all seven companies would be approximately 53 million tons, a 1.2-fold increase over their 2015 carbon footprint. In looking to maximize sales, manufacturers have tended to sell heavier vehicles, given the correlation between weight and price. Thus, their carbon footprint based on the optimal sales patterns has not decreased compared to the 2015 baseline value. One important finding in this study is that automobile manufacturer behavior—striving to achieve CAFE standards with the goal of maximizing car sales—will increase their carbon footprint and actually worsen the environment. It is therefore concluded that it is necessary for automobile manufacturers to mitigate the carbon footprint associated with vehicle lifecycle under the CAFE standards.

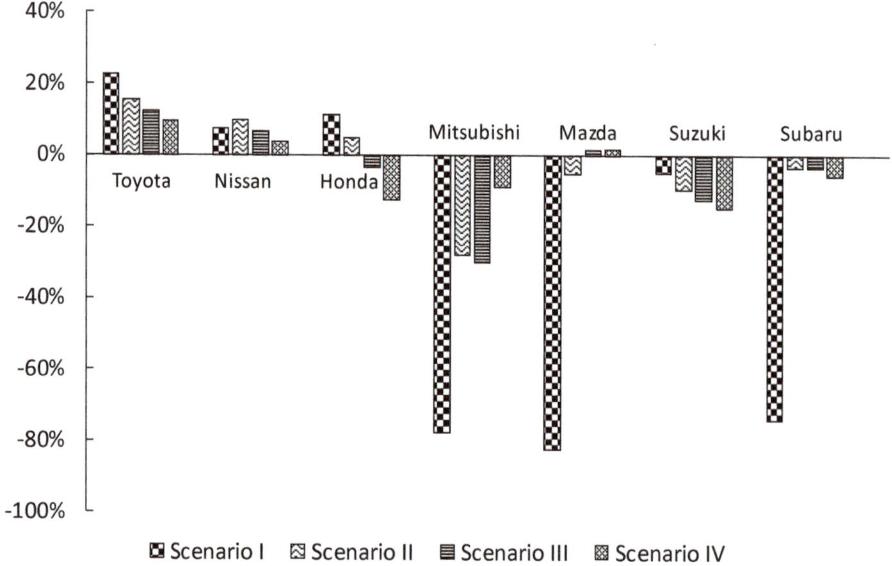

Figure 2. Percentage changes in lifecycle CO_2 emissions under optimal scenarios I–IV relative to the actual emissions.

5. Conclusions and Policy Implications

This study estimated the CAFEs and CAFE targets of seven Japanese automobile manufacturers, and identified the manufacturers that met their CAFE targets and those that did not. It was clearly observed that the manufacturers that met their CAFE targets were of two distinct types: a company that offered a wide range of vehicle models with good fuel economy (Honda) and a company that focused on selling vehicle models with exceptionally good fuel economy (Toyota).

This study further proposed an optimization problem with an objective function of maximizing the profit under constraints with respect to both car sales and CAFE standards, and addressed the question of how the optimized car sales of each company differ from the actual car sales and what the effect of meeting the CAFE standards would have on reduction in lifecycle CO_2 emissions under the optimized car sales.

Our main findings were as follows:

(1) Automobile manufacturers can maximize their sales under the constraints of the CAFE standards, but vehicle sales plans based on sales maximization will lower their CAFE standard scores and could cause a moral hazard among automobile manufacturers.

(2) Economically optimal automobile manufacturer behavior—striving to achieve CAFE standards while maximizing sales—will increase the manufacturers' overall carbon footprint and actually worsen the environment.

Toyota published an environmental report [30] concluding that "In the United States, Toyota's model year 2013 fleet achieved the required U.S. Corporate Average Fuel Economy (CAFE) standards and Toyota met the required greenhouse gas standards in both the United States and Canada". Although it is important to communicate environmental outcomes to the public, it seems that the relationship between CAFE and GHG emissions is still unclear, because the GHG emissions reported by Toyota took into consideration only CO_2 emissions generated by fuel consumption in a defined distance; the 2013 report did not assess how a strategy to achieve the CAFE standards would affect the overall CO_2 emissions through the automobile lifecycle.

Although one of the objectives of the Japanese CAFE standards is to promote more flexible motor vehicle sales by companies [5], the standard ignores an important aspect of life cycle CO_2 emissions. This paper suggests that automakers should pay more attention to the corporate life cycle CO_2 emissions and publish a more comprehensive sustainability report including answers to the questions of how meeting the CAFE standards would affect the corporate lifecycle CO_2 emissions, and what strategy can be effective for reducing the corporate lifecycle CO_2 emissions under the CAFE standards. This study demonstrates that the CAFE analysis framework proposed in this paper is powerful for addressing the above questions. In addition, the results reveal that Japanese automakers can significantly reduce CO_2 emissions under the CAFE standards.

It is also important to note that automobile manufactures that violate the CAFE standards in Japan will be fined one million Japanese yen after implementation of the CAFE standards, thus the fine under the Japanese CAFE standards will be much less than those in the U.S.A. and European countries [5]. To strengthen these currently weak regulations, the Japanese government should monitor the achievement status of all automobile manufactures and obligate the Japanese automobile manufactures to submit comprehensive sustainability reports as described above to the government. Such sustainability reports including the results estimated using the analysis framework proposed in this study can be practically useful for policy makers in arguing how the CAFE standards can contribute to reducing societal CO_2 emissions, and what might be a more effective policy centered around automobile lifecycle management under the CAFE standards.

Supplementary Materials: The following are available online at http://www.mdpi.com/1996-1073/12/4/677/s1.

Funding: This research received no external funding.

Acknowledgments: The author gratefully acknowledges the helpful comments of Shigemi Kagawa, Keisuke Nansai, Shogo Eguchi, Shunichi Hienuki, and two anonymous referees.

Conflicts of Interest: The author declares no conflict of interest.

Appendix A

Red circle: Vehicle model of achieving the target fuel efficiency value for a predefined passenger vehicle weight category.

Blue circle: Vehicle model of not achieving the target fuel efficiency value for a predefined passenger vehicle weight category.

Table A1. Weight categories.

Class	Vehicle Weight (kg)	Target Fuel Economy (km/L)
1	0–740	24.6
2	741–855	24.5
3	856–970	23.7
4	971–1080	23.4
5	1081–1195	21.8
6	1196–1310	20.3
7	1311–1420	19.0
8	1421–1530	17.6
9	1531–1650	16.5
10	1651–1760	15.4
11	1761–1870	14.4
12	1871–1990	13.5
13	1991–2100	12.7
14	2101–2270	11.9
15	2271–2600	10.6

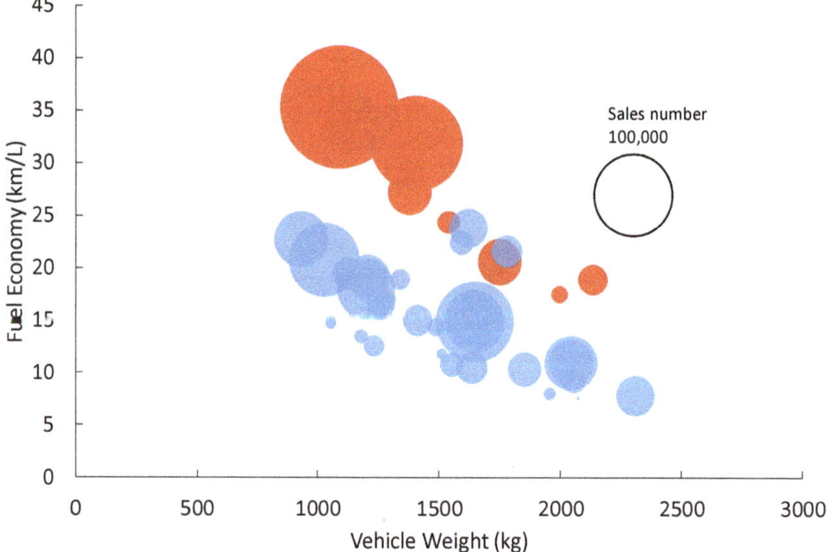

Figure A1. The relationships between fuel efficiency by vehicle model, vehicle weight, and the number of vehicle sales by model for Toyota.

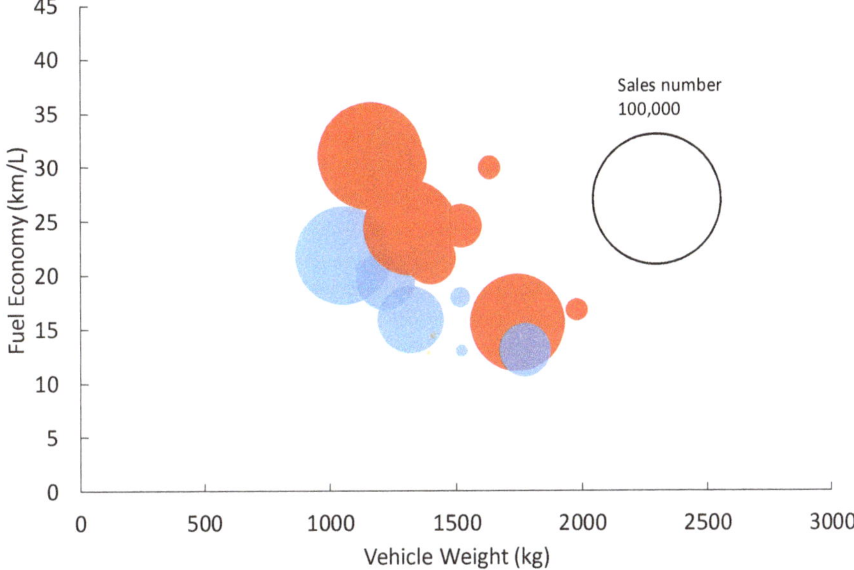

Figure A2. The relationships between fuel efficiency by vehicle model, vehicle weight, and the number of vehicle sales by model for Honda.

References

1. United National Framework Convention Climate Change (UNFCCC). 2015. Available online: http://unfccc.int/paris_agreement/items/9485.php (accessed on 10 December 2018).
2. International Energy Agency. Recent Trends in the OECD: Energy and CO_2 Emissions. 2016. Available online: http://www.iea.org/media/statistics/Recent_Trends_in_the_OECD.pdf (accessed on 10 December 2018).
3. Ministry of Land, Infrastructure, Transport and Tourism. CO_2 Emissions in the Transport Sector in Japan. 2017. Available online: http://www.mlit.go.jp/sogoseisaku/environment/sosei_environment_tk_000007.html (accessed on 10 December 2018).
4. National Highway Traffic Safety Administration (NHTSA), U.S.A. 2016. Available online: https://www.nhtsa.gov/laws-regulations/corporate-average-fuel-economy (accessed on 10 December 2018).
5. Ministry of Land, Infrastructure, Transport and Tourism. Report on Fuel Economy Standard in Japan in 2020. 2011. Available online: http://www.mlit.go.jp/common/000170128.pdf (accessed on 10 April 2018).
6. Kagawa, S.; Hubacek, K.; Nansai, K. Better cars or older cars? Assessing CO_2 emission reduction potential of passenger vehicle replacement programs. *Glob. Environ. Chang.* **2013**, *23*, 1807–1818. [CrossRef]
7. Whitefoot, K.; Skerlos, S. Design incentives to increase vehicle size created from the U.S. footprint-based fuel economy standards. *Energy Policy* **2012**, *41*, 402–411. [CrossRef]
8. Luk, J.; Saville, B.A.; MacLean, H.L. Vehicle attribute trade-offs to meet the 2025 CAFE fuel economy target. *Transp. Res. Part D: Transp. Environ.* **2016**, *49*, 154–171. [CrossRef]
9. Kiso, T. Evaluating New Policy Instruments of the Corporate Average Fuel Economy Standards: Footprint, Credit Transferring, and Credit Trading. *Environ. Resour. Econ.* **2007**, *72*, 1–32. [CrossRef]
10. Levinson, A. Environmental Protectionism: The Case of CAFE. Available online: https://www.sciencedirect.com/science/article/pii/S0165176517303427 (accessed on 12 February 2019).
11. Goldberg, P.K. The Effects of the Corporate Average Fuel Economy Standards in the U.S. *J. Ind. Econ.* **1998**, *46*, 1–33. [CrossRef]
12. Kleit, A.N. Impact of long-range increases in the fuel economy (CAFE) standard. *Econ. Inquiry* **2004**, *42*, 279–294. [CrossRef]
13. Bento, B.; Goulder, L.; Jacobsen, M. Distributional and Efficiency Impacts of Increased US Gasoline Taxes. *Am. Econ. Rev.* **2009**, *99*, 667–699. [CrossRef]

14. Jacobsen, M. Evaluating US fuel economy standards in a model with producer and household heterogeneity. *Am. Econ. J. Econ. Policy* **2013**, *5*, 148–187. [CrossRef]

15. Parry, I.W.H.; Walls, M.; Harrington, W. Automobile externalities and policies. *J. Econ. Lit.* **2007**, *45*, 373–399. [CrossRef]

16. Austin, D.; Dinan, T. Clearing the air: The costs and consequences of higher CAFE standards and increased gasoline taxes. *J. Environ. Econ. Manag.* **2005**, *50*, 562–582. [CrossRef]

17. Woo, J.; Moon, H.; Lee, M.; Jo, M.; Lee, J. Ex-ante impact evaluation of Corporate Average Fuel Economy standards on energy consumption and the environment in South Korea. *Transp. Res. Part D: Transp. Environ.* **2017**, *53*, 321–333. [CrossRef]

18. Jenn, A.; Azevedo, I.; Michalek, J. Alternative Fuel Vehicle Adoption Increases Fleet Gasoline Consumption and Greenhouse Gas Emissions under United States Corporate Average Fuel Economy Policy and Greenhouse Gas Emissions Standards. *Environ. Sci. Technol.* **2016**, *50*, 2165–2174. [CrossRef] [PubMed]

19. Guinee, J.B.; Heijungs, R.; Huppes, G.; Zamagni, A.; Masoni, P.; Buonamici, R.; Ekvall, T.; Rydberg, T. Life cycle assessment: Past, present, and future. *Environ. Sci. Technol.* **2011**, *45*, 90–96. [CrossRef] [PubMed]

20. Samaras, C.; Meisterling, K. Life cycle assessment of greenhouse gas emissions from plug-in hybrid vehicles: Implications for policy. *Environ. Sci. Technol.* **2008**, *42*, 3170–3176. [CrossRef] [PubMed]

21. Thomas, C.E. Fuel cell and battery electric vehicles compared. *Int. J. Hydrog. Energy* **2009**, *34*, 6005–6020. [CrossRef]

22. Hawkins, T.R.; Singh, B.; Majeau-Bettez, G.; Strømman, A.H. Comparative Environmental Life Cycle Assessment of Conventional and Electric Vehicles. *J. Ind. Ecol.* **2013**, *17*, 53–64. [CrossRef]

23. Bauer, C.; Hofer, J.; Althaus, H.; Del Duce, A.; Simons, A. The environmental performance of current and future passenger vehicles: Life cycle assessment based on a novel scenario analysis framework. *Appl. Energy* **2015**, *157*, 871–883. [CrossRef]

24. Nansai, K.; Moriguchi, Y. Embodied Energy and Emission Intensity Data for Japan Using Input–Output Tables (3EID): For 2005 IO Table, CGER, National Institute for Environmental Studies, Japan. 2012. Available online: http://www.cger.nies.go.jp/publications/report/d031/index.html (accessed on 9 February 2019).

25. Japan Automobile Dealers Association. New Car Sales in Japan 2016. 2016. Available online: http://www.jada.or.jp/contents/data/index.html (accessed on 10 December 2018).

26. Ministry of Land, Infrastructure, Transport and Tourism. Inventory Data of Fuel Economies of New Cars in Japan in 2015. 2016. Available online: http://www.mlit.go.jp/jidosha/Jidosha_fr10_000027.html (accessed on 10 December 2018).

27. Autoc-One. 2018. Available online: http://autoc-one.jp/ (accessed on 10 December 2018).

28. Ministry of Environment. Japan's National Greenhouse Gas Emissions in Fiscal Year 2015 (Final Figures). 2017. Available online: https://www.env.go.jp/en/headline/2309.html (accessed on 10 December 2018).

29. Ministry of Finance. Financial Statements Statistics of Corporations by Industry, Annually. 2016. Available online: https://www.mof.go.jp/pri/publication/zaikin_geppo/hyou/g774/774.htm (accessed on 10 December 2018).

30. TOYOTA. North American Environmental Report. 2013. Available online: https://www.toyota.com/usa/environmentreport2013/ (accessed on 10 December 2018).

Article

Sources of China's Fossil Energy-Use Change

Yawen Han [1,2,3], Shigemi Kagawa [4,*], Fumiya Nagashima [2] and Keisuke Nansai [5]

[1] School of Earth Sciences and Resources, China University of Geosciences, Beijing. 29 Xueyuan Road, Haidian District, Beijing 100083, China; yawenhancugb@foxmail.com
[2] Graduate School of Economics, Kyushu University, 744 Motooka, Nishi-ku, Fukuoka 819-0395, Japan; sbhfmb1122@yahoo.co.jp
[3] Research Center for Strategy of Global Mineral Resources, Chinese Academy of Geological Sciences, 26 Baiwanzhuang Street, Xicheng District, Beijing 100037, China
[4] Faculty of Economics, Kyushu University, 744 Motooka, Nishi-ku, Fukuoka 819-0395, Japan
[5] National Institute for Environmental Studies of Japan, 16-2 Onokawa, Tsukuba, Ibaraki 305-0053, Japan; nansai.keisuke@nies.go.jp
* Correspondence: kagawa@econ.kyushu-u.ac.jp

Received: 16 January 2019; Accepted: 16 February 2019; Published: 21 February 2019

Abstract: Technology improvement related to energy conservation and energy mix low-carbonization is a critical approach for tackling global warming in China. Therefore, we attempt to identify the technology factors of China's energy consumption change between 2007 and 2012, when China's economy started slowing. This study proposes a new refined structural decomposition analysis (SDA) based on a hybrid multi-regional input–output (MRIO) model. The technology factors are expressed through the energy input level effect, energy composition effect, and non-energy input effect. We find that the energy level effect was the primary driver for energy reduction, saving 1205 million tonnes of standard coal equivalent (Mtce) of energy, while 520 Mtce was offset by energy composition and non-energy input effects. The sector analysis shows that the energy input level, energy composition, and non-energy input effects of electricity, the chemical industry, and metallurgy are noteworthy. In addition, the sector contribution to energy-use change, by province, related to the three effects, is also studied. From these results, we propose policy suggestions for further energy saving, in order to achieve China's energy target through technology improvements by the higher priority contributors identified.

Keywords: hybrid MRIO; SDA; energy saving; energy composition; China

---

1. Introduction

Fossil energy consumption is one of the main reasons for global warming [1,2]. As a major greenhouse gas (GHG) emitter, China aims to keep energy consumption within six billion tonnes of standard coal equivalent by 2030 for GHG abatement. Meanwhile, natural gas and non-fossil energy are targeted to have a 35% share of China's energy pie [3]. However, it has been announced that the commitment to the Paris Agreement of each country is not enough to achieve the 1.5 degree Celsius goal [4], which undoubtedly poses additional challenges to national climate policy. Therefore, it is important to quantify the driving factors related to energy saving and energy mix low-carbonization in order to shed light on the achievement of the pursuit of a higher GHG reduction. Also, the realization of national goals requires the joint efforts of all of the provinces. Given the provincial disparities, the provincial contributions to the energy saving and energy mix change are also worth estimating, for a policy tailored to each province.

Structural decomposition analysis (SDA) is usually used to evaluate the driving factors behind energy consumption or environment load change. An advantage of SDA is that a wide variety of the driving factors can be identified considering the economic linkages between sectors. However, the empirical studies are restricted by the availability of the monetary input–output tables of nations [5]. The energy consumption changes or environment load changes are often decomposed into three factors, namely: efficiency, the Leontief inverse effect, and the final demand effect. The final demand effect is further decomposed into commodity shares of final demand, destination share, per-capita total final demand, and population [6].

For the SDA studies focusing on energy change, Weber (2009) [7] used SDA to decompose energy growth between 1997 and 2002 in the United States, and found that rising populations and household consumption were the two drivers of energy demand growth, while being offset by the considerable structural change within the economy. A study of Italy from 1999 to 2006 showed that the final demand for goods and services was the main factor increasing energy consumption, but energy consumption was offset by the energy intensity and production structure [8]. In the case of Japan, the total energy demand dramatically increased, mainly as a result of the growth in the non-energy final demand from 1985 to 1990 [9]. In Brazil, from 1970 to 1996, the primary accelerator of energy use was population growth, and after 1980, the increasing demand for energy-intensive products also drove the energy demand increase [10].

Since China experienced an alarming average annual growth rate of energy consumption of 13% from 2002 to 2007 [11], many researchers have focused on an energy SDA analysis. Xie (2014) [12] examined the driving factors of China's energy use from 1992 to 2010. The study proved that after 2002, when China joined the WTO (World Trade Organization), 38% of its energy growth came from the increasing exports, 36% from the increasing fixed capital formation, while the contribution of household demand growth was only 15%. This was quite different from the household consumption-driven energy growth in developed countries such as the United States or Italy. However, between 2007 and 2010, the energy change induced by exports decreased because of the financial crisis. Fixed capital formation acted as the main reason for energy growth, accounting for 75% of energy growth, because of the "package plan" proposed by the Chinese government. The results of Mi et al. (2018) [13] agreed with those of Xie (2014) [12].

Moreover, Zhang and Lahr (2014), and Zhang et al. (2016) [14,15] quantified the energy change related to the regional demand in China. Meanwhile, efficiency played a critical role in offsetting energy consumption. However, from 2002 to 2007, the efficiency of construction, abnormally increasing the total energy use, resulted in a weakened energy saving effect of efficiency [12–14].

From the above-mentioned literature, we found that the decomposition of the final demand factor has already been well discussed, while the decomposition of technology factors was seldom deeply studied. The present research is intended to fill this gap. A novelty of this study is the development of a new refined decomposition framework to identify the technology role of each sector of each province in the change of energy-use in China.

Specifically, we propose an energy SDA based on a "hybrid" multiregional input–output (MRIO) table, expressed in both monetary and physical units. To the best of our knowledge, this is the first attempt to apply a hybrid MRIO approach to the provincial primary energy consumptions of China. It is well known that a hybrid input–output approach is superior to a monetary input–output approach, because energy sectors are frequently characterized by different pricing for different sectors, whereas the analysis based on monetary input–output tables assumes uniform pricing across all sectors [16–18]. However, the SDA described in previous studies is fit for the monetary approach, but not the hybrid approach [16]. Our new SDA framework applies to both hybrid and monetary cases.

In doing the hybrid analysis, we focus on the primary fossil energy (i.e., coal, oil, and natural gas) consumed in China. Hydropower, nuclear power, renewable energy, and others are not considered, because the proportion of non-fossil energy consumption in China was less than 10% during our study period [19]. The proposed energy SDA method is applied to the Chinese hybrid MRIO tables of 2007 and 2012, and the important driving forces of China's energy growth are identified at a sector and province level. Finally, we identify important stakeholder sectors and provinces that are important for energy saving in China, and suggest that they should be environmentally monitored. Policy makers should commit to making important stakeholder sectors and provinces the highest priority.

This paper is organized as follows: Section 2 describes the methodology proposed in this study, Section 3 explains the data used in this study, Section 4 provides the results and discussion, and Section 5 concludes this paper.

2. Methodology

2.1. Primary Fossil Energy Consumed in China

The structure of the MRIO model can be expressed as $\mathbf{x} = \mathbf{Ax} + \mathbf{f}$, where $\mathbf{x} = (x_i^r)$ $(r = 1, \ldots, R; i = 1, \ldots, N)$ is the vector of outputs of N sectors of R provinces, and $\mathbf{f} = (f_i^r)$ is the vector of the total final demand for sector i of province r, including rural household consumption, urban household consumption, government consumption, fixed capital formation, and exports. Furthermore, $\mathbf{A} = (z_{ij}^{rs} / x_j^s) = (a_{ij}^{rs})$ $(r, s = 1, \ldots, R; i, j = 1, \ldots, N)$ is the input coefficient matrix, where z_{ij}^{rs} denotes the intermediate deliveries from sector i of province r to sector j of province s.

In the mixed-units approach (i.e., hybrid approach), if i is an energy sector, x_i^r, z_{ij}^{rs}, and f_i^r are measured in energy units (tonnes of coal equivalent in this study). As in the previous studies [12,16,17], the solution of the hybrid MRIO model is $\mathbf{x} = \mathbf{Lf}$, where $\mathbf{L} = (\mathbf{I} - \mathbf{A})^{-1}$ is the Leontief inverse matrix. It should be noted that the hybrid MRIO database of China describes both domestically produced energy and imported energy. The domestically produced primary fossil energy consumption in China triggered by the final demand can be estimated as follows:

$$\mathbf{d} = \hat{\mathbf{e}}\mathbf{Lf} \tag{1}$$

where $\mathbf{d} = (d_i^r)$ represents the fossil energy consumptions provided by energy sector k of province r, triggered by the domestic final demand of China and its exports (when $i = k$). It should be noted that $\hat{\mathbf{e}}$ is a diagonal matrix with diagonal elements of 0 and 1. When i is a fossil energy sector (coal mining, crude oil, and natural gas) of province r, $e_i^r = 1$, and otherwise, $e_i^r = 0$.

The physical energy import triggered by the domestic final demand of China and its exports can also be estimated as follows:

$$\mathbf{h} = \mathbf{MLf} \tag{2}$$

where $\mathbf{h} = (h_k)$ represents the imported energy (k) consumed in China, and $\mathbf{M} = (m_{kj}^s)$ is the energy import coefficient matrix expressing the physical amount of energy (k) imported to produce one unit of output of sector j of province s (see the Supporting Information).

The fossil energy (k) consumed in China triggered by total final demand can be expressed as follows:

$$P_k = \sum_{r=1}^{R} d_k^r + h_k \tag{3}$$

2.2. Structural Decomposition Analysis (SDA)

As the primary fossil energy consumptions formulated in Equation (3) include both domestically produced and imported energy, the SDA technique can be performed, focusing on either of these two sources. From Equation (1), the change in the consumption of domestically produced energy in China between 2007 and 2012 can be calculated by the following:

$$\Delta \mathbf{d} = \mathbf{d}_{12} - \mathbf{d}_{07} = \hat{\mathbf{e}}\mathbf{L}_{12}\mathbf{f}_{12} - \hat{\mathbf{e}}\mathbf{L}_{07}\mathbf{f}_{07} \tag{4}$$

The driving factors can be found as the final demand effect $(\Delta \mathbf{d}_F)$ and the Leontief inverse effect $(\Delta \mathbf{d}_L)$, respectively. The final demand effect refers to the energy-use change induced by the shifts of the final demand while holding the Leontief inverse constant. The Leontief inverse effect represents the energy-use change generated from the Leontief inverse shift, with a given final demand. We take the average of all of the possible decompositions in order to qualify each effect. Dietzenbacher and Los (1998), and Hoekstra and Van Den Bergh (2002) provide the theoretical details in their studies [20,21].

$$\Delta \mathbf{d}_F = \frac{1}{2}\left(\hat{\mathbf{e}}\mathbf{L}_{07}\Delta \mathbf{f} + \hat{\mathbf{e}}\mathbf{L}_{12}\Delta \mathbf{f} \right) \tag{5}$$

$$\Delta \mathbf{d}_L = \frac{1}{2}\left(\hat{\mathbf{e}}\Delta \mathbf{L}\mathbf{f}_{07} + \hat{\mathbf{e}}\Delta \mathbf{L}\mathbf{f}_{12} \right) \tag{6}$$

where Δ represents the change of a factor.

We further decompose the Leontief inverse effect ($\Delta \mathbf{d}_L$) into trade effect ($\Delta \mathbf{d}_T$) and technology effect ($\Delta \mathbf{d}_B$). Moreover, the technology effect is identified along three dimensions, as the energy-use change associated with shifts in the non-energy input, energy composition, and energy input level. Trade effect refers to the energy change induced by the shifts of the input factors' sourcing locations. Technology effect refers to the combined effect of the non-energy input, energy composition, and energy input level effects. Firstly, the non-energy input effect ($\Delta \mathbf{d}_G$) is the energy-use change generated from the change in non-energy inputs in production, when holding the other factors constant. Secondly, energy composition effect ($\Delta \mathbf{d}_C$) refers to the energy-use change induced by the energy mix shifts in the production process. Thirdly, the energy input level effect ($\Delta \mathbf{d}_E$) is the energy-use change brought about by the sum of all kinds of energy consumption variation per unit of output (i.e., energy efficiency).

Furthermore, we have the following matrix decomposition of the Leontief inverse matrix [22].

$$\Delta \mathbf{L} = \mathbf{L}_{12} - \mathbf{L}_{07} = \frac{1}{2}(\mathbf{L}_{12}\Delta \mathbf{A}\mathbf{L}_{07} + \mathbf{L}_{07}\Delta \mathbf{A}\mathbf{L}_{12}) \tag{7}$$

In this study, we re-defined the input coefficient matrices of 2007 and 2012, as follows:

$$\mathbf{A}_{07} = \mathbf{T}_{07} \circ \mathbf{B}_{07} \tag{8}$$

$$\mathbf{A}_{12} = \mathbf{T}_{12} \circ \mathbf{B}_{12} \tag{9}$$

Here, $\mathbf{T}$ is the inter-provincial trade coefficient matrix of China, showing the proportion supplied from province r of the total intermediate delivery from sector i, required for producing one unit of sector j of province s, and it can be defined as follows:

$$\mathbf{T} = (t_{ij}^{rs}) = \left(\frac{a_{ij}^{rs}}{\sum_{r=1}^{R} a_{ij}^{rs}} \right) = \begin{bmatrix} \mathbf{T}^{11} & \cdots & \mathbf{T}^{1R} \\ \vdots & \ddots & \vdots \\ \mathbf{T}^{R1} & \cdots & \mathbf{T}^{RR} \end{bmatrix} \tag{10}$$

where $\mathbf{T}^{rs}(r, s = 1, \ldots, R)$ is the inter-provincial trade coefficient submatrix for the goods and services flowing from province r to s. $\mathbf{B}$ in the right-hand sides of Equations (8) and (9) is the technical coefficient matrix of the Chinese provinces, and it can be written as follows:

$$\mathbf{B} = \begin{bmatrix} \mathbf{B}^1 & \cdots & \mathbf{B}^R \\ \vdots & \ddots & \vdots \\ \mathbf{B}^1 & \cdots & \mathbf{B}^R \end{bmatrix} \tag{11}$$

where $\mathbf{B}^s = (b_{ij}^s) = (\sum_{r=1}^{R} a_{ij}^{rs})$ is the technical coefficient submatrix of a specific province (s), showing the total intermediate delivery from sector i required for producing one unit of sector j of province s. It should be noted that $\circ$ denotes the Hadamard product.

We further propose decomposing the technical coefficient submatrix of a specific province (s), $\mathbf{B}^s$ ($s = 1, \ldots, R$), as follows:

$$
\mathbf{B}^s = (b_{ij}^s) =
\begin{matrix}
\text{Energy sector} \\ \vdots \\ \text{Energy sector} \\ \text{Non} - \text{energy sector} \\ \vdots \\ \text{Non} - \text{energy sector}
\end{matrix}
\underbrace{
\begin{bmatrix}
0 & \cdots & 0 \\
\vdots & \ddots & \vdots \\
0 & \cdots & 0 \\
b_{K+1,1}^s & \cdots & b_{K+1,N}^s \\
\vdots & \ddots & \vdots \\
b_{N1}^s & \cdots & b_{NN}^s
\end{bmatrix}}_{\mathbf{G}^s}
$$

$$
+
\underbrace{
\begin{bmatrix}
\dfrac{b_{11}^s}{\sum_{i=1}^{K} b_{i1}^s} & \cdots & \dfrac{b_{1N}^s}{\sum_{i=1}^{K} b_{iN}^s} \\
\vdots & \cdots & \vdots \\
\dfrac{b_{K1}^s}{\sum_{i=1}^{K} b_{i1}^s} & \cdots & \dfrac{b_{KN}^s}{\sum_{i=1}^{K} b_{iN}^s} \\
0 & \cdots & 0 \\
\vdots & \ddots & \vdots \\
0 & \cdots & 0
\end{bmatrix}}_{\mathbf{C}^s}
\underbrace{
\begin{bmatrix}
\sum_{i=1}^{K} b_{i1}^s & & & \\
& \sum_{i=1}^{K} b_{i2}^s & & \\
& & \ddots & \\
& & & \sum_{i=1}^{K} b_{iN}^s
\end{bmatrix}}_{\mathbf{E}^s}
\tag{12}
$$

where $\mathbf{G}^s$ is the technical coefficient submatrix for the non-energy sectors, $\mathbf{C}^s$ is the energy composition matrix showing the energy mix information of each sector, and $\mathbf{E}^s$ is the diagonal matrix with the overall energy input level (energy intensity in this study) of each sector. K is the total number of energy sectors.

Using the refined decomposition factors of Equations (10), (11), and (12), the input coefficient matrices of 2007 and 2012 can be formulated, respectively, as follows:

$$
\mathbf{A}_{07} = \mathbf{T}_{07} \circ \mathbf{B}_{07} = \mathbf{T}_{07} \circ (\mathbf{G}_{07} + \mathbf{C}_{07}\mathbf{E}_{07})
\tag{13}
$$

$$
\mathbf{A}_{12} = \mathbf{T}_{12} \circ \mathbf{B}_{12} = \mathbf{T}_{12} \circ (\mathbf{G}_{12} + \mathbf{C}_{12}\mathbf{E}_{12})
\tag{14}
$$

Accordingly, ΔL in Equation (6) can be further decomposed into the changes in the structural factors of $\mathbf{T}$, $\mathbf{G}$, $\mathbf{C}$, and $\mathbf{E}$, as follows:

$$
\Delta \mathbf{L}_T = \frac{1}{4}[\mathbf{L}_{12}\{\Delta\mathbf{T} \circ (\mathbf{B}_{07} + \mathbf{B}_{12})\}\mathbf{L}_{07} + \mathbf{L}_{07}\{\Delta\mathbf{T} \circ (\mathbf{B}_{07} + \mathbf{B}_{12})\}\mathbf{L}_{12}]
\tag{15}
$$

$$
\Delta \mathbf{L}_G = \frac{1}{4}[\mathbf{L}_{12}\{(\mathbf{T}_{07} + \mathbf{T}_{12}) \circ \Delta\mathbf{G}\}\mathbf{L}_{07} + \mathbf{L}_{07}\{(\mathbf{T}_{07} + \mathbf{T}_{12}) \circ \Delta\mathbf{G}\}\mathbf{L}_{12}]
\tag{16}
$$

$$
\Delta \mathbf{L}_C = \frac{1}{8}[\mathbf{L}_{12}\{(\mathbf{T}_{07} + \mathbf{T}_{12}) \circ \Delta\mathbf{C}(\mathbf{E}_{07} + \mathbf{E}_{12})\}\mathbf{L}_{07} + \mathbf{L}_{07}\{(\mathbf{T}_{07} + \mathbf{T}_{12}) \circ \Delta\mathbf{C}(\mathbf{E}_{07} + \mathbf{E}_{12})\}\mathbf{L}_{12}]
\tag{17}
$$

$$
\Delta \mathbf{L}_E = \frac{1}{8}[\mathbf{L}_{12}\{(\mathbf{T}_{07} + \mathbf{T}_{12}) \circ (\mathbf{C}_{07} + \mathbf{C}_{12})\Delta\mathbf{E}\}\mathbf{L}_{07} + \mathbf{L}_{07}\{(\mathbf{T}_{07} + \mathbf{T}_{12}) \circ (\mathbf{C}_{07} + \mathbf{C}_{12})\Delta\mathbf{E}\}\mathbf{L}_{12}]
\tag{18}
$$

We can finally quantify the Leontief inverse effect-driven energy-use change as the following four factors.

$$
\begin{aligned}
\Delta \mathbf{d}_L = {} & \tfrac{1}{2}\left(\hat{\mathbf{e}}\Delta \mathbf{L}\mathbf{f}_{07} + \hat{\mathbf{e}}\Delta \mathbf{L}\mathbf{f}_{12} \right) = \tfrac{1}{2}\left(\hat{\mathbf{e}}\Delta \mathbf{L}_T\mathbf{f}_{07} + \hat{\mathbf{e}}\Delta \mathbf{L}_T\mathbf{f}_{12} \right) \text{ (Trade effect)} \\
& + \tfrac{1}{2}\left(\hat{\mathbf{e}}\Delta \mathbf{L}_G\mathbf{f}_{07} + \hat{\mathbf{e}}\Delta \mathbf{L}_G\mathbf{f}_{12} \right) \text{ (Non-energy input effect)} \\
& + \tfrac{1}{2}\left(\hat{\mathbf{e}}\Delta \mathbf{L}_C\mathbf{f}_{07} + \hat{\mathbf{e}}\Delta \mathbf{L}_C\mathbf{f}_{12} \right) \text{ (Energy composition effect)} \\
& + \tfrac{1}{2}\left(\hat{\mathbf{e}}\Delta \mathbf{L}_E\mathbf{f}_{07} + \hat{\mathbf{e}}\Delta \mathbf{L}_E\mathbf{f}_{12} \right) \text{ (Energy input level effect)}
\end{aligned}
\left.\rule{0pt}{6em}\right\} \text{Pure technology effect}
\tag{19}
$$

Similarly, we decomposed the changes in the imported energy of China between 2007 and 2012 (see Supporting Information).

3. Data Sources

This study uses the monetary MRIO tables of China from 2007 and 2012 [23,24], and the energy-use data of the individual sectors by province [25], as well as the energy export, import, and inventory change of each province [26,27]. The "monetary" MRIO tables of 2007 and 2012 are of 30 sectors [23,24], while the energy-use data of the individual sectors by province is of 45 industrial sectors and 2 residential sectors [25]. For consistency, this study consolidates the industrial sectors in the different databases [23–25] into the same 31 sectors, with 8 energy sectors and 23 non-energy sectors. The energy types in the energy-use data are classified into eight groups corresponding to the eight energy sectors (see Table S1 and Table S2 in the Supporting Information for sector classification and energy grouping). Based on the data above, we compile "hybrid" MRIO tables of the years 2007 and 2012, whose energy inputs are described in physical terms (tonnes of coal equivalent in this study), and non-energy inputs are described in monetary terms (Chinese yuan in this study). A detailed description of constructing the hybrid MRIO tables for China can also be found in the Supporting Information.

The double deflation method is used to deflate the monetary flows in the hybrid MRIO table for the 2012 to 2007 constant prices [28] for SDA application. The price index of each sector is obtained from the National Statistical Yearbook [29,30] (see Table S3 for price deflators).

4. Results and Discussion

4.1. Nationwide SDA of China

We estimated the total primary fossil energy consumptions in China by using Equation (3). The total primary fossil energy consumed in China increased by 61%, from 2682 million tonnes of standard coal equivalent (Mtce) in 2007, to 4319 Mtce in 2012. In particular, coal consumption made up the most substantial fraction, growing from 2162 Mtce to 3399 Mtce (Figure 1). It is important to determine what the sources were of the remarkable increase in the fossil energy consumptions during the recent five years, between 2007 and 2012.

From the SDA results, we found that final demand shifts were the dominating sources that pushed each energy-use upward, by 2259 Mtce in total (84% of the energy consumption in 2007) (Figure 2). It is also important to identify the role of technology changes in primary energy use. Looking at the technology effect, which is a combined effect of the energy input level, energy composition, and non-energy input effects in Figure 2, technological changes have contributed to decreasing fossil energy use by 684 Mtce in total. Thus, expanding consumptions lead to economic growth and fossil energy growth in China, while the Chinese government can expect that technological changes can help to partially offset fossil energy growth.

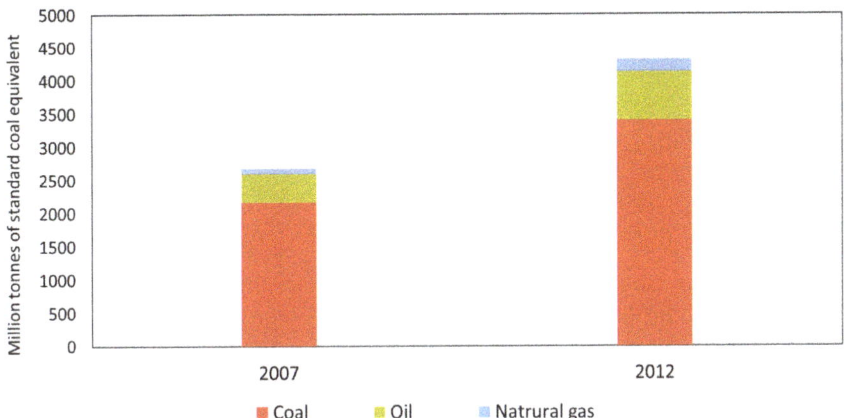

Figure 1. Total primary fossil energy consumed in China in 2007 and 2012.

Using the SDA technique developed in this study, we can further decompose the technology effect into the following three effects: energy input level, energy composition, and non-energy input effects. Looking at these three effects in detail, using Figure 2, between 2007 and 2012, the energy input level effect was the main driver offsetting the increase of all fossil energy types, helping to reduce the energy consumption by 1205 Mtce (45% of the energy consumption in 2007). On the other hand, the energy composition effect was a driver of increasing the consumptions of coal and natural gas by 198 and 12 Mtce, respectively, while reducing oil consumption by 79 Mtce, leading to a 131 Mtce energy growth overall (Figure 2). The non-energy input effect was a driver of increasing the fossil energy consumption by 389 Mtce (15% of energy consumption in 2007) (Figure 2). Thus, the energy saving effect through technological changes was weakened by both energy composition and non-energy input effects; therefore, the Chinese government should put effort into maximizing the technology-induced energy saving effect through an improvement in energy mix and non-energy input.

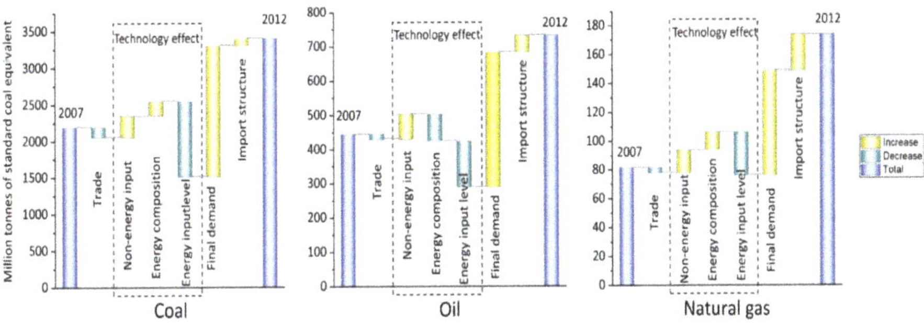

Figure 2. Nationwide structural decomposition analysis (SDA) for each fossil-energy type in China.

4.2. Decomposition of Technology Effects by Province and Sector

4.2.1. Energy Input Level Effect

We computed the aggregated energy input level effect of each sector by summing the energy input level effect of each sector over 30 provinces. Figure 3 shows that the aggregated energy input level effects of 22 sectors in China contributed to a coal reduction of 1281 Mtce, and the top three sectors were electricity and hot water production and supply (Electricity), metallurgy, and the chemical industry, contributing 71% of that 1281 Mtce coal reduction. The reason lies in the dramatic improvements in the energy efficiencies of these sectors [31,32].

Unfortunately, the other nine sectors had a total rise in coal consumption of 243 Mtce, due to the energy input level effect (Figure 3). Growth in the sectors of coal mining and coking products stand out the most (Figure 3). With the reduction of easy-to-mine and high-quality coal resources, the increasing difficulty in coal mining and the enlarging need of coal refinement promoted unit energy consumption in coal mining [33,34]. As coal and electricity were the main energy types consumed in coal mining, according to matrix **C** in Equation (13), the unit energy consumption change majorly affected the coal consumption. The installation of environmental protection equipment and the instigation of long-term low-load operation in coking products led to an increase of the unit of energy consumption in this sector [35,36]. For example, wastewater facilities were installed for environment protection, and the unit energy consumption of coking products rose [36]. For the coking facilities, a lower processing amount (a lower load rate) results in a higher allocated unit energy consumption [36]. Coal was the main material for coking. Therefore, the increased unit energy consumption mainly led to coal growth.

Concerning oil consumption, the energy input level effects of 24 sectors acted as drivers to reduce oil by 166 Mtce, with the top sector being transport and storage, which had a reduction of 83 Mtce (Figure 3). The energy input level effect of petroleum refining causing oil use to grow by 23 Mtce cannot be ignored, as this amount was significantly higher than those of the next highest sectors (Figure 3). The increasing share of high-sulfur crude oil and low-load operation owing to excess capacity, induced a unit energy consumption growth in petroleum refining [37,38]. High-sulfur crude oil not only increased the processing difficulty of refineries, but also expanded the scale of environmental protection facilities, such as desulfurization [39]. As a result, the unit energy consumption of refineries increased, which led to oil growth, because oil was the main raw material of refineries. In addition, the energy input level effect on natural gas use change was quite small (Figure 3).

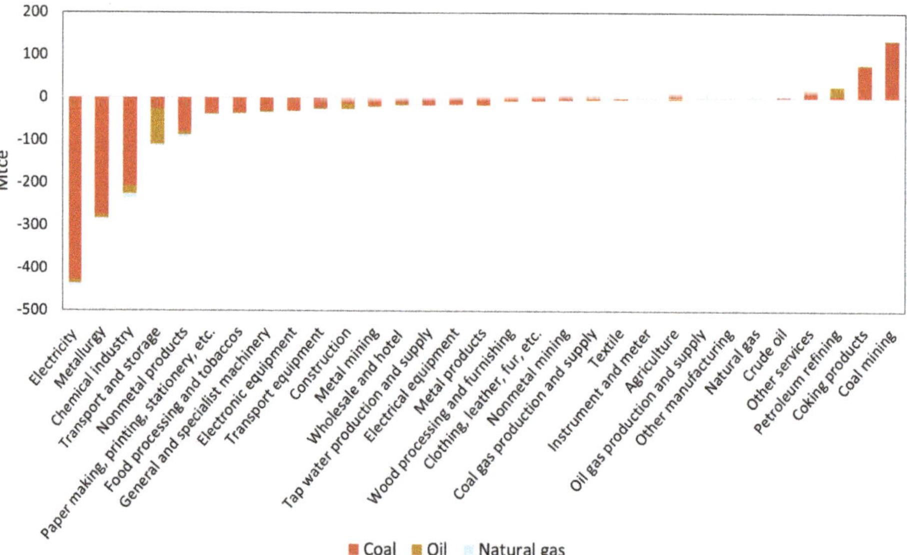

Figure 3. Energy input level effect by sector.

Subsequently, we ranked the disaggregated energy input level effects of 31 sectors of 30 provinces; the top ten and the bottom ten are listed in Figure 4. The coking products of Shanxi had the most energy growth among the 930 contributors (31 sectors × 30 provinces), at 58 Mtce (Figure 4). From the bottom ten, we were surprised to find that the energy input level effect of electricity was positive at 47 and 20 Mtce in Inner Mongolia and Shanxi, respectively, even though the local governments announced an update of equipment at the power stations during the study period [40]. The problems of these sectors lie in the fact that the low-load operation of power generation units, as well as the poor equipment management levels in Inner Mongolia and Shanxi, resulted in the ineffectiveness of the power generating units with large capacities and a high parameter [41–43]. The other contributors of the bottom ten were mainly related to coking products, coal mining, and petroleum refining, as expected (Figure 4).

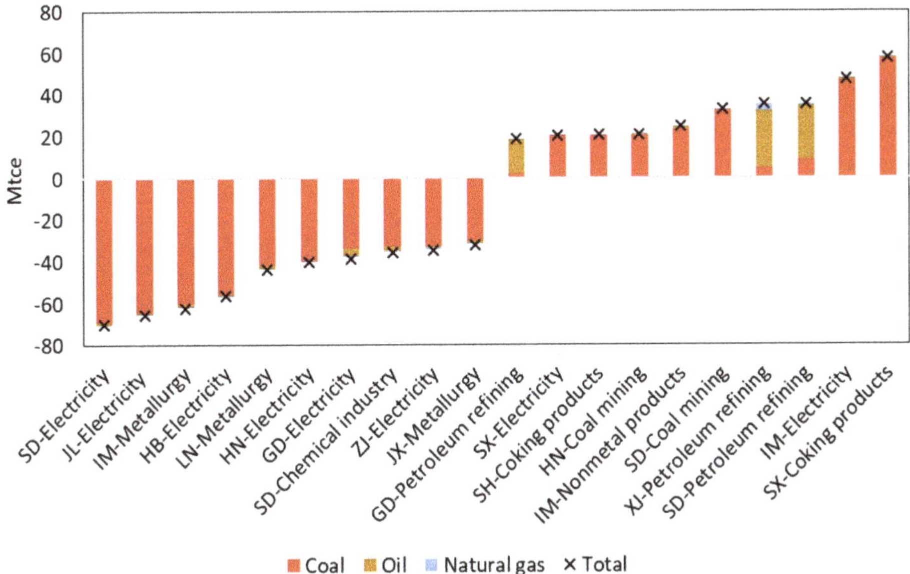

Figure 4. Top ten and bottom ten contributors for energy saving related to energy input level effect. SD: Shandong; JL: Jilin; IM: Inner Mongolia; HB: Hebei; LN: Liaoning; HN: Henan; GD: Guangdong; ZJ: Zhejiang; JX: Jiangxi; SX: Shanxi; SH: Shanghai; XJ: Xinjiang.

Accordingly, further energy efficiency improvement in electricity, the chemical industry, metallurgy, coal mining, coking products, transport and storage, and petroleum refining would dramatically reduce energy use. However, the fact is that, by 2017, when these sectors encountered excess capacity, some of the technologies of sectors such as electricity, the chemical industry, metallurgy, and coking products were already close to or at the advanced level in the world [31,32]. That being the case, what counts is not replacing the equipment frequently, but promoting an energy saving approach under a low-load operation, and phasing out inferior production capacities. As for other sectors, such as coal mining and petroleum refining, both the lag in technology and the over-capacity should be addressed.

4.2.2. Energy Composition Effect

A positive energy composition effect means increasing fossil energy consumptions through energy mix changes. It should be noted that even if the energy input level effect of the specific sector is negative as a result of the decreasing energy intensity as defined by the diagonal element of the matrix **E**, the fossil energy consumptions of China may increase as a result of the energy composition change of the sector. Specifically, the energy composition effects of electricity and the chemical industry distinctly stimulated the energy growth (Figure 5), although the most significant energy saving was due to the energy input level effects of the two sectors (Figure 3).

As for electricity, the energy composition effect caused fossil energy growth by 60 Mtce. Therein, the coal consumption increased by 76 Mtce, while the oil and natural gas went down by 16 and 0.5 Mtce (Figure 5). As for the chemical industry, the energy composition effect was positive at 30 Mtce. The coal consumption grew by 34 Mtce, with oil consumption and natural gas decreasing by 3 and 1 Mtce, respectively (Figure 5). The matrix $\Delta \mathbf{C}$ in Equation (17) gives us the source of the energy composition effect of a specific sector of a specific province. First of all, the increased coal and electricity proportion in the energy input promoted coal growth. Secondly, fuel oil saving at coal plants made the proportion of oil consumption for electricity, which was already the smallest, even smaller [44,45]. Policy factors [46] and the price disadvantage of oil impeded the use of oil as a material in the chemical industry [47]. Thirdly, an increasing natural gas price and tight natural gas supply hindered the application of natural gas for both electricity and the chemical industry [48,49]. In addition, the Natural Gas Utilization Policy set selective restrictions on natural gas use for electricity and the chemical industry [47,50]. The factors described above influenced the energy structure for electricity and the chemical industry.

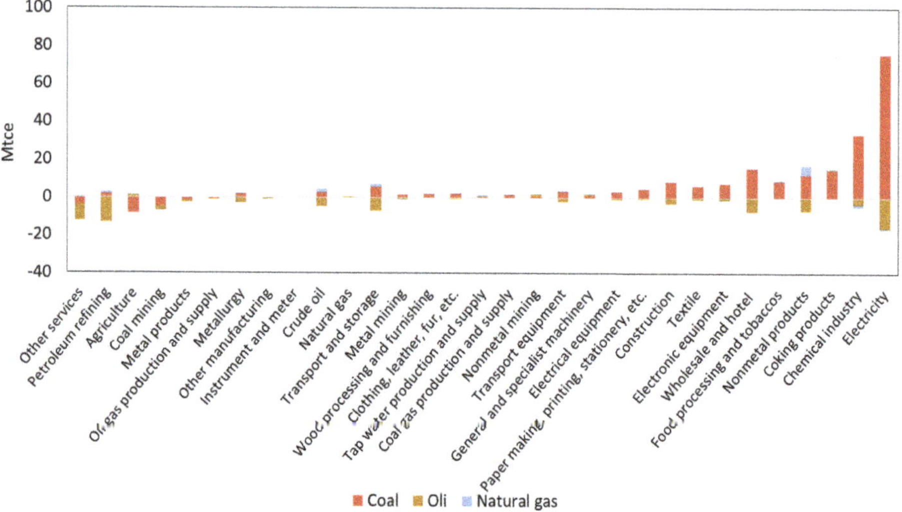

Figure 5. Energy composition effect by sector.

Figure 6 shows the top ten and the bottom ten sectors regarding the disaggregated energy composition effects of 31 sectors of 30 provinces. We found that the energy composition effect of metallurgy, which ranked second among the top ten for energy saving related to the energy input level effect, also had a clear impact on energy use, although the aggregated energy composition effect was negligible at −0.5 Mtce (Figure 5). Here, the energy composition effects of metallurgy of Liaoning, Jiangxi, Shanxi, and Hunan, ranked in the bottom ten for energy saving, triggering a fossil energy growth of 44 Mtce in total, due to the increased proportion of coking products' input to metallurgy in these provinces (Figure 6). However, the energy composition effects of five of the provinces ranking in the top ten helped to reduce fossil energy by 37 Mtce (Figure 6), because the secondary energy (coke oven gas, converter gas, etc.) recovery improvement substantially reduced the coal proportion in the metallurgy of these provinces [51]. In addition, the other growth in the bottom ten was related to electricity, the chemical industry, and coking products, as expected (Figure 6).

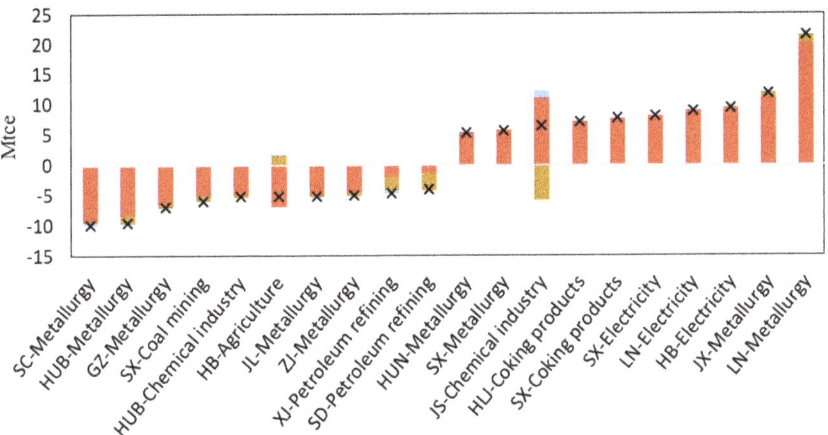

Figure 6. Top ten and bottom ten contributors for energy saving related to the energy composition effect. SC: Sichuan; HUB: Hubei; GZ: Guizhou; SX: Shanxi; HB: Hebei; JL: Jilin; ZJ: Zhejiang; XJ: Xinjiang; SD: Shandong; HUN: Hunan; JS: Jiangsu; HLJ: Heilongjiang; LN: Liaoning; JX: Jiangxi.

We found that, compared with the other sectors, the energy composition effects of electricity, the chemical industry, and metallurgy noticeably impacted the fossil energy-use change, especially coal use change. The large coal fraction in the change was because the energy composition change was mainly impacted by a coal-related technology change and a growing electricity share. If electricity generation remains dominated by coal, the wide application of electricity in end-use will trigger more coal consumption, like the coal growth from the chemical industry related to the energy composition effect. Thus, the energy input mix low-carbonization of electricity fundamentally plays an important and beneficial role in the changes in China's primary energy structure. In addition, the significant natural gas growth from the energy composition effect that we expected did not appear in the individual sectors of the individual provinces, as a result of a natural gas supply shortage and price increases. Thus, ensuring the natural gas supply and reasonable prices are key to the development of natural gas.

4.2.3. Non-Energy Input Effect

Figure 7 shows that the aggregated non-energy input effect of 24 sectors contributed to fossil energy growth. Note that the non-energy input effects of electricity, metallurgy, and the chemical industry weakened their energy savings related to the energy input level effect. The non-energy input effects of the chemical industry and metallurgy generated noticeable fossil energy growth at 65 and 34 Mtce, respectively, although that of electricity was insignificant at 7 Mtce (Figure 7). Thus, the −441 Mtce energy input level effect of electricity was offset by 68 Mtce, as a result of the energy growth from the energy composition effect and the non-energy input effect, resulting in a −373 Mtce of a combined effect from the three factors (i.e., the technology effect). Similarly, the −286 Mtce energy input level effect of metallurgy was offset by 34 Mtce, leading to a −252 Mtce technology effect. The −237 Mtce energy input level effect of the chemical industry was offset by 94 Mtce, generating a −143 Mtce technology effect overall.

In addition to the above, the non-energy input effect of construction triggered the most significant energy growth, leading the fossil energy to increase by 123 Mtce (Figure 7). The other consequential sectors that contributed to energy growth were coal mining and electronic equipment (Figure 7). From $\Delta \mathbf{G}$ in Equation (16), we found that the energy growth induced by the non-energy input effect of construction and metallurgy mainly came from the increased metallurgy products' input. Similarly, the growing chemical industry products input and electronic equipment input were mainly responsible for the energy growth of the chemical industry and the electronic equipment related to the non-energy input effect, respectively. Particularly, the principal input factor that was responsible for the fossil energy growth from the non-energy input effect of coal mining varied province by province, but mainly included growth in the chemical industry products, metallurgy products, electrical equipment, and general and specialist machinery input in the sector of coal mining.

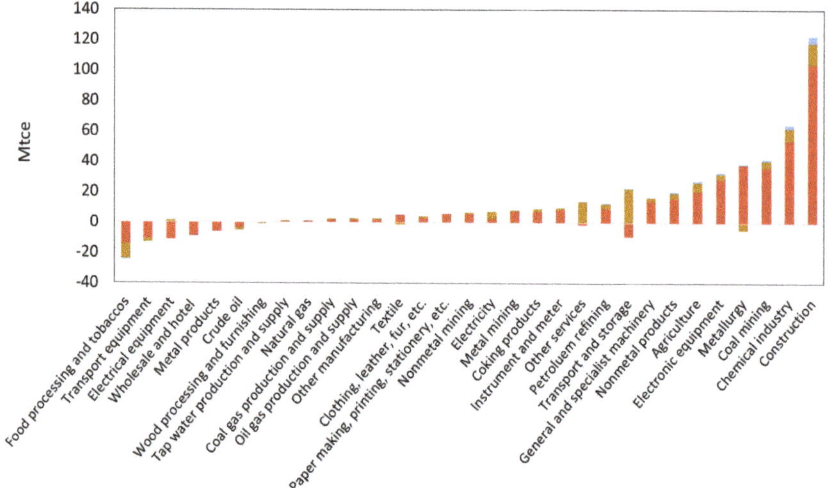

Figure 7. Non-energy input effect of each sector.

Figure 8 shows the most influential disaggregated contributors to energy-use related to the non-energy input effect. The energy growth generated from the non-energy input effect of coal mining in Shanxi was the highest, mainly owning to the growing metallurgy products input (Figure 8). The other contributors in the bottom ten were related to construction, metallurgy, the chemical industry, and electronic equipment, not surprisingly (Figure 8).

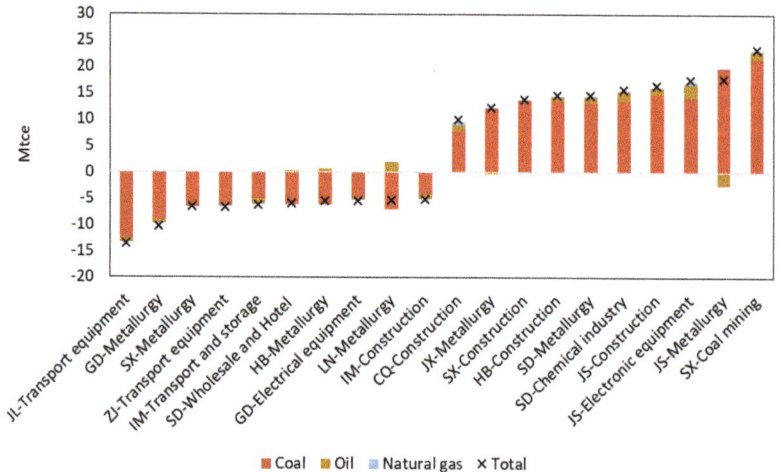

Figure 8. Top ten and bottom ten contributors for energy saving related to non-energy input effect. JL: Jilin; GD: Guangdong; SX: Shanxi; ZJ: Zhejiang; IM: Inner Mongolia; SD: Shandong; HB: Hebei; LN: Liaoning; CQ: Chongqing; JX: Jiangxi; JS: Jiangsu.

We consider that reducing the non-energy input should be taken seriously. Metallurgy products' utilization in construction, metallurgy, and coal mining; chemical industry products' utilization in the chemical industry; and electronic equipment products' utilization in electronic equipment should be given priority attention in order to save energy indirectly.

4.3. Analysis Combining the Present and Previous Studies

We collected five seminal peer-reviewed articles [12–14,22,52] relevant to our research, and classified the driving forces behind China's energy-use change into a production technology effect, and a final demand effect. We computed the "average" decomposition effects of different studies in the same period, and Figure 9 shows the percentage contribution of each driving factor to China's energy-use change during the specific five-year period. Importantly, we found that the energy saving effect of the production technology was weaker between 2002 and 2012 than between 1981 and 2002. If the production technology effect becomes smaller over time, then the climate mitigation goal of China will probably not be achieved because of the expanding final demand of goods and services in China (see the larger final demand effects in Figure 9).

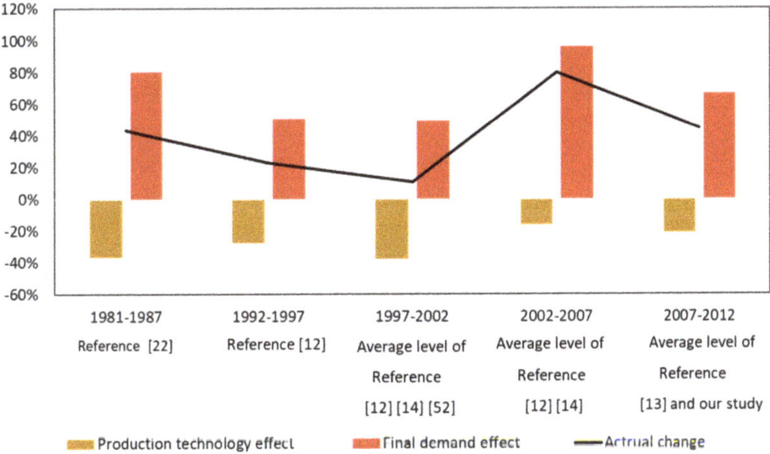

Figure 9. Energy consumption change from 1981 to 2012.

5. Conclusions and Policy Implication

The main conclusions are summarized as follows:

Between 2007 and 2012, technology changes affected fossil energy consumption in China in three aspects. The energy input level effect was the most influential factor offsetting the primary fossil energy growth, by 1205 Mtce, in China. Unfortunately, although the energy composition effect saved 79 Mtce of oil, it promoted coal and natural gas growth by 198 and 12 Mtce, respectively, ultimately resulting in an energy growth of 131 Mtce. The non-energy input effect indirectly caused the fossil energy consumption to increase by 389 Mtce.

The role of the energy input level effect was in opposition to the energy composition and non-energy input effects in most sectors. The energy input level effects of electricity, metallurgy, and the chemical industry stood out the most at −936 Mtce in total, but the noteworthy energy composition effects and non-energy input effects of the three sectors weakened the energy input level effect, triggering a fossil energy growth of 194 Mtce.

The other main sources of energy growth were as follows. For the energy input level effect, coal mining, coking products, and petroleum refining were the primary source sectors promoting energy growth, by 243 Mtce overall. For the non-energy input effect, construction, coal mining, and electronic equipment generated energy growths of 198 Mtce overall.

We suggest that the energy growth of the important stakeholder sectors identified in this study should be given a higher priority of attention by policy makers.

Given the tougher climate challenges facing all countries, we propose the following policy, based on our analysis, for further energy conservation in China. The energy inputs of the contributors listed in Table 1 should be the main monitoring points for energy saving. Annual reports for the main monitoring points should be compiled by the relevant governments. We also suggest that this contributors list be updated every five years for dynamic monitoring of the major sources of energy growth related to technology change. Therefore, the focus of technology improvement for further fossil energy saving should always be clear.

Table 1. Main energy growth sources by sector and by province related to technology effects.

Factor	Source sector	Source province
Energy level effect	Coking products	Shanxi and Shanghai
	Coal mining	Shandong and Henan
	Petroleum refining	Xinjiang, Shandong and Guangdong
	Electricity	Inner Mongolia and Shanxi
	Non-metal products	Inner Mongolia
Energy composition effect	Electricity	Hebei, Liaoning, and Shanxi
	Chemical industry	Jiangsu
	Metallurgy	Liaoning, Jiangxi, Hunan, and Shanxi
	Coking products	Shanxi, Heilongjiang
Non-energy input effect	Construction	Jiangsu, Hebei, Shanxi, and Chongqing
	Metallurgy	Jiangsu, Shandong, and Jiangxi
	Coal mining	Shanxi
	Chemical industry	Shandong
	Electronic equipment	Jiangsu

Supplementary Materials: The following are available online at http://www.mdpi.com/1996-1073/12/4/699/s1, Table S1: Results of industrial sectors classification, Table S2: Energy grouping, Table S3: Price deflaters for sectors, Table S4: Abbreviation of each province, method S1: Decomposition of the changes in the imported energy of China during 2007 to 2012, method S2: Constructing hybrid MRIO tables of China.

Author Contributions: S.K. and Y.H. designed the research and developed the method. Y.H. and F.N. analyzed the data. Y.H. wrote the manuscript with inputs from all authors.

Funding: This research was funded by the Grant-in-Aid for Grant-in-Aid for Scientific Research (A) [16H01797].

Acknowledgments: We are grateful to the two anonymous referees for their helpful comments and suggestions. We also sincerely appreciate the support from the China Scholarship Council. All errors are our own.

Conflicts of Interest: The authors declare no conflict of interest.

References

1. Sher, F.; Pans, M.A.; Sun, C.; Snape, C.; Liu, H. Oxy-fuel combustion study of biomass fuels in a 20 kWth fluidized bed combustor. *Fuel* **2018**, *215*, 778–786. [CrossRef]
2. Sher, F.; Pans, M.A.; Afilaka, D.T.; Sun, C.; Liu, H. Experimental investigation of woody and non-woody biomass combustion in a bubbling fluidised bed combustor focusing on gaseous emissions and temperature profiles. *Energy* **2017**, *141*, 2069–2080. [CrossRef]
3. NDRC (China National Development and Reform Commission). *Revolutionary Strategy for Energy Production and Consumption (2016–2030)*; NDRC: Beijing, China, 2016.
4. IPCC (Intergovernmental Panel on Climate Change). *Global Warming of 1.5 °C*; IPCC: Geneva, Switzerland, 2018.
5. Su, B.; Ang, B.W. Structural decomposition analysis applied to energy and emissions: Some methodological developments. *Energy Econ.* **2012**, *34*, 177–188. [CrossRef]

6. Lenzen, M. Structural analyses of energy use and carbon emissions—An overview. *Econ. Syst. Res.* **2016**, *28*, 119–132. [CrossRef]

7. Weber, C.L. Measuring structural change and energy use: Decomposition of the US economy from 1997 to 2002. *Energy Policy* **2009**, *101*, 115–126. [CrossRef]

8. Cellura, M.; Longo, S.; Mistretta, M. Application of the Structural Decomposition Analysis to assess the indirect energy consumption and air emission changes related to Italian households consumption. *Renew. Sustain. Energy Rev.* **2012**, *16*, 1135–1145. [CrossRef]

9. Kagawa, S.; Inamura, H. A structural decomposition of energy consumption based on a hybrid rectangular input-output framework: Japan's case. *Econ. Syst. Res.* **2001**, *13*, 339–364. [CrossRef]

10. Wachsmann, U.; Wood, R.; Lenzen, M.; Schaeffer, R. Structural decomposition of energy use in Brazil from 1970 to 1996. *Appl. Energy* **2009**, *86*, 578–587. [CrossRef]

11. BP. *Statistical Review of World Energy*; BP: London, UK, 2018.

12. Xie, S.C. The driving forces of China's energy use from 1992 to 2010: An empirical study of input-output and structural decomposition analysis. *Energy Policy* **2014**, *73*, 401–415. [CrossRef]

13. Mi, Z.; Li, Z.; Meng, J.; Shan, Y.; Zheng, H.; Ou, J.; Guan, D.; Wei, Y. China's Energy Consumption in the New Normal. *Earth's Future* **2018**, *6*, 1007–1016. [CrossRef]

14. Zhang, H.; Lahr, M.L. China's energy consumption change from 1987 to 2007: A multi-regional structural decomposition analysis. *Energy Policy* **2014**, *67*, 682–693. [CrossRef]

15. Zhang, B.; Qiao, H.; Chen, Z.M.; Chen, B. Growth in embodied energy transfers via China's domestic trade: Evidence from multi-regional input–output analysis. *Appl. Energy* **2016**, *184*, 1093–1105. [CrossRef]

16. Dietzenbacher, E.; Stage, J. Mixing oil and water? Using hybrid input-output tables in a structural decomposition analysis. *Econ. Syst. Res.* **2006**, *18*, 85–95. [CrossRef]

17. Miller, R.E.; Blair, P.D. *Input-Output Analysis: Foundations and Extensions*, 2nd ed.; Cambridge University Press: Cambridge, UK, 2009; ISBN 9780511626982.

18. Weinzettel, J.; Steen-Olsen, K.; Hertwich, E.G.; Borucke, M.; Galli, A. Ecological footprint of nations: Comparison of process analysis, and standard and hybrid multiregional input-output analysis. *Ecol. Econ.* **2014**, *101*, 115–126. [CrossRef]

19. NBS (China National Bureau of Statistics). Total Consumption of Energy and Its Composition. Available online: http://www.stats.gov.cn/tjsj/ndsj/2018/indexeh.htm (accessed on 23 December 2018).

20. Hoekstra, R.; Van Den Bergh, J.C.J.M. Structural decomposition analysis of physical flows in the economy. *Environ. Resour. Econ.* **2002**, *23*, 357–378. [CrossRef]

21. Dietzenbacher, E.; Los, B. Structural Decomposition Techniques: Sense and Sensitivity. *Econ. Syst. Res.* **1998**, *10*, 307–324. [CrossRef]

22. Lin, X.; Polenske, K.R. Input-Output Anatomy of China's Energy Use Changes in the 1980s. *Econ. Syst. Res.* **1995**, *7*, 67–84. [CrossRef]

23. Mi, Z.; Meng, J.; Guan, D.; Shan, Y.; Song, M.; Wei, Y.-M.; Liu, Z.; Hubacek, K. Chinese CO_2 emission flows have reversed since the global financial crisis. *Nat. Commun.* **2017**, *8*, 1712. [CrossRef]

24. Liu, W.; Chen, J.; Tang, Z.; Liu, H.; Han, D.; Li, F. *Theories and Practice of Constructing China's Interregional Input-Output Tables between 30 Provinces in 2007*; China Statistics Press: Beijing, China, 2012.

25. CEADs (China Emission Accounts and Datasets). Provincial Energy Inventory. Available online: http://www.ceads.net/data/energy-inventory/ (accessed on 19 February 2019).

26. NBS (China National Bureau of Statistics). Regional energy balance sheet. In *China Energy Statistics Yearbook*; China Statistics Press: Beijing, China, 2013.

27. NBS (China National Bureau of Statistics). Regional energy balance sheet of. In *China Energy Statistics Yearbook*; China Statistics Press: Beijing, China, 2008.

28. Dietzenbacher, E.; Hoen, A.R. Deflation of input-output tables from the user's point of view: A heuristic approach. *Rev. Income Wealth* **1998**, *44*, 111–122. [CrossRef]

29. NBS (China National Bureau of Statistics). Price Index. Available online: http://www.stats.gov.cn/tjsj/ndsj/2013/indexch.htm (accessed on 19 February 2019).

30. NBS (China National Bureau of Statistics). Price Index. Available online: http://www.stats.gov.cn/tjsj/ndsj/2010/indexch.htm (accessed on 19 February 2019).

31. CEC (China Electricity Council). *China Coal Power Clean Development Report*; CEC: Beijing, China, 2017.

32. RITE (The Research Institute of Innovative Technology for the Earth). *Estimation of Energy Consumption per unit in 2015 (Steel Department-Converter Steel)*; RITE: Kyoto, Japan, 2015.

33. Gu, Y.; Yu, W.; Ma, X. Comparative study on current situation of coal preparation and policy at home and abroad. *Coal Prep. Technol.* **2012**, *4*, 110–116.

34. Guan, X. Study on Reinforcement Design for Open-Pit Slope under the Underground Mining. Master's Thesis, Qingdao Technological University, Qingdao, China, 2010.

35. Wu, D.; Zhang, T.; Zhao, Z. Private Enterprises in Shanxi Coking Explore Energy Conversation Issues. *Sci. Technol. Ind.* **2010**, *10*, 67–71.

36. Wang, M.; Deng, S.; Jiang, Y.; Xiong, C.; Song, H. 12th 5 year Energy Saving Study of Coke Industry in Shanxi Province. *Shanxi Energy Conserv.* **2010**, *5*, 57–60.

37. Wang, M. Discussions on Energy Saving Measures for Continuous Catalytic Reforming Unit at Low Load. *Qilu Petrochem. Technol.* **2009**, *37*, 1–3.

38. NPCPI (China National Petroleum and Chemical Planning Institute). *Current Situation and Development Suggestion of Refining Industry in China*; NPCPI: Beijing, China, 2013.

39. Zhuang, J.; Hou, K.; Yan, C.; Li, Z. Energy Consumption and Analysis on Energy Saving Measures at SINOPEC's Large Refineries. *China Pet. Process. Petrochem. Technol.* **2007**, *3*, 1–8.

40. The People's Government of Shanxi. *The 12th Five-Year Plan of Shanxi Electric Power Industry Development*; The People's Government of Shanxi: Taiyuan, China, 2012.

41. Ma, Y. Analysis on development of thermal power industry in Inner Mongolia Autonomous Region. *Dev. Environ.* **2017**, *10*, 14.

42. Liu, G.; Jiang, X.; Li, Z. Investigation on Affects of Generation Load on Coal Consumption Rate in Fossil Power Plant. *Power Syst. Eng.* **2008**, *24*, 47–49.

43. Shi, X.; Wang, X.; Pang, L. A survey of coal-fired power generation in Shanxi. *Shanxi Electr. Power* **2015**, *6*, 51–54.

44. World Bank. Electricity Production from Oil Sources. Available online: https://data.worldbank.org.cn/indicator/EG.ELC.PETR.ZS?locations=CN (accessed on 15 November 2018).

45. NEA (China National Energy Administration). Development Status of Key Technologies for Energy Conservation in China's Power Industry. Available online: http://www.nea.gov.cn/2012-01/12/c_131355690_2.htm (accessed on 15 November 2018).

46. NDRC (China National Development and Reform Commission). *China's Energy 11th 5 Year Plan*; NDRC: Beijing, China, 2007.

47. EIA (U.S. Energy Information Administration). Spot Prices. Available online: https://www.eia.gov/dnav/pet/pet_pri_spt_s1_d.htm (accessed on 15 November 2018).

48. Ren, H. Future Development Situation Analysis of Natural Gas Power Industry. *East China Electr. Power* **2014**, *42*, 1457–1459.

49. Wang, S.; Ran, W.; Chen, F. Discuss on the energy-saving measures of low-load operation of natural gas purification device. *Pet. Nat. Gas Chem. Ind.* **2013**, *42*, 447–456.

50. NDRC (China National Development and Reform Commission). *Natural Gas Utilization Policy*; NDRC: Beijing, China, 2007.

51. MPI (China Metallurgical Industry Planning and Research Institute). Review of 11th five year and prospect 12th five year development of circular economy in China's iron and steel industry. *Metall. Econ. Manag.* **2012**, *2*, 8–12.

52. Liang, J.; Zheng, W.; Cai, J. The decomposition of Energy consumption growth in China based on input-output model. *J. Nat. Resour.* **2007**, *22*, 853–864.

Article

How Does Information and Communication Technology Capital Affect Productivity in the Energy Sector? New Evidence from 14 Countries, Considering the Transition to Renewable Energy Systems

Hidemichi Fujii [1],*, Akihiko Shinozaki [1], Shigemi Kagawa [1] and Shunsuke Managi [2]

[1] Faculty of Economics, Kyushu University, 744 Motooka, Nishi-ku, Fukuoka 819-0395, Japan;
   shinozaki@kyudai.jp (A.S.); kagawa@econ.kyushu-u.ac.jp (S.K.)
[2] Urban Institute & Faculty of Engineering, Kyushu University, 744 Motooka, Nishi-ku, Fukuoka 819-0395,
   Japan; managi.s@gmail.com
* Correspondence: hidemichifujii@econ.kyushu-u.ac.jp; Tel.: +81-92-802-5494

Received: 1 April 2019; Accepted: 7 May 2019; Published: 10 May 2019

Abstract: By focusing on a distributed energy system that has been widely diffused for efficient utilization of renewable energy generation in recent years, this paper investigates the relationship between productivity growth and information and communications technology capital in the energy sector. Information and communications technology is a key factor in operating distributed energy systems in a way that balances energy supply and demand in order to minimize energy loss and to enhance capacity utilization. The objective of this study is to clarify the determining factors that affect productivity growth, focusing on three different information and communications technologies: information technology capital, communication technology capital and software capital. Our estimation sample covers energy sectors in 14 countries from 2000 to 2014. The results show that information technology and software capital contribute to increasing material productivity and capital productivity in the energy sector, respectively. Meanwhile, communication technology capital negatively affects these two productivity indicators.

Keywords: information and communications technology; productivity; renewable energy; energy sector; distributed energy system

---

1. Introduction

1.1. ICT and Productivity

As a general-purpose technology, information and communications technology (ICT) can play an important role in productivity growth, which is the main driver of the wealth of nations and market competitiveness [1,2]. Investment in ICT enables new technologies to enter the production process and is viewed as a key factor in efficiency gains in ICT using industry [3,4]. The OECD [5] has noted that developments in ICT, combined with internationally fragmented production processes, are making business services increasingly dynamic, transportable and tradeable. According to Miller and Atkinson [6], approximately two-thirds of U.S. growth in total factor productivity (TFP) between 1995 and 2004 was due to ICT, and ICT has contributed roughly one-third of growth ever since. Kvochko [7] identified five common economic effects of ICT: direct job creation, contribution to GDP growth, the emergence of new services and industries, workforce transformation, and business innovation. Thus, ICT contributes to economic development through multiple pathways in various sectors.

The energy sector (e.g., electricity and gas supply) is one of the key industries investing in ICT to deliver cost savings and efficiency gains [8]. In particular, distributed energy systems have been widely diffused to efficiently utilize renewable energy generation in recent years. Figure 1 shows the investment trend related to the energy sectors. Figure 1a shows that investment in the global power sector has shifted from fossil fuel to renewable energy and networks in the past decade. Another important trend is that the ICT sector's investment in new energy technology companies has rapidly increased in recent years (see Figure 1b)). The rapid growth in investment by the ICT sector has helped diffuse distributed energy systems with renewable energy generation in networked environments such as smart grids [9].

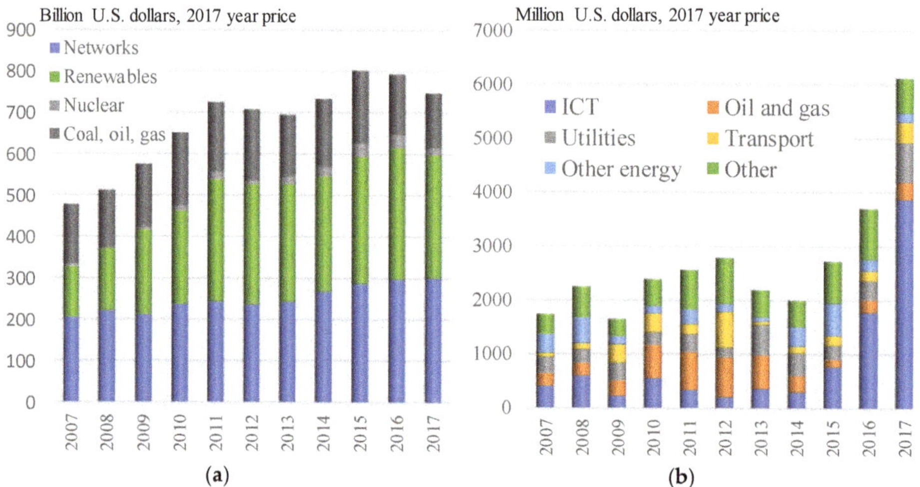

Figure 1. The investment trend related to the energy sector. (**a**) Investment by global power sector; (**b**) Corporate investment in new energy technology companies, by sector of investing company. (Source: IEA World Energy Investment 2018).

According to the Information Technology Industry Council [10], ICT can be used to improve the reliability, resiliency, and efficiency of grids' transmission, storage and distribution infrastructure through better real-time monitoring and control of the grid systems under increasingly complex energy grids. Nagai et al. [11] explained that ICT can be used in energy infrastructure in three ways: (1) as a system for cost-based analysis of operational efficiency, (2) as a support system for optimizing operation and maintenance, and (3) as a visualization tool for the management of key performance indicators and risks. They also noted that ICT has the advantage of developing an autonomous decentralized energy system, which is essential for controlling large-scale and various types of renewable energy supplies. According to the World Energy Council [12], ICT, especially software tools, can provide data and information on how to better configure the various elements of an energy generation system so as to optimize its overall performance in a cost-effective manner.

1.2. Literature Review and Novelty of This Study

Many previous studies have analyzed how ICT contributes to productivity growth [13–15]. According to Polák [16], more than 70 articles in the last 20 years have investigated the contribution of ICT. For example, Edquist and Henrekson [17] examined the effect of ICT on the change in TFP in 50 industries in Sweden from 1993 to 2013. They concluded that ICT capital growth is not significantly associated with TFP growth. Strobel [18] compared the contribution of ICT capital to TFP and its spillover effect in thirteen manufacturing industries between the U.S. and Germany from 1991 to 2005. Strobel [18] clarified that ICT has a different function in affecting productivity growth. Regarding cross-country analysis, Ceccobelli et al. [19] investigated the impact of ICT capital on labor productivity (LP) using data on 14 countries from 1995 to 2005 and a nonparametric approach to estimate productivity change.

However, most previous studies focus on national-level activities or manufacturing sectors, with few studies addressing the energy sector (e.g., the German energy sector [20]). Additionally, many previous studies cast a spotlight on the differences between ICT capital data and non-ICT capital data, and most of them use ICT gross capital stock to investigate the impact of ICT capital on productivity. ICT capital is composed of several different types of capital, including information technology (IT) capital, communication technology (CT) capital, and software capital.

To clarify the details on the relationship between ICT capital and productive performance in the energy sector, the differences in the characteristics of ICT capital must be considered. Notably, not all ICTs contribute equally to improved productive performance in the energy sector. Certain ICTs directly contribute to reducing labor costs, such as smart meters and sensors for remote measuring, whereas others contribute to improving efficiency because they improve the grid management system. Therefore, the incentives for the energy sector to invest in ICT vary depending on the type of technology considered. A determinant analysis of productive performance that focuses on the characteristics of each type of ICT is important for suggesting effective policies to encourage development and to induce activities in such technology in the energy sector.

Another important contribution of this study is that it focuses on the period from 2000 to 2014. As explained above, previous studies on the effect of ICT capital mainly focus on the period from 1990 to 2005. Meanwhile, innovative ICT utilization, such as the internet of things, has dramatically advanced in recent years [21]. Considering recent ICT innovations is important for proposing policy. Energy strategies, especially with regard to nuclear power and renewable energy diffusion, have been strongly affected by the Fukushima Daiichi nuclear disaster on 11 March 2011 [22]. Investigating the relationship between ICT capital and productivity change using a recent dataset on the energy sector can provide key information for strategy building for future energy systems and ICT investments.

Based on this background, this study investigates the effect of ICT capital stock share as a determining factor in market competitiveness using both production efficiency as a performance evaluation method and an econometric approach for the analysis of determinants. The objective of this study is to clarify the ICT capital effect on productivity in the energy sector, focusing on the characteristics of each type of ICT.

1.3. Research Framework

To focus on the characteristics and effects of ICT capital stock, this study clarifies how each aspect of ICT capital concentration affects the productive performance in the energy sector. As explained above, ICT contributes to the performance of the energy sector through various pathways [10,11]. Thus, the contribution of ICT should be investigated not only from a one-dimensional perspective but also from a multidimensional perspective using multiple indicators. To investigate this relationship, this study applies four productive performance evaluation indicators: LP, capital productivity (CP), material productivity (MP), and TFP.

Figure 2 shows the research framework for this study. The methodological approach involves two steps. First, this study evaluates the performance of the energy sector using four indicators. A performance evaluation indicator can be applied as a proxy for market competitiveness in the industrial sector, and the four indicators allow us to perform a multidimensional evaluation of productive performance in the energy sector.

Second, this study tries to explain the differences in the productive performance indicators among countries based on three multidimensional factors: (1) the ICT capital share, (2) the renewable energy share, and (3) the electricity price and research and development (R&D) capital share. The ICT capital share focuses on the concentration of IT capital, CT capital, and software capital within the gross capital stock. Next, this study investigates the effect of a distributed electricity system on the performance of the energy sector, focusing in particular on solar photovoltaic and wind power generation.

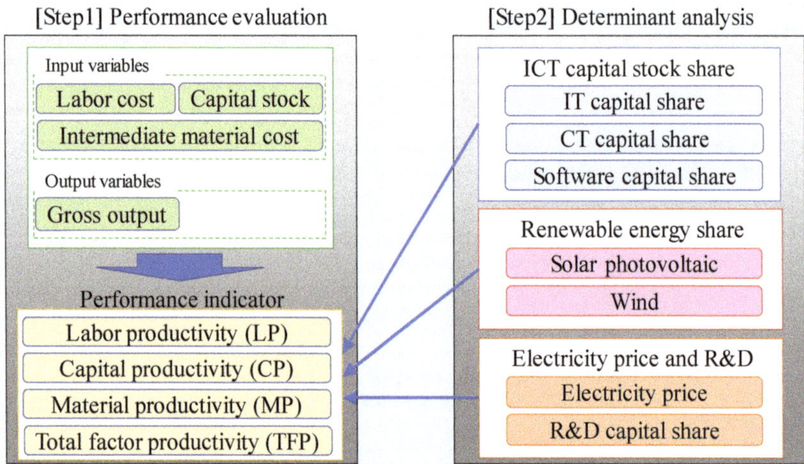

Figure 2. The research framework for this study.

Additionally, this study applies electricity price and R&D capital share as the control variables in the determinant analysis. According to the OECD [23], investment in intellectual property products, such as R&D, not only contributes to expanding the technological frontier but also enhances the ability of firms to adopt existing technologies, playing an important role in productivity performance. For this reason, the R&D share is selected as the control variable in the 2nd step of the analysis.

This paper contributes by seeking to explain productivity changes using econometric techniques, with a specific focus on the effect of ICT capital stock composition. This study investigates the effect of different types of ICT capital formations on measures of performance of the energy sector. More specifically, this study investigates the role of ICT capital in the energy sector, differentiating this type of capital according to factors such as renewable energy management systems.

2. Materials and Methods

2.1. Performance Evaluation Indicators

2.1.1. Labor Productivity (LP)

LP is defined as the desirable output (e.g., sales, production amount) per labor input (e.g., labor cost, hours worked) [24]. LP can be increased by reducing the labor input while maintaining the same amount of production or by increasing the production amount with the same labor input. In other words, LP is the inverted score of the labor input per unit of production, which represents the production of scale-adjusted labor input. In this study, LP is estimated by the gross output divided by labor compensation.

LP growth, in which ICT capital is typically used, was generally much higher and more volatile between 1995 and 2013 [4]. ICT capital contributes to productivity growth in labor-intensive industries because transaction costs, including information sharing among laborers, can be decreased due to ICT capital utilization [25]. However, the energy sector, which is a typical capital-intensive industry, has not been investigated with regard to the relationship between LP and ICT capital stock. In general, productivity in the energy sector is strongly related to energy efficiency, which is determined by the technological level of equipment. Thus, this study assumes that the contribution of ICT capital to LP growth in the energy sector is limited.

2.1.2. Capital Productivity (CP)

CP is measured as the ratio between the volume of output and the volume of capital input, defined as the flow of productive services that the capital delivers in production [24]. CP can be increased by reducing the capital input while maintaining the same amount of production or by increasing the production amount with the same capital input. Therefore, CP reflects how efficiently capital is used to produce output [24]. In this study, CP is estimated by gross output divided by capital stock.

ICT capital utilization has an important role in increasing CP in the energy sector. One reason is that renewable energy systems, especially solar photovoltaic and wind power, are widely diffused worldwide. ICT capital is an important factor in the efficient use of renewable energy generation in distributed energy systems [26]. In particular, the control system for balancing energy supply and demand requires ICT capital to minimize energy loss and to enhance capacity utilization [27]. Based on this background, this study assumes that ICT capital stock contributes to CP growth and that this contribution effect is stronger in countries that achieve a high share of distributed energy (e.g., solar photovoltaic and wind power) in their total energy supply.

2.1.3. Material Productivity (MP)

MP is defined as the desirable output per intermediate input (e.g., natural resources). MP can be increased by reducing the intermediate input while maintaining the same amount of production or by increasing the production amount with the same intermediate input. Therefore, MP reflects the efficiency of intermediate input utilization [23]. Thus, in addition to LP and CP, MP is an important indicator for evaluating the energy sector from the resource efficiency perspective.

Notably, the definition of the material is often different between the economics and engineering research fields. Baptist and Hepburn [28] explain that engineers tend to define materials to mean physical inputs (e.g., iron ore), while economists often do not differentiate between materials and other intermediate inputs because it can be difficult to distinguish raw materials. Because of the data constraints on identifying raw materials, this study evaluates MP using monetary-based intermediate input data, following the idea in the economics research field. In this study, MP is estimated by gross output divided by intermediate material cost.

This study assumes that the IT and software capital contribute to increasing MP due to the automated control of resource inputs in the production process. Additionally, sensing technology can reduce the risk of resource waste, such as leakages of electricity and gas.

2.1.4. Total Factor Productivity (TFP)

TFP is defined as the portion of output not explained by the number of inputs used in production [29]. TFP is also interpreted as a proxy for advancements in production technology [30]. TFP can be increased by reducing the input factors while maintaining the same amount of production or by increasing the production amount with the same input factors. Therefore, TFP reflects the overall production technology, which is a key factor in gaining market competitiveness.

This study measures TFP change by examining the relative productivity among the energy sectors of 42 countries using a directional distance function (DDF) model. The formula for calculating the distance function for country k can be computed using the following optimization problem:

$$\vec{D}\left(x_k^l, y_k^m, g_{x^l}, g_{y^m}\right) = \text{Maximize} \beta_k \tag{1}$$

$$\text{s.t.} \sum_{i=1}^{N} \lambda_i x_i^l \le x_k^l + \beta_k g_{x^l} \qquad l = 1, \cdots, L \tag{2}$$

$$\sum_{i=1}^{N} \lambda_i y_i^m \geq y_k^m + \beta_k g_{ym} \qquad m = 1, \cdots, M \tag{3}$$

$$\lambda_i \geq 0, \qquad (i = 1, \cdots, N) \tag{4}$$

where β_k is the production inefficiency score of country k, and i is the country name. λ_i is the weight variable used to identify the reference point on the production frontier line. l and m are the input and output variable names, respectively; x is the production input factor in the $L \times N$ input factor matrix; and y is the output in the $M \times N$ output factor matrix. In addition, g_x is the directional vector of the input factor, and g_y is the directional vector of the output factors. To estimate the production inefficiency score of all countries, a model calculation must be applied independently N times for each country.

To estimate the productivity change indicators, this study sets the directional vector $= \left(g_{xl}, g_{ym} \right) = \left(x_k^l, y_k^m \right)$. This type of directional vector assumes that an inefficient firm can decrease its productive inefficiency while increasing its desirable outputs and that it can decrease its inputs in proportion to the initial combination of actual outputs. Under this directional vector setting and the selection of data variables in Figure 2, the following equation can be obtained:

$$\vec{D}\left(x_k^l, y_k^m, g_{xl}, g_{ym} \right) = \text{Maximize} \beta_k \tag{5}$$

$$\sum_{i=1}^{N} \lambda_i x_i^l \leq (1 - \beta_k) x_k^l \qquad l = \text{Labor, Capital stock, material} \tag{6}$$

$$\sum_{i=1}^{N} \lambda_i y_i^m \geq (1 + \beta_k) y_k^m \qquad m = \text{gross output} \tag{7}$$

$$\lambda_i \geq 0 \qquad i = 1, \cdots, N \tag{8}$$

This study employs the Luenberger productivity indicator (LPI) as a TFP measure because the LPI is believed to be more robust than the widely used Malmquist indicator [31]. The LPI is computed with the results of the DDF model and is derived as follows [31,32]:

$$\text{TFP}_t^{t+1} = \frac{1}{2} \left\{ \vec{D}^{t+1}(x_t, y_t) - \vec{D}^{t+1}(x_{t+1}, y_{t+1}) + \vec{D}^{t}(x_t, y_t) - \vec{D}^{t}(x_{t+1}, y_{t+1}) \right\} \tag{9}$$

where x_t is the input for year t, x_{t+1} is the input for year $t + 1$, y_t is the desired output for year t, and y_{t+1} is the desired output for year $t + 1$. $\vec{D}^{t}(x_t, y_t)$ is the inefficiency score of year t based on the frontier curve in year t. Similarly, $\vec{D}^{t+1}(x_t, y_t)$ is the inefficiency score of year $t + 1$ based on the frontier curve in year $t + 1$.

2.2. Determinant Factor Analysis

To investigate the effect of ICT capital share on productive performance indicators in the energy sector, panel regression analysis is applied. Four performance indicators are regressed on the seven determinant factors (see Figure 2). In addition, the interaction terms of the ICT capital share and the renewable energy share to investigate the CP improvement effect. The specification for the regression is assumed to be that in Equation (2):

$$\text{Performance}_{it}^{j} = \sum_{k} \beta_1^k ICT_{it}^k + \sum_{k} \beta_2^k (ICT_{it}^k \times Renewable_{it}) + \mathbf{X}\beta + \mu_t + \delta_i + \varepsilon_{it} \tag{10}$$

The subscripts i, t, j, and k represent the country, time, type of performance indicator, and type of *ICT* capital, respectively, whereas β_1, β_2, and β are the coefficient parameters. To capture the characteristics of the energy sector in each country, the control variable vector **X** was incorporated into the models. μ_t and δ_i are unobserved time- and country-specific fixed effects, respectively. ε_{it} is an idiosyncratic error term.

2.3. Data

For the productivity analysis in the 1st step, this study uses four data variables in energy sector data from 42 countries between 2000 and 2014 (Table 1). In this study, the energy sector is defined as the electricity, gas, steam and air conditioning supply sectors following the WIOD and the United Nations Statistics Division [33].

Table 1. The description of data variables for the 1st step of estimation (productivity analysis).

Data Category	Data Variable	Unit	Mean Value	Std. dev.	Min.	Max.
Output variable	Gross output	MillionU.S. $	57,453	108,921	321	999,835
Input variable	Labor compensation	MillionU.S. $	5,921	11,841	37	77,550
	Capital stock	MillionU.S. $	145,438	313,294	844	2,050,479
	Intermediate material cost	MillionU.S. $	36,331	75,793	147	796,417
Countries (42)	Australia, Austria, Belgium, Brazil, Bulgaria, Canada, China, Croatia, Cyprus, Czech Republic, Denmark, Estonia, Finland, France, Germany, Greece, Hungary, India, Indonesia, Ireland, Italy, Japan, Korea, Latvia, Lithuania, Luxembourg, Mexico, Netherlands, Norway, Poland, Portugal, Romania, Russia, Slovakia, Slovenia, Spain, Sweden, Switzerland, Turkey, Taiwan, United Kingdom, United States					

Source: Figure created by the author using the World Input-Output Database (WIOD) [34].

The analysis includes observations on gross output, labor compensation, capital stock, and intermediate input data from the World Input-Output Database (WIOD) [34]. This study uses the following four data variables from WIOD: (1) the gross output by industry at current basic price (in millions of national currency), (2) intermediate inputs at current purchasers' price (in millions of national currency), (3) compensation of employees (in millions of national currency), and (4) nominal capital stock (in millions of national currency).

All financial data are in 2010 dollars ($ U.S.), applying the currency exchange and price deflation factors from the WIOD. Using the dataset for the 1st step of the analysis, the four productive performance indicators are calculated. It should be noted that the DDF model for TFP estimation requires a large sample size to identify the production frontier line [35]. To estimate TFP change using a large dataset, the data of 42 countries are included in the 1st step of the analysis.

For the 2nd step of the analysis, this study uses four productivity indicators estimated as dependent variables and seven data variables as independent variables (Table 2). Seven independent variable datasets are obtained from three different databases. The first database is the EU KLEMS database, which provides capital stock data by type of usage [36]. Data on 14 countries from 2000 to 2014 are obtained from the EU KLEMS database. The data variables include IT capital stock, CT capital stock, software capital stock, R&D capital stock, and gross capital stock. This study estimates each capital stock share using the gross capital stock as the denominator.

The second database is the Renewable Energy Information 2017 published by the International Energy Agency (IEA). To investigate the ICT capital effect on distributed energy systems, energy production by solar photovoltaic and wind generation are used to estimate data on the share of renewable energy. Additionally, the total data on all energy sources are used as the denominator.

Finally, the electricity price index is obtained from the Energy Price and Taxes 2018 database published by the IEA. The electricity price index is used as the proxy of the market environment in the energy sector (e.g., a feed-in tariff policy makes the electricity price increase).

This study combines two datasets: the financial dataset for productivity analysis and the dataset for the determinant analysis. Data on 14 countries are available from both datasets; thus, these data are used for the 2nd step of the analysis. Table 2 shows the average value of the data variables for the 14 countries in the determinant analysis.

Table 2. The description of data variables for the 2nd step of estimation (determinant analysis).

Data Category	Data Variable (Code)	Unit	Mean	Std. dev.	Min.	Max.
Dependent variable (Productivity indicator)	Labor productivity (LP)	$/$	10.369	3.713	4.240	21.470
	Capital productivity (CP)	$/$	0.511	0.282	0.180	1.340
	Material productivity (MP)	$/$	1.830	0.509	1.180	3.510
	Total factor productivity (TFP)	-	0.001	0.020	-0.068	0.065
Independent variable	ICT capital stock share (ICT)	%	1.744	1.155	0.400	4.920
	IT capital stock share (IT)	%	0.276	0.204	0.020	0.960
	CT capital stock share (CT)	%	0.813	1.025	0.010	4.030
	Software capital stock share (Soft)	%	0.656	0.565	0.050	3.040
	R&D capital stock share (R&D)	%	0.924	2.400	0.000	14.610
	Share of solar photovoltaic and wind power generation (Renewable)	%	1.963	3.060	0.000	15.840
	Electricity price index (Price)	-	80.364	16.743	42.400	115.200
Countries (14)	Austria, Czech Republic, Denmark, Finland, France, Germany, Italy, Luxembourg, Netherlands, Slovenia, Spain, Sweden, United Kingdom, United States					

According to Stiroh [15], the share of ICT capital stock in total capital stock is the preferred way to measure ICT capital intensity. Thus, this study estimates the share of each type of capital stock in gross capital stock. Notably, the share of each capital type of stock in gross capital stock can reflect the relative priority of the accumulation of capital stock compared with other types of capital stock, including non-ICT capital. To conduct the determinant analysis of productive performance with ICT capital shares, this study clarifies the impact of ICT capital stock on productive performance.

This research uses three types of capital stock (IT capital, CT capital, and software capital) as data on ICT capital. This categorization follows the definition of ICT investments reported by the OECD. According to the OECD [24], ICT investment is defined as the acquisition of equipment and computer software, and ICT has three components: IT equipment (e.g., computers and related hardware), CT equipment (e.g., telecommunications equipment), and software (e.g., packaged software and customized software).

Notably, the sector integration method of the EU KLEMS database is different from that of the WIOD. The EU KLEMS database provides data only on the utility sector; such data integrate data on the energy sector and on the water supply sector. Therefore, it is difficult to distinguish the ICT capital data in the energy sector from the EU KLEMS database, which is a limitation of this research. To overcome this limitation, this study assumes that the ICT capital share in the energy sector is broadly similar to that in the utility sector. This assumption is based on the fact that the capital stock data on the energy sector are much higher than those on the water supply sector based on the WIOD database. For example, in 2014, the energy sector accounted for a 92% share and an 80% share of the capital stock in the utility sector in the U.S. and Italy, respectively. This evidence supports our assumption that the trend of capital stock formation between the energy sector and the utility sector is similar.

Tables 1 and 2 describe the countries and the variables in the 1st and 2nd steps of the analysis. Because of the limited availability of data on ICT capital stock from the EU KLEMS database, the data sample was decreased from 42 countries in the 1st step of the analysis to 14 countries in the 2nd step of the analysis.

It should be noted that ICT capital utilization is just one dimension of the productive performance improvement in the energy sector; there are other ways to promote this improvement (e.g., fossil fuel combustion efficiency and distribution efficiency). One limitation of this study is that the data on R&D capital stock are limited to the total value and do not reveal the type of technology. Thus, this study assumes that technological innovation related to resource utilization (e.g., fuel combustion and distribution technology) is reflected in the R&D capital stock value. Based on this assumption, the R&D capital stock is applied as an innovation factor of resource utilization technology in the 2nd step of the analysis.

3. Results

3.1. Bivariate Analysis of Productive Performance and ICT Capital

Figures 3–6 present the relationships between the four performance indicators and the share of ICT capital stock. Each dot indicates pooled data on the energy sector in 14 countries from 2000 to 2014. The vertical axis shows the performance indicator and the horizontal axis shows the share of each type of ICT capital stock.

To compare the relationship between productive performance and ICT capital stock share among countries with different economic scales, this study divides the 14 countries into two groups. The first group comprises countries with a large economic scale. France, Germany, Italy, the U.K., and the U.S. are selected for this group. The other group comprises countries with a medium or small economic scale. Austria, the Czech Republic, Denmark, Finland, Luxembourg, the Netherlands, Slovenia, Spain, and Sweden are included in this group. These two groups are distinguished from one another by different colors in the scatter plot figure. Grey color is used to indicate the large-economic-scale group in each figure.

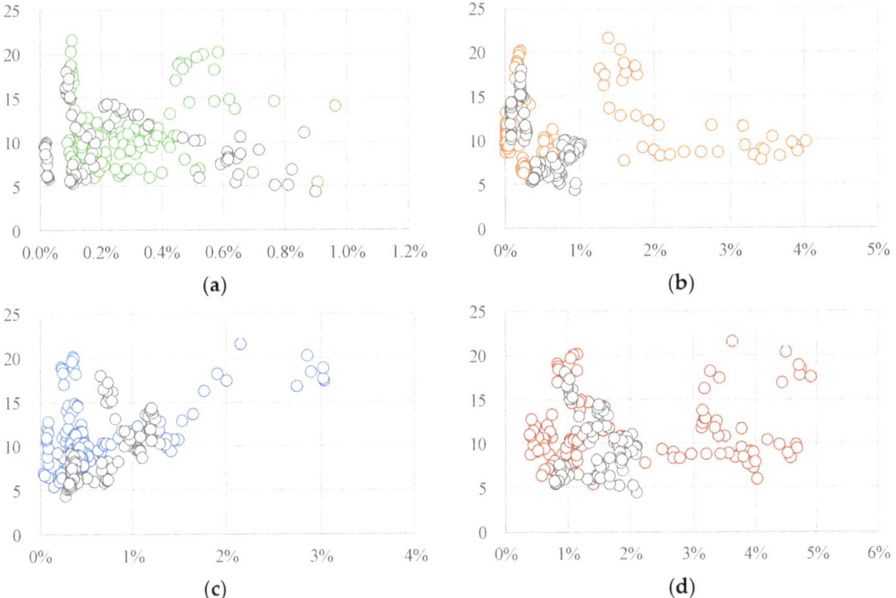

Figure 3. The scatter plot of labor productivity (LP) and information and communications technology (ICT) capital share. (**a**). Information Technology (IT) capital share; (**b**). Communication technology (CT) capital share; (**c**). Software capital share; (**d**). Share of total ICT capital. **Note:** The vertical axis shows the LP and the horizontal axis shows the capital share in gross capital stock. Grey color represents countries with a large economic scale.

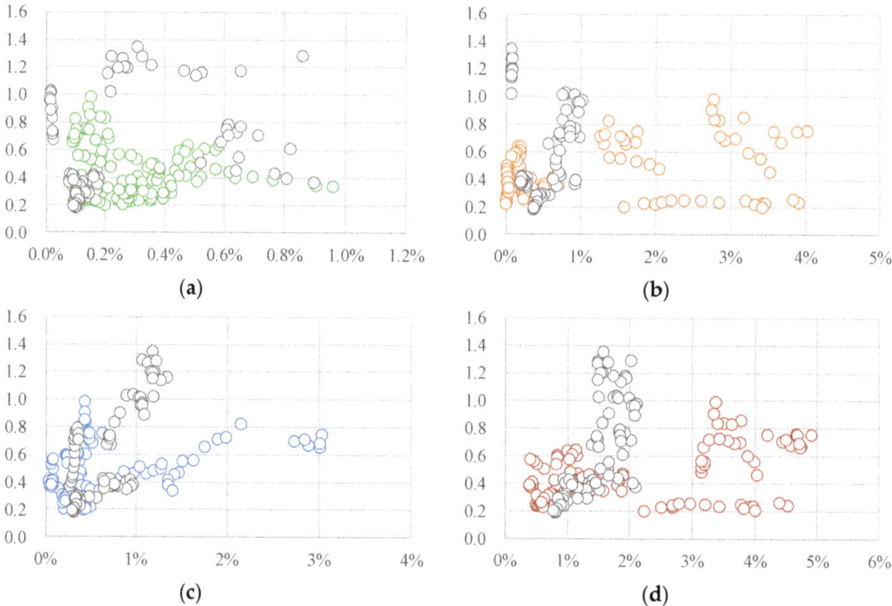

Figure 4. The scatter plot of capital productivity (CP) and ICT capital share. (**a**). IT capital share; (**b**). CT capital share; (**c**). Software capital share; (**d**). Share of total ICT capital. **Note:** The vertical axis shows the CP and the horizontal axis shows the capital share in gross capital stock. Grey color represents countries with a large economic scale.

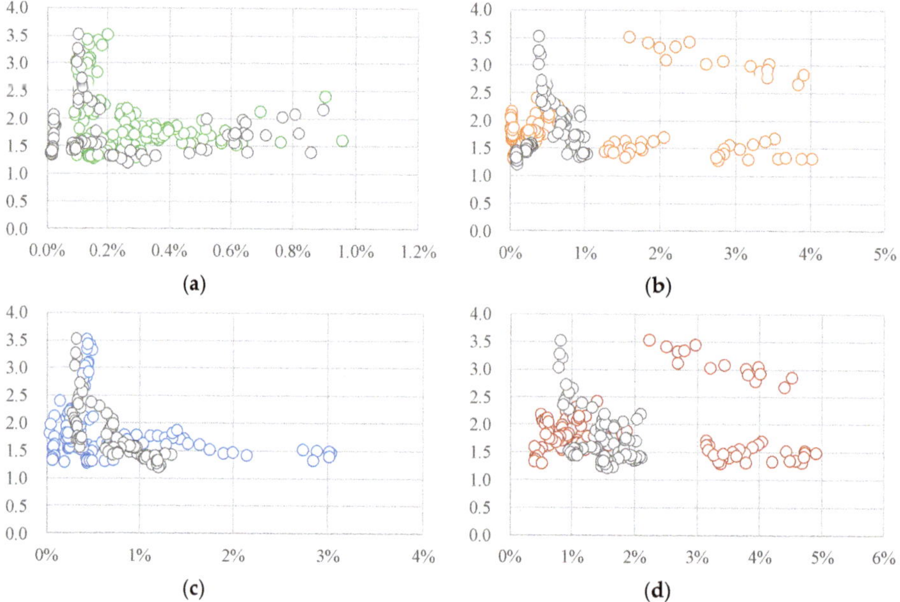

Figure 5. The scatter plot of material productivity (MP) and ICT capital share. (**a**). IT capital share; (**b**). CT capital share; (**c**). Software capital share; (**d**). Share of total ICT capital. **Note:** The vertical axis shows the MP and the horizontal axis shows the capital share in gross capital stock. Grey color represents countries with a large economic scale.

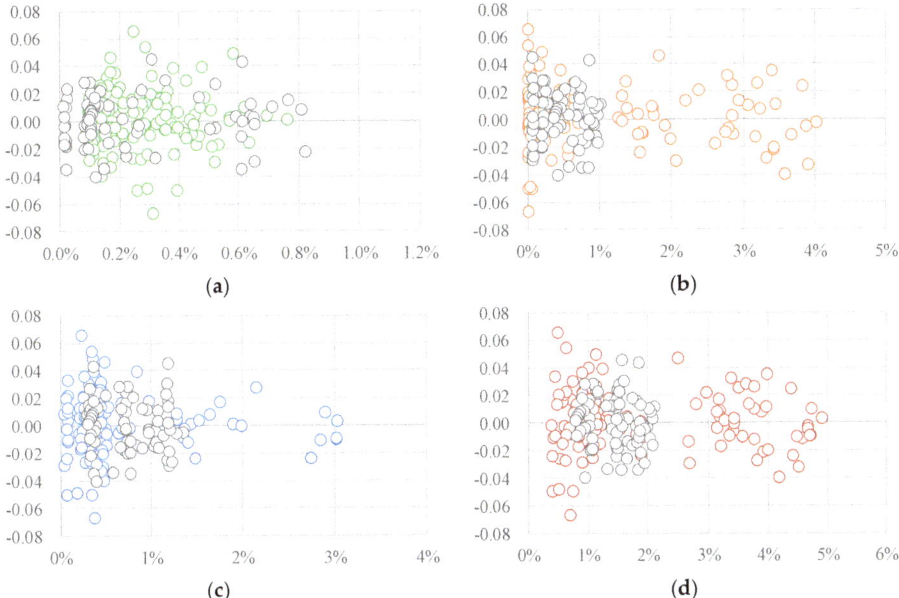

Figure 6. The scatter plot of the total factor productivity (TFP) change and ICT capital share. (**a**). IT capital share; (**b**). CT capital share; (**c**). Software capital share; (**d**). Share of total ICT capital. **Note:** The vertical axis shows the TFP change and the horizontal axis shows the capital share in gross capital stock. Grey color represents countries with a large economic scale.

Figure 3 shows that the relationship between ICT capital and LP differs based on the type of technology. Figure 3a,b imply that there are negative relationships between LP and the shares of IT and CT capital. Meanwhile, Figure 3c implies that the share of software capital has a positive relationship with LP. These relationships are similar in both economic scale groups. Finally, Figure 3d indicates an ambiguous relationship between LP and the share of total ICT capital. These results indicate the importance of using not only total ICT capital data but also specific ICT capital data.

The ambiguous relationship between LP and total ICT capital should be investigated in more detail using specific ICT capital data because there are several possible explanations for it. One possibility is that each ICT capital stock has an ambiguous relationship with LP. Another possibility is that the effect of each type of ICT capital on LP is canceled out if the ICT capital data are integrated. In the former situation, there is an ambiguous relationship between LP and ICT capital. In the latter situation, the relationship between LP and each ICT capital share should be considered carefully. Otherwise, the estimation results might lead to a misleading discussion and policy implications.In addition to LP, CP is observed to have different relationships based on each type of ICT capital share. Figure 4 shows that there is a positive relationship between CP and software capital share (see Figure 4c), even though there is an ambiguous relationship with total ICT capital share (Figure 4d). Finally, Figures 5 and 6 show the relationship of ICT capital with MP and TFP, respectively. In contrast to Figures 3 and 4, there are similar trends in the four figures in Figures 5 and 6, which show that there are diverse effects of ICT capital among the performance indicators. Additionally, there is no large difference between the economic scale groups. Based on these findings, this study further investigates the relationship between the performance indicators and ICT capital share using an econometric approach.

3.2. Determinant Analysis of the Productive Performance Indicators

Tables 3–5 present the results of the determinant analysis, focusing on the impact of ICT capital share on the productive performance indicators. Table 3 indicates the results of the determinant analysis that does not expressly consider the differences in the specific types of ICT capital. Table 4 shows the results of the determinant analysis that applies three ICT capital shares separately as determinant variables to consider the differences in specific ICT capital characteristics. In addition to the two models, this study applies the interaction term of each ICT capital share and the renewable energy share to investigate the hypothesis that the impact of ICT capital is different due to the degree of renewable energy diffusion (see Table 5).

Table 3. The results of the determinant analysis using integrated information and communications technology (ICT) capital data.

Dependent Variable	LP		CP		MP		TFP	
	Coef.	Sig	Coef.	Sig	Coef.	Sig	Coef.	Sig
ICT	−90.60	***	−0.63	-	−5.49	-	−0.01	-
R&D	−27.23	***	−2.08	***	4.30	***	−0.07	-
Price	0.04	***	0.00	**	−0.00	-	−0.00	***
Renewable	35.37	***	0.77	**	−1.89	**	−0.04	-
Constant	8.12	***	0.44	***	2.06	***	0.02	***
Observation	210		210		210		196	
Within	0.463		0.224		0.203		0.060	
Between	0.040		0.149		0.000		0.222	
Overall	0.113		0.047		0.012		0.061	
Wald chi2	166.990		51.620		48.410		12.490	
Prob > chi2	0.000		0.000		0.000		0.014	

Note: *** and ** indicate significance at the 1 and 5% levels, respectively. The random effect model is applied for all estimations.

Table 4. The results of the determinant analysis using individual ICT capital data.

Dependent Variable	LP		CP		MP		TFP	
	Coef.	Sig	Coef.	Sig	Coef.	Sig	Coef.	Sig
IT	−350.15	***	−9.56	-	71.38	***	−0.54	-
CT	−119.79	***	−4.19	**	−23.47	***	0.00	-
Software	56.01	-	9.41	***	−3.09	-	−0.12	-
R&D	−25.11	***	−1.90	***	4.55	***	−0.09	-
Price	0.04	***	0.00	***	0.00	-	−0.00	***
Renewable	22.42	***	−0.15	-	−2.40	***	−0.04	-
Constant	8.69	***	0.44	***	1.70	***	0.03	***
Observation	210		210		210		196	
Within	0.492		0.277		0.307		0.066	
Between	0.159		0.001		0.054		0.170	
Overall	0.229		0.011		0.018		0.064	
Wald chi2 / F-value	188.210		70.390		14.010		12.850	
Prob > chi2 / Prob > F	0.000		0.000		0.000		0.045	

Note: *** and ** indicate significance at the 1 and 5% levels, respectively. The fixed effect model is applied for the model with the MP.

The 2nd stage of the analysis includes the preferred specification from fixed effects or random effects based on the results of a Hausman test.

First, this study compares the impacts of specific ICT capital shares and the total ICT capital share on the productive performance indicators in Tables 3 and 4. From Table 3, a significant effect of the total ICT capital share on CP and MP is not observed. Meanwhile, Table 4 shows that CT and software capital shares significantly affect CP, with different signs. Additionally, IT and CT capital shares significantly affect MP, with different signs. These results imply that the total ICT capital share does not significantly affect CP and MP because the effects of specific ICT capital shares are canceled out. This finding can be clarified if and only if specific ICT capital shares are applied separately to consider the differences in ICT capital characteristics, which is necessary to precisely understand the impact of ICT capital.

Table 5. The results of the determinant analysis using individual ICT capital with interaction terms.

Dependent Variable	LP		CP		MP		TFP	
	Coef.	Sig	Coef.	Sig	Coef.	Sig	Coef.	Sig
IT	−245.66	*	−15.80	**	64.35	***	−0.84	-
CT	−128.26	***	−9.73	***	−22.02	***	0.09	-
Software	83.95	-	10.68	**	−27.52	***	−0.34	-
R&D	−23.35	***	−2.13	***	3.65	***	−0.09	-
Price	0.04	***	0.00	-	0.00	*	−0.00	***
Renewable	36.33	***	−1.12	*	−5.44	***	−0.16	-
IT*Renewable	−7,540.18	**	598.92	***	329.89	-	28.52	-
CT*Renewable	776.87	-	203.60	***	−81.32	-	−7.87	-
Software*Renewable	648.79	-	−101.97	***	259.97	***	9.45	-
Constant	8.02	***	0.52	***	1.87	***	0.03	***
Observation	210		210		210		196	
Within	0.497		0.404		0.347		0.077	
Between	0.261		0.001		0.010		0.168	
Overall	0.307		0.015		0.000		0.070	
Wald chi2 / F-value	180.110		14.110		11.030		14.050	
Prob > chi2 / Prob > F	0.000		0.000		0.000		0.121	

Note: ***, **, and * indicate significance at the 1, 5, and 10% levels, respectively. The fixed effect model is applied for the model with CP and MP.

4. Discussion

This study discusses the impact of specific ICT capital shares on each productive performance indicator. Table 4 shows that IT and CT capital shares negatively affect LP. According to Biagi and Falk [25], IT capital and CT capital contribute to increasing LP in labor-intensive industries due to the reduction in transaction costs (e.g., smooth communication between employees). Meanwhile, the energy sector is a typical capital-intensive industry. Thus, the differences in industrial characteristics are one interpretation of the different results from those of the previous research.

Another finding is that the software capital share contributes to increasing the CP indicator even though the CT capital share negatively affects it (Table 4). According to Nagai et al. [11] and Paiho et al. [37], the software system contributes to improving CP by optimizing control and efficient capital utilization in the energy sector. Our results are consistent with those of these previous studies.

IT capital share contributes to increasing MP, while CT capital negatively affects MP. One interpretation of the positive contribution of IT capital share to MP is the increased incineration efficiency of fossil fuels due to the optimal control of resource utilization by sensing technology.

Finally, this study discusses the results of the determinant analysis model with the interaction terms in Table 5. From Table 5, the interaction terms of renewable energy with IT and CT capital shares significantly contribute to increasing CP. This result implies that the contribution effects of IT capital and CT capital are stronger in countries that achieve a high share of distributed energy systems. These findings have been introduced in previous results as case studies (e.g., References [26,27]), and our results empirically support this relationship using a panel dataset in 14 countries.

Notably, the determinant analysis with the interaction terms provides different information from the models without the interaction terms. From Table 4, a significant effect of the IT capital share on CP is not observed. Additionally, this study observes a significantly negative effect of the CT capital share on CP. These results mislead us to believe that IT and CT capital shares do not contribute to increasing CP in the energy sector if interaction terms are not applied. In recent years, the diffusion of renewable energy systems has become increasingly important to mitigate issues associated with climate change. To evaluate the impact of ICT capital under a widely diffused distributed energy system, a research framework with an interaction term between ICT capital and the renewable energy share is important.

In other words, these results indicate that ICT capital has an important role in managing distributed energy systems to increase CP. In particular, the interaction term of renewable energy and CT capital contributes to increasing CP even though the CT capital share negatively affects the three productive performance indicators. This result implies that the CT capital contributes more if an energy system is distributed.

An interpretation of this result is that decreasing the capital-labor ratio due to renewable energy penetration contributes to improving CP. To confirm this relationship, this study estimates the correlation between the capital-labor ratio and the share of solar photovoltaic and wind power generation using data on 14 countries from 2000 to 2014. The correlation score is 0.110 (p-value = 0.1130), which implies that there is no statistically significant relationship between renewable energy penetration and the capital-labor ratio in our dataset. Therefore, this study considers that renewable energy penetration contributes to improving productivity through the synergy effect with ICT utilization but does not decrease the capital-labor ratio.

Finally, this study does not observe a significant effect of ICT capital share on TFP in any of the estimation models. One interpretation of this result is that the main driver of technological progress in the energy sector is energy efficiency, which is determined by the field of engineering technology [38]. Therefore, the contribution of ICT, which supports technology for energy system management, is limited with regard to enhancing technological progress in the energy sector.

5. Conclusions

This study investigates the impact of information and communication technology in several ways using multiple productive performance indicators and data on three types of information and communication technology capital. The main results are summarized as follows.

Total information and communication technology capital do not significantly affect capital productivity and material productivity. Meanwhile, information technology and software capital contribute to increasing material productivity and capital productivity in the energy sector, respectively. On the other hand, communication technology capital negatively affects these two indicators. These results imply that the effects of specific types of information and communication technology capital are canceled out.

Another important finding is that the interaction term of renewable energy share with the information technology and communication technology capital shares significantly contributes to improving capital productivity. Meanwhile, information technology capital and communication technology capital negatively affect capital productivity when renewable energy diffusion is not considered. This result indicates the importance of a research framework that assumes that the impacts of information and communication technology capital on productive performance differ according to the degree to which there are renewable energy systems.

The limitation of this research is the detailed data (e.g., cost and investment) on specific information and communication technology capital. Further analysis with more detailed data regarding each technology is expected to compare the cost-effectiveness of different technologies in increasing the productivity of the energy sector.

Author Contributions: Conceptualization, H.F. and A.S.; methodology, H.F.; investigation, H.F.; writing—original draft preparation, H.F.; writing—review and editing, H.F., A.S., S.K., and S.M.

Funding: This research was funded by a Grant-in-Aid for Young Scientists (B) [JP17K12858] from the Ministry of Education, Culture, Sports, Science and Technology (MEXT), Japan.

Conflicts of Interest: The authors declare no conflict of interest.

References

1. Basu, S.; Fernald, J. Information and communications technology as a general-purpose technology: Evidence from US industry data. *Ger. Econ. Rev.* **2007**, *8*, 146–173. [CrossRef]
2. Kretschmer, T. *Information and Communication Technologies and Productivity Growth: A Survey of the Literature*; OECD Digital Economy Papers, No. 195; OECD Publishing: Paris, France, 2012.
3. Bassanini, A.; Scarpetta, S. Growth, technological change, and ICT diffusion: Recent evidence from OECD countries. *Oxf. Rev. Econ. Policy* **2002**, *18*, 324–344. [CrossRef]
4. OECD. *OECD Compendium of Productivity Indicators 2015*; OECD Publishing: Paris, France, 2015.
5. OECD. *OECD Compendium of Productivity Indicators 2018*; OECD Publishing: Paris, France, 2018.
6. Miller, B.; Atkinson, R.D. *Raising European Productivity Growth through ICT*; The Information Technology and Innovation Foundation: Washington, DC, USA, 2014.
7. Kvochko, E. *Five Ways Technology Can Help the Economy*; World Economic Forum: Cologny, Switzerland, 2013.
8. International Telecommunication Union. *ICT for Energy: Telecom and Energy Working together for Sustainable Development*; Telecommunication Development Bureau: Geneva, Switzerland, 2017.
9. International Energy Agency. *World Energy Investment 2018*; IEA Publications: Paris, France, 2018.
10. Information Technology Industry Council. *Benefits of Information Communications Technology to Energy Infrastructure*; Information Technology Industry Council: Washington, DC, USA, 2014.
11. Nagai, K.; Murakami, M.; Oowada, K.; Fukumoto, T.; Sato, Y.; Matsubara, T. Energy infrastructure and ICT systems. *Hitachi Rev.* **2016**, *65*, 20–25.
12. World Energy Council. *The Role of ICT in Energy Efficiency Management*; World Energy Council: London, UK, 2018.
13. David, P.A. The dynamo and the computer: An historical perspective on the modern productivity paradox. *Am. Econ. Rev.* **1990**, *80*, 355–361.
14. Jorgenson, D.W. Information technology and the U.S. Economy. *Am. Econ. Rev.* **2001**, *91*, 1–32. [CrossRef]
15. Stiroh, K.J. Information technology and the US productivity revival: What do the industry data say? *Am. Econ. Rev.* **2002**, *92*, 1559–1576. [CrossRef]
16. Polák, P. The productivity paradox: A meta-analysis. *Inf. Econ. Policy* **2017**, *38*, 38–54. [CrossRef]
17. Edquist, H.; Henrekson, M. Do R&D and ICT affect total factor productivity growth differently? *Telecommun. Policy* **2017**, *41*, 106–119. [CrossRef]
18. Strobel, T. ICT intermediates and productivity spillovers-evidence from German and US manufacturing sectors. *Struct. Chang. Econ. Dyn.* **2016**, *37*, 147–163. [CrossRef]
19. Ceccobelli, M.; Gitto, S.; Mancuso, P. ICT capital and labour productivity growth: A non-parametric analysis of 14 OECD countries. *Telecommun. Policy* **2012**, *36*, 282–292. [CrossRef]
20. Wissner, M. ICT, growth and productivity in the German energy sector-on the way to a smart grid? *Util. Policy* **2011**, *19*, 14–19. [CrossRef]
21. Naveed, K.; Watanabe, C.; Neittaanmäki, P. The transformative direction of innovation toward an IoT-based society-increasing dependency on uncaptured GDP in global ICT firms. *Technol. Soc.* **2018**, *53*, 23–46. [CrossRef]
22. Managi, S.; Guan, D. Multiple disasters management: Lessons from the Fukushima triple events. *Econ. Anal. Policy* **2017**, *53*, 114–122. [CrossRef]
23. OECD. *Green Growth Indicators 2017*; OECD Publishing: Paris, France, 2017.
24. OECD. *OECD Digital Economy Outlook 2015*; OECD Publishing: Paris, France, 2015.
25. Biagi, F.; Falk, M. The impact of ICT and e-commerce on employment in Europe. *J. Policy Model.* **2017**, *39*, 1–18. [CrossRef]
26. Trillo-Montero, D.; Santiago, I.; Luna-Rodriguez, J.J.; Real-Calvo, R. Development of a software application to evaluate the performance and energy losses of grid-connected photovoltaic systems. *Energy Convers. Manag.* **2014**, *81*, 144–159. [CrossRef]
27. Nijhuis, M.; Gibescu, M.; Cobben, J.F.G. Assessment of the impacts of the renewable energy and ICT driven energy transition on distribution networks. *Renew. Sustain. Energy Rev.* **2015**, *52*, 1003–1014. [CrossRef]
28. Baptist, S.; Hepburn, C. Intermediate inputs and economic productivity. *Philos. Trans. A Math. Phys. Eng. Sci.* **2013**, *371*, 20110565. [CrossRef]
29. Comin, D. Total factor productivity. In *The New Palgrave Dictionary of Economics*; Macmillan, P., Ed.; Palgram Macmillan: London, UK, 2008; pp. 1806–1877.

30. Elgin, C.; Çakır, S. Technological progress and scientific indicators: A panel data analysis. *Econ. Innov. New Technol.* **2015**, *24*, 263–281. [CrossRef]

31. Chambers, R.G.; Chung, Y.; Färe, R. Profit, directional distance functions, and nerlovian efficiency. *J. Optim. Theory Appl.* **1998**, *98*, 351–364. [CrossRef]

32. Luenberger, D.G. Benefit functions and duality. *J. Math. Econ.* **1992**, *21*, 461–481. [CrossRef]

33. United Nations Statistics Division. *International Standard Industrial Classification of All Economic Activities Revision 4*; United Nations Statistics Division: New York, NY, USA, 2008.

34. Timmer, M.P.; Dietzenbacher, E.; Los, B.; Stehrer, R.; de Vries, G.J. An illustrated user guide to the world input–output database: The case of global automotive production. *Rev. Int. Econ.* **2015**, *23*, 575–605. [CrossRef]

35. Banker, R.D.; Charnes, A.; Cooper, W.W.; Swarts, J.; Thomas, D. An introduction to data envelopment analysis with some of its models and their uses. *Res. Gov. Nonprofit Account.* **1989**, *5*, 125–163.

36. Jäger, K. *EU KLEMS Growth and Productivity Accounts 2017 Release—Description of Methodology and General Notes*; The Conference Board: New York, NY, USA, 2018.

37. Paiho, S.; Saastamoinen, H.; Hakkarainen, E.; Similä, L.; Pasonen, R.; Ikäheimo, J.; Rämä, M.; Tuovinen, M.; Horsmanheimo, S. Increasing flexibility of Finnish energy systems-a review of potential technologies and means. *Sustain. Cities Soc.* **2018**, *43*, 509–523. [CrossRef]

38. Johnstone, N.; Managi, S.; Rodríguez, M.C.; Haščič, I.; Fujii, H.; Souchier, M. Environmental policy design, innovation and efficiency gains in electricity generation. *Energy Econ.* **2017**, *63*, 106–115. [CrossRef]

Article

Resource Security Strategies and Their Environmental and Economic Implications: A Case Study of Copper Production in Japan

Ran Motoori [1,*], **Benjamin McLellan** [1], **Andrew Chapman** [2] **and Tetsuo Tezuka** [1]

[1]   Graduate School of Energy Science, Kyoto University, Kyoto 6068501, Japan
[2]   International Institute for Carbon Neutral Energy Research, Kyushu University, Fukuoka 8190395, Japan
*   Correspondence: motoori.ran.45m@st.kyoto-u.ac.jp

Received: 25 June 2019; Accepted: 3 August 2019; Published: 6 August 2019

Abstract: Japan is a nation which is highly dependent on the import of raw materials to supply its manufacturing industry, notable among them copper. When extracting copper from ore, a large amount of energy is required, typically leading to high levels of CO_2 emissions due to the fossil fuel-dominated energy mix. Moreover, maintaining security of raw material supply is difficult if imports are the only source utilized. This study examines the environmental and economic impacts of domestic mineral production from the recycling of end-of-life products and deep ocean mining as strategies to reduce CO_2 emissions and enhance security of raw material supplies. The results indicate that under the given assumptions, recycling, which is typically considered to be less CO_2 intensive, produces higher domestic emissions than current copper processing, although across the whole supply chain shows promise. As the total quantity of domestic resources from deep ocean ores are much smaller than the potential from recycling, it is possible that recycling could become a mainstream supply alternative, while deep ocean mining is more likely to be a niche supply source. Implications of a progressively aging society and flow-on impacts for the recycling sector are discussed.

Keywords: resource security; domestic mineral production; input-output analysis; environmental assessment; transition

1. Introduction

As nations undergo a low-carbon energy transition, including large-scale electrification and a shift toward a more efficient transportation system, the need for mineral raw materials (ore and concentrates), some of which are critical, are expected to increase [1,2]. Japan is a nation which is highly dependent on the import of such raw materials for industry, and the need to conserve limited resources and maintaining resource security are considered important to the Japanese public [3]. Japan is completely dependent on imports of raw materials such as for ferrous and non-ferrous metal ores and concentrates (e.g., iron, copper, lead and zinc), which makes it vulnerable to potential global supply disruptions. Such disruptions may occur for various political, economic and environmental reasons. Recent discourse around resource security has been focused on critical materials, essential for industry, civil life and military applications for which alternative materials are not available and which use minerals whose probability of supply restriction is elevated or, in some cases, whose environmental implications of supply are high [4]. "Critical materials" can be broadly defined as relating to materials that fulfil an important (or vital) role in society and for which there is a high possibility of supply restriction, which could lead to higher prices (economic scarcity) or physical unavailability (physical scarcity). A variety of factors are typically included in the evaluation of "criticality", which includes factors of economic importance or reliance (vulnerability to supply restriction, including the number of people using, the importance to economic sectors and the potential for substitution) and supply risk (factors that affect this might include the level of monopolization of supply chains, environmental intensity of production and governance in producing countries). Taking copper for example, it is vital to some applications such as electric wire but can be substituted with aluminum with some loss of performance; at the same time, the global production of copper ore is quite diversified, giving a potentially lower supply risk. Due to increasing demand and potential for lagging supply, ore grade decline in terrestrial mines and potential subsequent cost increases per ton of metal, it may face both economic and physical scarcity.

The specific definition of a critical material differs from nation to nation, depending on the applied evaluation criteria although materials such as rare earth elements (REE) and Platinum group metals (PGM) are common among materials considered as critical in most nations. In Japan, as a country that is almost entirely dependent on imports of raw materials, apart from the concept of "criticality", other metals are considered "strategic" due to their economic importance (regardless of the likelihood of global supply disruptions which is included in criticality), including copper, lead, nickel and zinc [5–7]. This reasoning underpins their consideration as strategic materials in Japan.

Resource importing countries have severe limits on their ability to exert control over political, economic or environmental issues which are occurring in resource exporting countries. Cognizant of this limitation and, as metals play an important role in achieving a low-carbon society [8], many import-reliant countries have seen the need to develop a resource acquisition or resource security strategy. Application of these strategic metals to renewable energy includes, for example, photovoltaic panel requiring copper, indium, gallium and selenium and wind turbines requiring nickel, molybdenum, neodymium and boron as functional materials [8]. As renewable energy installation increases, battery demand will also likely increase. Lithium-ion batteries are one key storage option in renewable energy societies, and these also include potentially critical minerals such as cobalt and lithium [8]. In order to improve low levels of raw-material self-sufficiency, each country needs to decide their own approach. For example, the USA drew up the Critical Materials Strategy [7], the EU developed the Critical raw materials for the EU [5], while the Japanese government outlined their strategy in the Strategic Energy Plan [9]. This plan includes a raft of measures, such as, supply diversification, recycling of end-of-life products and research and development for extraction from unestablished or alternative materials [6]. Of these measures, all are dependent on other nations, except for research and development and domestic recycling of end-of-life products. In order to ameliorate this issue, domestic mineral production needs to be considered alongside domestic recycling in order to achieve resource security, critical to the low-carbon energy transition. The circular economy is one tightly-linked concept to maximize resource efficiency and minimize waste production, within the context of sustainable economic and social development [10]. Promoting recycling is one of the key components of the circular economy, as well as a key strategy for critical material resource security.

This research focuses on the effects of domestic copper production and recycling in Japan. Copper is selected from among the group of critical materials as it has many applications and demand is expected to increase due to expanded renewable energy deployment such as CIGS photovoltaic panel and CIS (copper (C), indium (I) and selenium (S)) photovoltaic panel [11]. Recently, the United States published a federal strategy, which focuses on the improvement of critical mineral supply by: Identifying new sources of critical minerals, enhancing activity at all levels of the supply chain, seeking to stimulate private sector investment and growth of domestic downstream value-added processing and manufacturing, ensuring that miners, producers, and land managers have access to the most advanced mapping data; and outlining a path to streamline leasing and permitting processes in a safe and environmentally responsible manner [12]. Resource acquisition options for an import-reliant country include (1) on-land mining, (2) imports of raw materials, (3) recycling, and (4) unconventional resource exploration. In the context of Japan, because of its lack of on-land mines, option (1) is unlikely. In terms of improving resource security, option (2) does not relieve the vulnerability to export restrictions by exporting countries. Lacking known deposits on land or near shore, development of domestic primary production is focused on ore production by means of deep ocean seafloor massive sulfide (SMS) mining. Thus, recycling and deep sea mining (which is a subset of option (4)) are the options to be discussed in this study.

This study aims to assess the constraints of energy, material, economic and labor under domestic mineral production.

2. Background and Literature Review

In recent history, mineral resource security policy in Japan has been highly dependent on the acquisition of mining rights in resource exporting countries [13]. For example, Japan Oil, Gas and Metals National Corporation (JOGMEC), the organization in charge of contributing to a stable supply of petroleum, natural gas and mineral products, undertakes joint exploration for copper and gold in Australia. Through this project JOGMEC expects to contribute toward resource security in Japan [14]. Although Japan's defined self-sufficiency rate reflects the acquisition of mining rights in resource exporting countries (so that the apparent self-sufficiency rate of base metals in Japan is about 50%) [9], for most metals there is no domestic mining, and there is no policy promoting the extraction of minerals in Japan.

With regards to copper in Japan, 100% of mineral concentrate is imported, and the annual production of copper in Japan in 2017 was 1488 kt, split between the end uses of wire (64%), brass (35%), and casting (1%) [15]. The recycling rate for 2017 was reported as 32% [15]. The recycling rate reflects the percentage of copper produced utilizing copper scrap. In-process scrap, which is scrap generated during wire or brass production, is utilized at the rate of almost 100% and the wire and brass industry also utilizes end-of-life product scrap, which is generated from waste copper products, for production [16]. Scrap produced in Japan was distributed to the wire, brass, copper smelting and export industries [16]. Wire and brass industries use scrap so that they can reduce material input costs [16]. The smelting industry uses scrap to productively utilize the excess heat created by the converting reaction in converter furnaces [17]. Copper smelting and export industries currently process lower grade scrap, whereas high grade scrap was recycled in the wire and brass sectors [16]. China has consistently been a major importer of Japan's scrap—particularly scrap which required labor inputs in order to be economically recycled (such as plastic-coated cables). However, scrap imports, including not only copper scrap but also various kinds of metal scrap, slag and plastic imports were banned by China in 2018, in order to protect the environment [18]. This outcome means that Japan will need to process lower grade scrap domestically or find an alternative processing destination.

In the EU, following China's scrap import ban, waste paper and plastics which were anticipated to be shipped to China were shifted to other countries such as Vietnam, Thailand and India [19]. Copper scrap export data after China's policy change is not yet available, however it is expected that overflows of copper scrap will be sent to third parties.

Lack of an export market for scrap, with a shift to processing within Japan could potentially create new industry activities such as end-of-life product collection.

Considering the dual issues of resource security and environmental protection in both scrap producing and importing nations, recycling more copper within Japan could help ameliorate resource security and environmental issues.

One issue that arises from a strategy of greater domestic recycling is the downgrading of copper scrap due to impurities. Currently, the wire and brass industries both seek to prevent impurities and maintain a high-quality product, cognizant of the fact that no dedicated recycled copper refineries exist in Japan. According to the Metal Economic Research Institute Japan (MERIJ), two-thirds of brass makers maintain their own standards for scrap metals, 80% of which are stricter than the current Japanese Industrial Standard (JIS) [16]. MERIJ has also identified the issue of the increasing difficulty faced in securing employees for the typical techniques required to ensure high-quality copper scrap that are highly dependent on labor [16].

In addition to scrap recycling and direct imports, an alternative supply source for raw materials in Japan is provided by extraction of unconventional ores which lie on the seafloor using deep ocean mining. Deep ocean resources with Japan's exclusive economic zone (EEZ) such as SMS, manganese nodules and rare earth rich deep-sea muds have been widely studied [20]. The key reason for this is Japan's lack of commercial terrestrial domestic mining, which makes importing the only way for Japan to supply its mineral concentrate needs. Under this constraint, the current supply system is vulnerable to supply disruptions from external parties and this may engender significant economic damage to the manufacturing industry if key mineral inputs are affected. Although modern-day deep ocean mining technologies and the current deposit levels are not yet economically competitive with land-based mining, utilizing these resources in the future may ease the potential for untoward impacts should supply disruptions occur. The strength of deep ocean mining is its lower waste footprint. Land-based mining requires land clearing and, removal of overburden before extraction of the valuable ore. On the other hand, deep ocean mining does not need to remove much overburden for extraction. When deep ocean resources are extracted, waste rock is expected to be significantly lower than terrestrial mining [21]. Of course, there are many uncertainties around deep ocean mining, across all fronts—technical, environmental, economic and social. Notably, most studies consider the biodiversity and direct ecosystem impacts of deep ocean mining, which are difficult to compare adequately with land-based mining because of the lack of environmental baselines as well as the lack of commercial deep ocean mines as precedents. Thus, these aspects—which may be negative for deep ocean mining—are left out of the present examination.

One of Japan's well-studied potential deep ocean ore bodies lies approximately 110 km offshore from Okinawa Island [22]. It is reported that this deposit has 0.4% of copper (3000 kt), 1.4% of lead (10,000 kt), 5.8% of zinc (43,000 kt), 1.5g/t of gold (11 t), and 95.6 g/t of silver (1 kt) [23]. Although the total quantity of copper in this deposit is not large considering Japan's annual copper demand is approximately 1500 kt [15], combined production with other metals such as zinc, lead, gold and silver could contribute positively toward overall national resource security.

This study aims to contribute to the debate surrounding resource security for nations reliant on raw material imports, and to specifically investigate recycling approaches, deep ocean mining, and a combination of the two. This approach is applicable to not only copper but also other non-ferrous minerals. Copper has been chosen as a case study because of following reasons. Lead, which is used in applications such as automobile batteries or large-scale batteries, is already recycled effectively, giving only marginal potential for improvement. On the other hand, zinc is typically used in alloys, or for additives in galvanization or paints or as sacrificial anodes, which can be considered dissipative uses. Generally speaking, it is difficult to recover zinc metal by itself, but it is recovered in the steel recycling process. Unlike these metals, copper is used as the main material in products, in relatively pure form, and there is still room for improvement in recycling rates. Copper is therefore chosen as a case study, although this approach could be applied to these and other metals that are found in unconventional ore deposits.

3. Methodology

This study focuses on domestic mineral production in Japan as primary ore via deep ocean mining or as secondary material from recycling. In order to elucidate the energy, material, economic and labor flows, input-output analysis and material flow analysis will be employed in this study. The system boundary considered in this study is shown in Figure 1. Copper production is dependent on (A) the import of raw material, and domestic mineral production including: (B) Deep ocean mining and (C) end-of-life product recycling.

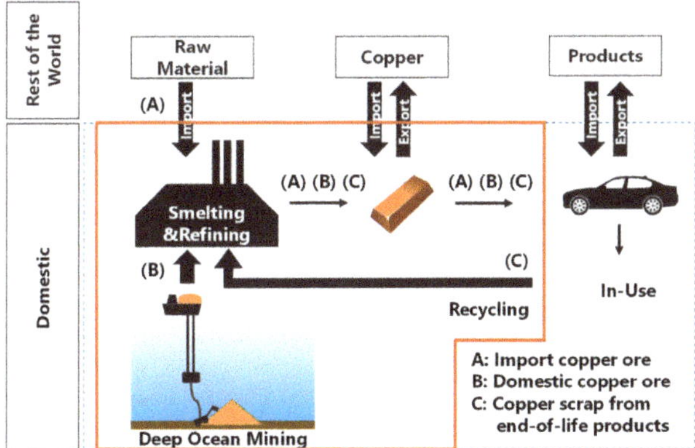

Figure 1. Copper material flow of domestic copper production system (bold box indicates the system boundary).

Detailed processes are shown in Figure 2. Due to the fact that both imported ores and domestic ores are sulfides, it is possible to utilize the same smelting facilities. As each production process will produce refined copper, the focus of this study, there are considered to be no quality issues that could prevent certain sectors from utilizing copper. The scope of this study is specific to the copper production process, while further processing for manufacturing electrical wire, brass or automobiles is not considered.

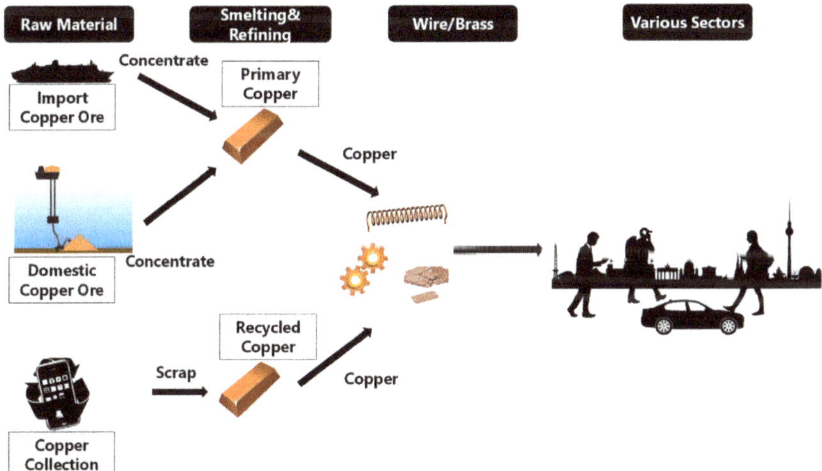

Figure 2. Material flows between sectors related to copper production in this study (boxes indicate sectors in the I-O table).

3.1. Input-Output Analysis

This study analyzes domestic mineral production in Japan utilizing the Japanese Input-Output (I-O) Tables from 2011, the latest version available at the time of writing, in which copper is differentiated from other minerals [24]. The I-O table is a statistical table showing inter-industry transactions of goods and services conducted in the domestic economy for a certain period (usually one year) in matrix form. This table has been prepared for the purpose of understanding the economic structure of a single region (country, prefecture or city level, although international I-O tables also exist). It is also possible to analyze economic ripple (multiplier) effects by using this table. Economic ripple effect can be defined as new economic production which is triggered to meet new demand. Currently, there are no industries dedicated specifically to deep ocean mining, secondary copper smelting/refining and end-of-life product collection in Japan. Deep ocean resource exploration requires mining and concentration processes which do not currently exist in Japan. Within the I-O table, the recycling sector is reflected, however this is not specific to copper recycling. Should domestic mineral production become mainstream, such sectors would likely emerge in the I-O table. However, it is impossible to investigate these sectors when employing the I-O table in its current form, and for this reason the Japanese 2011 I-O table has been extended in order to add these emerging sectors.

Although I-O analysis is widely used, we will present a brief description of the methods here. Figure 3 shows a conceptualized I-O table. Each row of the I-O table indicates the output of inter-industry transactions, while the columns of the I-O table show the inputs to each industry on a monetary basis. Intermediate demand sectors, which are sectors producing each good or service, perform production activities through the purchase of raw materials, services or energy and applying capital and labor. The final demand sector, which covers consumption, exports and imports, is mainly a buyer of consumer and capital goods as finished products. The intermediate input sector is a supplier of goods and services as intermediate goods. Each supply industry supplies goods and services to demand industries. The gross value added sector consists of a factor cost for production (such as capital and labor). Essentially, the subtotal of final demand and the subtotal of gross added value are balanced. The I-O table is balanced by Equations (1) and (2).

$$[x + F + Ex - Im = Y] \tag{1}$$

$$[x + V = Y] \tag{2}$$

x: Intermediate input, F: Domestic final demand, Ex: Export, Im: Import, Y: Domestic production, V: Value added.

Demand side		Intermediate demand				Final demand			Domestic production
		Industry 1	Industry 2	...	Industry n	Domestic final demand	Export	Import	
Intermediate input	Industry 1	$x_{1,1}$	$x_{1,2}$	...	$x_{1,n}$	F_1	Ex_1	Im_1	Y_1
	Industry 2	$x_{2,1}$	$x_{2,2}$	...	$x_{2,n}$	F_2	Ex_2	Im_2	Y_2
	...	...	...	...	...	...	...	...	...
	Industry n	$x_{n,1}$	$x_{n,2}$	...	$x_{n,n}$	F_n	Ex_n	Im_n	Y_n
Gross value added		V_1	V_2	...	V_n				
Domestic production		Y_1	Y_2	...	Y_n				

(Supply site label appears at the lower-left of the "Demand side" header cell.)

Figure 3. General conceptualized input-output table.

The input coefficient matrix, which is explained by Equation (3) and also known as the technology coefficient matrix, represents the raw material inputs required to produce 1 unit of the desired product. By utilizing the inverse matrix of the input coefficient matrix, the economic ripple effect can be estimated using Equation (4). The inverse matrix in Equation (4), $(I - A)^{-1}$, is known as the Leontief inverse matrix.

$$a_{ij} = x_{ij} / Y_j \tag{3}$$

$$\text{Economic ripple effect} = (I - A)^{-1} \Delta \tag{4}$$

$a_{i,j}$: input coefficient, $x_{i,j}$: intermediate input in sector ij, Y_j: Domestic production in sector j, A: Input coefficient matrix, I: Identity matrix.

As noted above, since the Japanese I-O table combines various mineral ore industries into one sector, copper ore should be separated from this sector to consider domestic mineral production. Also, the recycling sector does not clearly delineate copper flows. Thus, this study modifies the conventional I-O table to suit to domestic mineral production by using the procedures detailed below.

3.1.1. Copper Ore

The copper ore sector, which was separated from the mineral sector, is then separated into two sectors; imported copper ore and domestic copper ore. Both sectors only produce ore for the primary copper sector. The input structure (which is the materials and services for producing copper expressed in the columns of the I-O table), of domestic copper ore is approximated using the Chilean copper production structure [25].

3.1.2. Copper Collection

The end-of-life copper collection sector was added to the extended I-O table. This sector collects end-of-life copper products such as electronic home appliances or metal containing products, disassembles and sorts them and produces materials to be recycled by copper smelters, the ferrous sector and plastic sector. Note that the I-O table only allows a single output from a single sector, so, it is assumed that this sector produces raw materials only for the secondary copper sector. The input structure of this sector is approximated using life cycle assessment (LCA) data from a previous study [26].

3.1.3. Copper Recycling

When large amounts of scrap become the main raw material for copper production, primary copper smelters cannot be utilized due to technical limitations. This limitation necessitates the use of a furnace for recycling, which can process scrap materials [17]. To highlight the flow of recycled copper, this sector is newly added to the extended I-O table. The output structure of this sector is identical to primary copper smelting.

End-of-life products are collected and sorted into individual grades and provided to suitable sectors for modern day copper recycling. This means that copper scrap collected from end-users is not always recycled as copper and could be used for wire or brass production. This type of recycling, often called cascading may cause quality issues due to impurities. To prevent downgrading, the recycled copper sector and copper collection sectors are independent of each other, which means that collected copper scrap considered in this study is processed only to high-quality copper. Table 1 highlights the sectors and options considered in the extended I-O table.

Table 1. Sectors added to the extended I-O table.

Added Sector	Code	Notes
Import copper ore	CONV	Import copper concentrate; does not include concentrating processes in Japan.
Domestic copper ore	DOM	From mining to concentration. Does not include the distance from concentration site to smelter.
Recycled copper	REC	Separated from conventional primary copper processing. Does not use flash smelting.
Copper collection	REC	Collecting, disassembling and sorting of end-of-life products. Distance from collecting point to sorting point is included.

3.2. Assumptions

In addition to the augmentation of the I-O table, the following assumptions are made for this study.

3.2.1. Copper Production

This study assumes that copper production in any scenario is set at 1500 kt, equal to copper production at 2017 in Japan [15]. The gross domestic product (GDP), which is the total of the value added by all sectors, is also set to the same value as the 2011 I-O table in order to highlight the difference in final demand distribution among sectors, rather than overall GDP change.

3.2.2. End-of-Life Product Recycling

We assume that the copper collecting sector recovers end-of-life products from consumers, dismantles, sorts and separates them, and provides secondary materials for secondary copper production, iron and steel and plastic sectors. Here, it is assumed that there is a collection point of end-of-life products in the city, requiring consumers to bring end-of-life products for collectors to gather them. The "recovery" of the end-of-life copper collecting sector is transported from these collecting points to the copper recycling smelters [26].

As recycling progresses, the value of production in the recycled copper production sector and the copper recovery sector will increase and the intermediate input and import value of the imported copper ore sector will decrease. The copper recovery sector is assumed to be an industry that requires labor. Wages for this labor were substituted using values from the similarly structured waste treatment sector [27].

3.2.3. Extended I-O Table

Imported copper ore, domestic copper ore, primary copper, recycled copper, and the copper collecting sector were newly added to the 2011 I-O table. Copper production and GDP are assumed to remain constant at 2011 levels. When estimating final demand and domestic production for the progressing recycle rate, the input factor and ratio of the final demand sector to domestic production for each industry are used.

It was also assumed that the wage rates for estimating the labor force costs are the same as in the 2011 I-O table. In addition, wages are set in accordance with the employment table of the I-O Table.

This study applies both a bottom-up and top-down approach to estimating the input coefficients for extended industries. Figure 4 details the methodology used to extend the I-O table for this study. For example, for deep ocean mining (a non-existent industry in Japan) the input coefficient is substituted for that of land-based mining in Chile (a top-down estimate). On the other hand, the input coefficients for industries related to recycling are estimated through a bottom-up estimate approach, using LCA data [26]. LCA data is mass-based, and for the purposes of this study, these data were converted to a monetary basis utilizing the value and quantity table attached to the I-O table.

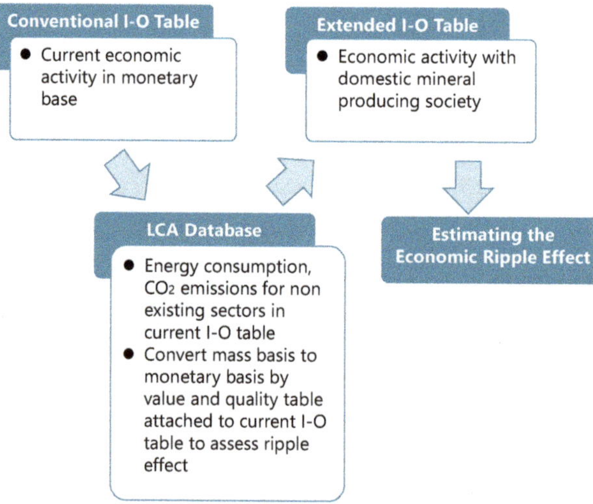

Figure 4. Methodology to extend the I-O table.

3.2.4. SMS Ore Production

As SMS deposits are polymetallic, deep ocean mining is not specifically targeted toward copper recovery and will result in multiple minerals being recovered—in fact, in many cases copper is more likely to be a coproduct. For example, the ore body considered here consists of copper, lead, zinc, gold and silver. The ore grade has been estimated at 0.4% (copper), 1.4% (lead) and 5.8% (zinc) [22]. To estimate environmental impact factors of SMS ore production, allocation by weight is usually adopted. Previous research revealed that multiple mineral production from SMS ore reduces energy consumption, CO_2 emissions and waste disposal amounts (due to their allocations to each material extracted) when compared to the production of copper ore alone [10]. This study applies the assumption that deep ocean mining will produce various minerals in addition to the target metal of copper for the estimation of environmental impacts.

According to current estimates, SMS reserves of ore are not sufficient to cover Japan's production [12], however there are large uncertainties as to the available resources, so this study assumes for the purpose of analysis that there is no limitation on this resource and that SMS ore can meet the demand.

3.2.5. Energy Consumption and CO_2 Emissions

The recycling of end-of-life products is considered to reduce energy consumption and CO_2 emissions when compared with processing ore, since recycling does not require an oxidization process. Additionally, since it is possible to utilize waste plastics as a reducing agent and energy source, we estimate that utilizing end-of-life products will assist in reducing CO_2.

Deep ocean mining is also able to reduce energy consumption and CO_2 emissions due to shorter material transportation distances. Mining, concentrating and transportation are invisible processes for resource importing countries since these processes are conducted only in resource exporting nations. Domestic mineral production by Japan may also have the flow on effect of reducing CO_2 emissions in resource exporting countries, although this is not considered in this study. To estimate Japanese CO_2 emission impacts, existing LCA data are used [26].

3.2.6. Labor

The I-O table contains supplemental information regarding labor and employment. An example portion of the employment table is shown in Figure 5. The inducement of employment by demand increase is estimated utilizing this data, focusing on all employed persons, including the employer as well as self-employed persons and family workers.

	Sector 1	Sector 2	Sector 3	$\cdots$	Sector n	
Employee	l_{11}	l_{12}	l_{13}	$\cdots$	l_{1n}	
Self-employed	l_{21}	l_{22}	l_{23}	$\cdots$	l_{2n}	Employment Table
Family worker	l_{31}	l_{32}	l_{33}	$\cdots$	l_{3n}	
$\vdots$	$\vdots$	$\vdots$	$\vdots$		$\vdots$	
Domestic production	X_1	X_2	X_3	$\cdots$	X_n	

Figure 5. Employment table.

The employment factor is estimated using Equation (5). As shown in Equation (6), by transforming employment factors to a diagonal matrix and by multiplying the domestic production of each sector, employment inducement due to 1 unit of demand increase is estimated. The scale of the inducement can be estimated by dividing the average of the sum of all industries.

$$p_j = L_j/X_j \tag{5}$$

p_j: employment coefficient L_j: employed person in sector j X_j: Domestic production in sector j.

$$Employment\ Inducement = \hat{L}\cdot(I - A)^{-1} \tag{6}$$

$\hat{L}$: *diagonal matrix of* p_j.

Annual income per person can also be estimated by the employment table, as the table includes average incomes along with the number of employed persons. Since copper collection does not exist as an activity within the current I-O table, incomes from the similar waste collection industry are substituted here.

4. Results

The results are expressed in three parts beginning with I-O analysis, final energy consumption and CO_2 emissions, followed by labor impacts.

4.1. I-O Analysis

I-O analysis reveals industries that will be affected by domestic mineral production. This study considers two extremes; 100% SMS production and 100% recycling. I-O analysis reveals that in a recycling society, 1 unit increase of final demand of recycled copper engenders 1.43 overall units. Table 2 shows the top five industries that will be affected.

Table 2. Economic ripple effect in recycled copper sector.

Industry	Ripple Effect
Recycled Copper	1.00
Copper Collection	0.26
Transportation	0.06
Service Industry	0.03
Electricity	0.02

On the other hand, an increase in deep ocean mining on final demand engenders a 1.75 unit increase, with the top 5 industries shown at Table 3.

Table 3. Economic ripple effect in domestic copper ore sector.

Industry	Ripple Effect
Domestic Copper Ore (Deep Ocean Mining)	1.14
Service Industry	0.19
Electricity	0.09
Mining (minerals and quarrying except copper ore)	0.09
Transportation	0.06

The estimated ripple effect in either case is smaller than for other industries, since the average ripple effect of all industries is 2.40. However, observing these impacts, deep ocean mining will relatively increase the output (on a monetary basis) of other industries more than copper recycling alone, as demand for copper increases.

4.2. Final Energy Consumption and CO_2 Emissions

Final energy consumption is estimated based on I-O and material flow analysis. In each process, multiple minerals are produced and this study allocates energy consumption and CO_2 emissions by the ratio of mass of final products.

Figure 6 indicates domestic energy consumption for the copper recycling scenario. A recycling rate of up to 10% only requires conventional copper smelting due to the relatively low capacity. For rates of 20% and above both conventional and recycled copper processing are required. As the recycling rate increases, the amount of energy required for transportation (collecting end-of-life products) increases rapidly.

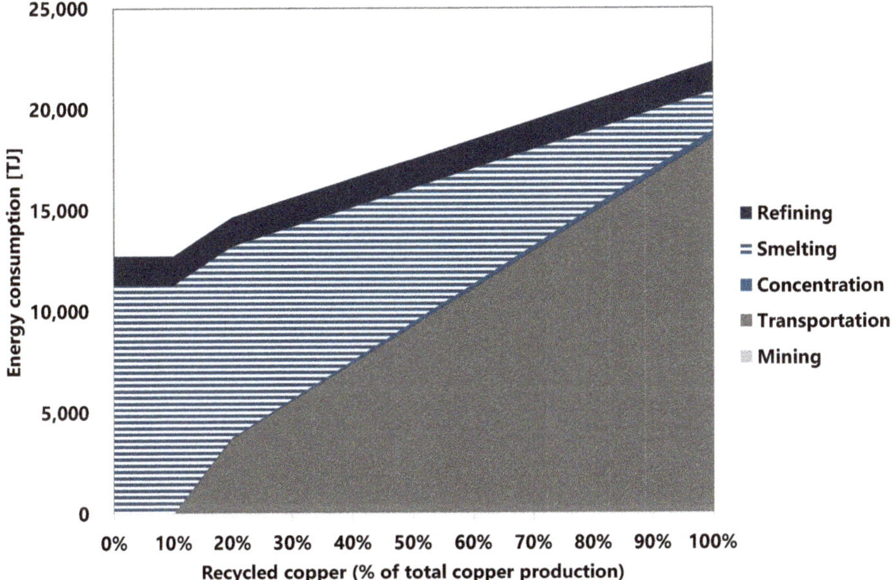

Figure 6. Final energy consumption in a recycling society.

Figure 7 shows energy consumption required for deep ocean mining. Although the reserves of deep ocean ore are very small relative to the total copper demand in Japan, this limitation is deliberately ignored here. Basically, the deep ocean mining scenario can be considered to be using conventional smelting and refining processes with added domestic mining and ore concentration. Thus, the energy required for mining and concentration processes increases as the supply provided by deep ocean ores increases.

According to this result, domestic mineral production will increase overall energy consumption. In the case of recycling, energy for smelting succeeds in reducing energy consumption due to the utilization of waste plastics and the absence of an oxidization process. However, transportation of end-of-life products consumes the most energy across all processes. In the case of deep ocean mining, mining and concentration processes are added to the current copper production system. The mining processes require more energy than concentration processes. Note that the current import-based system estimate does not consider mining, transportation and concentration processes because these are all currently conducted abroad. As our estimates only encompass the domestic impacts, actual total energy consumption will be greater than our estimate due to international energy consumption.

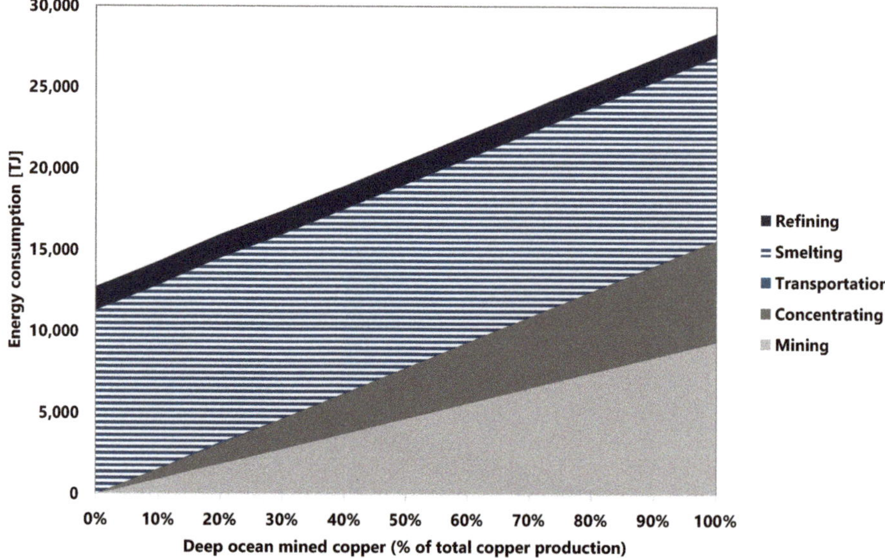

Figure 7. Final energy consumption for deep ocean mining.

Contrarily, CO_2 emissions in domestic mineral production may be less than for conventional copper processing. Figure 8 shows CO_2 emissions, incorporating those from overseas. It can be seen that deep ocean mining will contribute additional CO_2 emissions when compared with current copper production, however, considering the available reserves it is likely not a feasible method to supply Japan's copper ore needs solely from SMS deposits. It is more likely that domestic mineral production will utilize recycling, leading to a reduction in CO_2 emissions from copper production. Note that since environmental impact allocation is based on the mass balance, CO_2 emissions in mining and concentration of deep ocean mining process are smaller than those in conventional processes. A value-based allocation may change these outcomes.

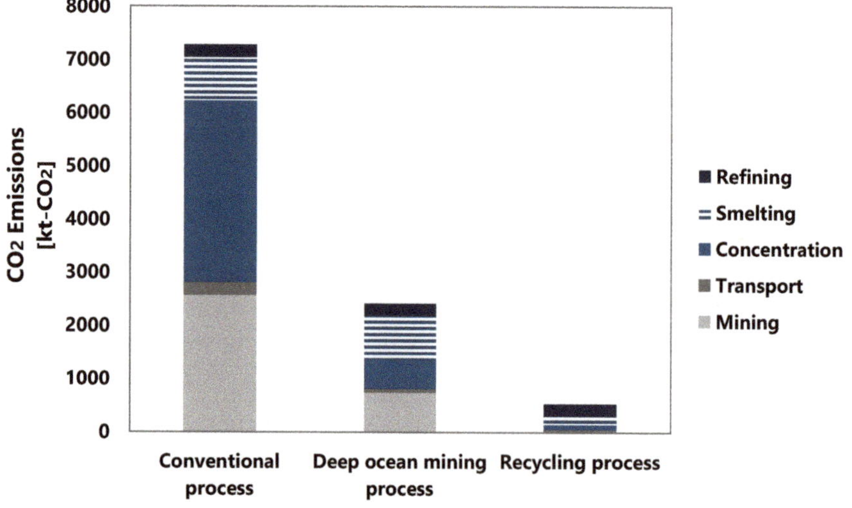

Figure 8. CO_2 emissions in copper production.

In addition, Japan's population is forecast to continue to decrease into the future. According to the National Institute of Population and Social Security Research, by 2055, the population in Japan will reduce to roughly 100 million people, approximately 80% of the current population [28].

Table 4 shows the population change observed and forecasted in Japan [28,29]. Following the assumption that copper consumption per person will remain constant, Japan's overall requirements for copper will decrease. Thus, by 2055 we anticipate that environmental impacts will be limited to those shown at an 80% DOM or recycling level in Figures 6 and 7, although alternatively exports could increase to take-up additional copper production. This study assumed that copper consumption per capita and GDP is constant. It has been shown that copper consumption is strongly correlated to GDP per capita [30]. Following the assumption that copper consumption per capita is stable, it is more important to maintain the GDP per capita when considering an aging society, which could lead to a decline in copper consumption commensurate with population decrease.

Table 4. Population change in Japan and copper consumption.

Year	Age Group 0–14 [Thousand People] [28]	Age Group 15–64 [Thousand People] [28]	Age Group 65–74 [Thousand People] [28]	Age Group More That 75 [Thousand People] [28]	Copper Consumption [kt] [28]
1995	19,400	87,000	12,000	7800	1300
2015	15,900	77,300	17,500	16,300	1300
2035	12,500	64,900	15,200	22,600	1200
2055	10,100	50,300	12,600	24,500	1000

4.3. Labor

Based on the I-O analysis, employment induced by recycling will make it the third largest sector next to textiles and construction. This indicates that the demand increase for recycled copper will lead to a larger labor demand.

When the final demand increased by 1 unit (1 million yen), 0.17 people are required as labor. Comparing to other industries under the same assumptions, recycling is the industry which requires labor force the most.

Average annual income per person in the recycling industry is estimated to be about three million yen, the second lowest income level in Japan. This industry may struggle to attract employees under these conditions

5. Discussion

In terms of economic ripple effect, it is estimated that copper ore production via deep ocean SMS mining will give a larger effect than that yielded by recycling alone. Demand increase for recycled copper leads to a commensurate demand increase for copper collection. It is also estimated that this demand increase for copper collection will also lead to a larger labor demand. It is unclear from the methodology applied in this study as to whether the recycling sector will increase the overall number of employees or simply extend the working hours of the current employee pool. This is one of the limitations of I-O analysis and an issue faced by Japan as a whole. Another limitation of the I-O table is the limitation of bottom-up estimation of input coefficients. As this study mainly employed bottom-up analysis based on LCA data, it does not consider industries which do not appear in the LCA for the given technologies (e.g., service industry). In that sense, energy consumption or CO_2 emissions can be considered as a conservative estimate. Regarding LCA, allocation in this study employed mass balance allocation assuming multiple mineral recovery the likelihood may be very low of its actual occurrence, but when only copper is recovered, the environmental impacts of deep ocean mining are much larger than the conventional process. Figure 9 shows the comparison of CO_2 emissions when only copper is recovered. CO_2 emissions from the recycling process are smaller than the conventional process though, unlike multiple mineral recovery, domestic emissions will become larger than what they are today. In the case of deep ocean mining, it will not be able to reduce CO_2 emissions across the entire copper supply chain.

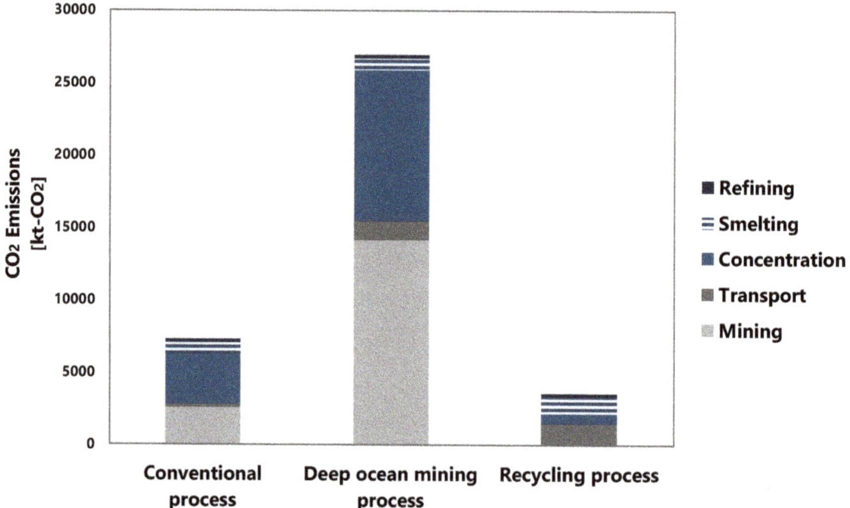

Figure 9. CO_2 emissions under the assumption that all emissions for deep ocean mining and recycling are allocated to copper.

As mentioned in the introduction, up until the end of 2018, Japan had been exporting lower grade copper scrap to China [31]. In the context of Japanese resource security, processing scrap in Japan could contribute to building a more stable supply chain and build a measure of resilience against inopportune external events including political, economic or environmental restrictions. A measure of independence from importing raw materials will result in resource security improvements. Also, according to the results of this study, a national approach to recycling or deep ocean mining, or a combination of the two can also contribute toward decreasing CO_2 emissions across the entire copper supply chain. Although some positive aspects are identified, we also find that the available working population may be a limitation for Japan's copper recycling capacity. In order to address these issues, we assessed three mitigatory measures, as follows.

Sensitivity Analysis

Process 1. Smelter Process End-of-Life Products Scraps

This study assumed that all scrap is dismantled and separated, and then processed in a smelting plant. Under this type of recycling regime, generally, high-quality separation is not required since impurities are treated at the smelting and refining process. Processed end-of-life copper scrap by smelters proposes that smelters disassemble end-of-life products and collect scrap by utilizing shredders and separation machinery. This type of recycling is expected to decrease energy requirements for concentration and labor when compared to the recycling option assessed in this study. The labor and energy needs of each facility is based on previous studies [32] and it is assumed that all pre-treatment processes such as crushing are conducted close to the smelting plant, so that energy for transportation is negligible.

Process 2. Producing Sector Processes Collected Waste

Since smelters play an important role in recycling in this study, it is assumed to be preferential to maintain smelters' profits under any future regime. One way to achieve this is to improve the recovery rate from scrap. Smelters' recycling process proposed that the smelter would be responsible for scrap collection, however, the mixture of materials will likely cause a lower recovery level of materials. Thus, in processed end-of-life products by producing sectors, producers of final products are made responsible for dismantling and sorting. This is achieved by producers installing mechanical dismantling facilities in their production factories. Metals and a certain amount of plastics will be provided to the recycled copper sector, with producers able to collect desirable materials. As the wire industry is able to reuse their end-of-life products, only scrap for brass production theoretically needs to be recycled. Generally, brass scrap contains about 60% copper. Copper demand for the brass industry was approximately 30% of total copper production in 2011 [15]. Thus, energy consumption for transportation is allocated at 60%, with smelting and refining at 30%. In this case zinc from brass scrap is also assumed to be recovered. Since this recycling process requires careful dismantling processes, the labor requirements are assumed to be the same as for the production stream. It is expected to improve the smelters' profit rate compared to the recycling process by smelters, but it may reduce overall profits due to a reduced quantity of available scrap. Moreover, this approach will increase system costs and energy for transportation. This approach also requires a large investment for producing industries, as it is assumed that the labor requirements for this process are the same amount as for the production process, i.e. securing the labor base is critical to its success. This type of recycling does not necessarily produce copper and may not increase the recycling rate according to the current Japanese definition [15].

Process 3. Local Recycling

The above two recycling process options both rely on hardware. However, this process encourages community level (including residential and commercial) recycling by consumers. This approach does not consider disassembling and sorting as full-time work. By consigning the disassembling and sorting process to consumers, the copper collection industry will not require large amounts of labor as is necessary for options (1) and (2). Finalizing scrap before the smelting process is a major function for the copper collection process within this process. According to the Ministry of Environment of Japan, copper accounts for 7% of end-of-life household electronic appliance materials after removing plastic and ferrous metals [33]. For this process, instead of hiring workers, the copper collection industry will provide incentives to consumers who separate their end-of-life products prior to recycling. Labor used for disassembling is not considered in the same way as options (1) and (2), but as a voluntary practice. Labor within this process only includes the workers in the copper finalizing processes. As we assume rapid aging in Japan, the retiring generations will likely play an important role in this proposed process. In addition, consumer driven recycling may provide non-traditional employment opportunities for those wishing to continue working past the retirement age. Table 5 illustrates a comparison among the three proposed mitigatory processes and the recycling of copper through traditional channels in Japan when 1 unit of final demand increase.

Table 5. Efficiency of processes proposed when compared with baseline copper recycling [corresponding to 1 unit change in copper recycling sector].

Process	Energy [GJ]	CO_2 Emissions [t-CO_2]	Labor [No. of People]
Process 1	+0.2	0	−0.02
Process 2	+0.5	0	+0.03
Process 3	−0.1	0	−0.02

Comparing these three process, we observe the existence of an energy-emissions-labor trade-off. Process 1 increase energy, however it is able to produce comparable amounts of copper with a relatively small labor force. Process 2 increase energy consumption and the estimated labor requirement the most, it may be impossible to realize under Japan's current demographic restrictions, i.e. an aging, shrinking society. Process 3 seems to be the ideal process approach with a moderate energy and emissions contribution and the smallest labor (actually hired) requirements of the three proposed approaches. Process 3 is unlikely to be realized unless consumers are willing to cooperate, as separating and sorting processes are dependent on consumers under this process. Note that CO_2 emissions in each process is less than 1 kg.

6. Conclusions

Japan is facing a number of challenges, among them, an aging shrinking population, and over-dependence on raw material imports from foreign nations. This study seeks to provide policy implications and recommendations for improving resource security in Japan, incorporating both recycling and indigenous deep ocean mining operations. Findings show that although deep ocean mining may have strong flow-on economic impacts, the overall quantity of copper extracted may not contribute strongly to Japan's resource security. On the other hand, I-O analysis of domestic copper production scenarios provides a new finding, that labor may be a constraint of realizing a recycling-centric society. Although producing copper by recycling eases environmental impacts, it requires significant manpower.

In order to overcome these shortcomings of individual resource security policies, this study provides three processes, complemented by economic, environmental and social analyses—considering employment, demographics and societal norms.

The analysis undertaken in this study enables a deeper understanding of the effect of demand increase for new industries under constant technological coefficients. However, I-O table-based analysis has some limitations. Since the I-O table is a snapshot of a single year, the effect of investment over the long-term is not recognized. A model to analyze the effect of longer-term investments, such as recycling plant construction which takes longer than a single I-O period will allow for the further development of this research

Author Contributions: Conceptualization, R.M. and B.M.; methodology, R.M.; software, R.M.; validation, R.M., B.M; formal analysis, R.M.; investigation, R.M.; resources, R.M.; data curation, R.M.; writing—original draft preparation, R.M.; writing—review and editing, A.C., B.M. and T.T.; visualization, R.M.; supervision, A.C., B.M. and T.T.;

Funding. This research received no external funding.

Conflicts of Interest: The authors declare no conflict of interest.

References

1. Watari, T.; McLellan, B.C.; Giurco, D.; Dominish, E.; Yamasue, E.; Nansai, K. Total material requirement for the global energy transition to 2050: A focus on transport and electricity. *Resour. Conserv. Recycl.* **2019**, *148*, 91–103. [CrossRef]

2. Månberger, A.; Stenqvist, B. Global metal flows in the renewable energy transition: Exploring the effects of substitutes, technological mix and development. *Energy Policy* **2018**, *119*, 226–241. [CrossRef]

3. Chapman, A.; Shigetomi, Y. Developing national frameworks for inclusive sustainable development incorporating lifestyle factor importance. *J. Clean. Prod.* **2018**, *200*, 39–47. [CrossRef]

4. National Research Council of the US. *Minerals, Critical Minerals, and the U.S. Economy*; National Academies Press: Washington, DC, USA, 2008; ISBN 978-0-309-11282-6.

5. European Commission. *Report on Critical Raw Materials for the EU, Report of the Ad Hoc Working Group on Defining Critical Raw Materials*; European Commission: Brussels, Belgium, 2014; p. 41.

6. Prime Minister's Office of Japan. *Shigen Kakuho Senryaku [Resource Securing Strategy]*; Prime Minister's Office of Japan: Tokyo, Japan, 2012; pp. 1–30.

7. U.S. Department of Energy. *Critical Materials Strategy*; U.S. Department of Energy: Washington, DC, USA, 2011; p. 191.

8. La Porta, D.; Hund, K.; McCormick, M.; Ningthoujam, J.; Drexhage, J. *The Growing Role of Minerals and Metals for a Low Carbon Future*; World Bank: Washington, DC, USA, 2017.

9. Government of Japan. *Strategic Energy Plan*; Government of Japan: Tokyo, Japan, 2018.

10. Hislop, H.; Hill, J. *Reinventing the Wheel: A Circular Economy for Resource Security*; Green Alliance: London, UK, 2011.

11. Elshkaki, A.; Graedel, T.E.; Ciacci, L.; Reck, B. Copper demand, supply, and associated energy use to 2050. *Glob. Environ. Chang.* **2016**, *39*, 305–315. [CrossRef]

12. U.S. Department of Commerce. *A Federal Strategy to Ensure Secure and Reliable Supplies of Critical Minerals*; U.S. Department of Commerce: Washington, DC, USA, 2019; p. 51.

13. Hatayama, H.; Tahara, K. Evaluating the sufficiency of Japan's mineral resource entitlements for supply risk mitigation. *Resour. Policy* **2015**, *44*, 72–80. [CrossRef]

14. JOGMEC. Signed a Joint Copper and Gold Exploration Contract in Australia. Available online: http://www.jogmec.go.jp/news/release/news_06_000271.html (accessed on 18 June 2019). (In Japanese).

15. JOGMEC. *Kobutsu Shigen Materiaru Furo 2017 [Mineral Resources Material Flow 2017]*; JOGMEC: Tokyo, Japan, 2017; pp. 1–15.

16. MERIJ. *Do Oyobi Dogokin Risaikurugenryo No Genjo to Anteikakuho Katsuyo Eno Kadai [Current Situation of Copper and Copper Alloy Recycled Materials and the Challenges to Its Stability Ensured and Utilization]*; MERIJ: Tokyo, Japan, 2015.

17. Mark, E.S.; King, M.J.; Sole, K.C.; Davenport, W.G. *Extractive Metallurgy of Copper*, 5th ed.; Elsevier: Amsterdam, The Netherlands, 2011; ISBN 9780080967899.

18. Ministry of Ecology and Environment of the People's Republic of China. *Announcement on Adjusting "Catalogue of Imported Waste Management"*; Ministry of Ecology and Environment of the People's Republic of China: Beijing, China, 2018.

19. Eurostat Recycling—Secondary Material Price Indicator. Available online: https://ec.europa.eu/eurostat/statistics-explained/index.php?title=Recycling_%E2%80%93_secondary_material_price_indicator&oldid=422150 (accessed on 18 June 2019).

20. Masuda, N. Challenges toward the sea-floor massive sulfide mining with more advanced technologies. In Proceedings of the 2011 IEEE Symposium on Underwater Technology and Workshop on Scientific Use of Submarine Cables and Related Technologies, Tokyo, Japan, 5–8 April 2011; pp. 1–4.

21. Hein, J.R.; Mizell, K.; Koschinsky, A.; Conrad, T.A. Deep-ocean mineral deposits as a source of critical metals for high- and green-technology applications: Comparison with land-based resources. *Ore Geol. Rev.* **2013**, *51*, 1–14. [CrossRef]

22. Agency for Natural Resource and Energy of Japan. Kaitei Nessui Kosho Kaihatsu Keikaku Dai Ikki Saishu Hyoka Hokokusho [1st Term Final Evaluation Report of Deep Ocean Hydrothermal Ore Development Project]. Available online: https://www.jogmec.go.jp/content/300110684.pdf (accessed on 11 November 2018).

23. Agency for Natural Resource and Energy of Japan; JOGMEC. *Kaitei Nessui Kosho Kaihatsu Sogo Hyoka Hokokusho [Report of the Comprehensive Assessment]*; Agency for Natural Resource and Energy of Japan: Tokyo, Japan, 2018; p. 158.

24. Ministry of Internal Affairs and Communications of Japan. 2011 Input-Output Tables for Japan. Available online: http://www.soumu.go.jp/english/dgpp_ss/data/io/io11.htm (accessed on 16 June 2019).

25. OECD. Input-Output Tables in Chile. 2011. Available online: https://stats.oecd.org/Index.aspx?DataSetCode=IOTSI4_2018 (accessed on 6 June 2019).

26. Motoori, R.; McLellan, B.C.; Tezuka, T. Environmental Implications of Resource Security Strategies for Critical Minerals: A Case Study of Copper in Japan. *Minerals* **2018**, *8*, 558. [CrossRef]

27. Ministry of Internal Affairs and Communications. Employment Matrix. Available online: https://www.e-stat.go.jp/en/stat-search/files?page=1&layout=datalist&toukei=00200603&tstat=000001073129&cycle=0 (accessed on 11 July 2019).

28. National Institute of Population and Social Security Research. *Population Projections for Japan: 2016–2065*; National Institute of Population and Social Security Research: Tokyo, Japan, 2017; pp. 1–385.

29. Statistics Breau of Japan. *Japan Statistical Yearbook 2019*; Ministry of Internal Affairs and Communications Japan: Tokyo, Japan, 2018; p. 781.

30. Kishita, Y.; Inoue, Y.; Kobayashi, H.; Fukushige, S.; Umeda, Y. Feasibility Assessment of Sustainability Scenarios Based on the Estimation of Metal Demand (Case Analysis of Long-Term Energy Scenarios Focusing on the Risk of Copper Depletion). *Trans. Japan Soc. Mech. Eng. Ser. C* **2013**, *79*, 3221–3233. [CrossRef]

31. Ministry of Finance of Japan. Trade Statistics of Japan. Available online: http://www.customs.go.jp/toukei/info/index_e.htm (accessed on 16 June 2019).

32. Coelho, A.; De Brito, J. Economic viability analysis of a construction and demolition waste recycling plant in Portugal—Part I: Location, materials, technology and economic analysis. *J. Clean. Prod.* **2013**, *39*, 338–352. [CrossRef]

33. Ministry of Environment of Japan. Kogata Denki Denshi Kiki Ni Fukumareru Yuyo Kinzoku Ganyuryo [List of Metals Contained in Small Electrical and Electronic Equipment (Board, Parts, Material Composition)]. Available online: http://www.env.go.jp/council/former2013/03haiki/y0324-05b.html (accessed on 11 November 2018).

Article

Banning Diesel Vehicles in London: Is 2040 Too Late?

Moayad Shammut [1], Mengqiu Cao [2,3,*], Yuerong Zhang [3,4,*], Claire Papaix [5], Yuqi Liu [6] and Xing Gao [3]

1   School of People, Environment and Planning, Massey University, Palmerston North 4442, New Zealand
2   School of Architecture and Cities, University of Westminster, London NW1 5LS, UK
3   Bartlett School of Planning, University College London, London WC1H 0NN, UK
4   Bartlett Centre for Advanced Spatial Analysis, University College London, London W1T 4TJ, UK
5   Department of Systems Management and Strategy, University of Greenwich, London SE10 9LS, UK
6   Department of Social Work and Social Administration, The University of Hong Kong, Hong Kong 999077, China
*   Correspondence: m.cao@westminster.ac.uk or mengqiu.cao.13@ucl.ac.uk (M.C.); yuerong.zhang.14@ucl.ac.uk (Y.Z.)

Received: 27 July 2019; Accepted: 6 September 2019; Published: 11 September 2019

Abstract: Air pollution contributes to 9400 deaths annually in London and diesel vehicles are considered a major source of lethal air pollutants. Consequently, the UK government announced its intention to ban diesel vehicles by 2040 to achieve a sustainable zero-carbon road transport system. Since no empirical studies have used a bottom-up approach to seek Londoners' views, it is therefore worth investigating the public opinion regarding this forthcoming ban. This paper aims to fill this research gap by taking London as a case study. A survey was designed, and fieldwork was conducted to distribute questionnaires to Londoners. Completed questionnaires were analysed using both quantitative and qualitative methods. The findings revealed that the majority of Londoners would be in favour of the ban if they were sufficiently exposed to the appropriate sources of information and were favourably disposed towards environmental protection measures. The results also showed that Londoners were more likely to switch to electric vehicles (EVs) if they were offered generous incentives and encouraged to use scrappage schemes. The present study makes a strong case for enforcing the ban well before 2040. The significance of this research is to provide clearer signals regarding the future of diesel vehicles, which in turn will strengthen the EV policy and uptake.

Keywords: low carbon technologies; low carbon transition; decarbonisation; zero carbon; air pollution; diesel ban; electric vehicles; transport policy; transport planning; London

---

1. Introduction

In recent years, European governments including the UK, have committed to addressing the adverse air pollution in their cities, by setting targets to decrease the carbon dioxide (CO_2) levels. Since then, the use of diesel vehicles has been incentivised and these are favoured over petrol vehicles in several European cities through the 1998 agreement between the European Automobile Manufacturers Association and the European Commission [1,2]. As a result, diesel vehicles now form a significantly increased share of the UK's licensed vehicle fleet. Since 1994, the proportion of diesel vehicles increased from 7% to about 40% in 2016, reaching a total of more than 12 million vehicles [3]. Although previous research has asserted that diesel vehicles are more CO_2 efficient than petrol vehicles [4,5], a very recent study revealed that diesel vehicles produce similar "real-world" CO_2 emissions to petrol vehicles [6]. In the same study, the authors indicated that diesel vehicles have not demonstrated significant advantages over petrol vehicles in terms of CO_2 emissions since 1995 in Europe. The policy of favouring diesel vehicles over petrol vehicles came at the cost of a considerable increase in air pollution in many cities, particularly in London, since diesel vehicles emit toxic pollutants such as nitrogen oxides (NO_x) and particulate matter (PM) [4], which are recognised as the most dangerous to the human health [4,7]. Thus, there is an urgent need to take action against the continuation of diesel vehicles.

Studies have shown that the transport sector emits the greatest amounts of greenhouse gas (GHG) emissions [8], whereby 95% of transport sector emissions are caused by the road transport [9]. In 2014, the road transport was responsible for 20% of all global CO_2 emissions [10]. Although the scope of this paper focuses primarily on road transport emissions due to the scale of their contribution to London's air pollution [2], it is imperative to point out that there are other sources of air pollution that originate from within the transport sector such as aviation, rail, and maritime activities as well as other sectors including industrial and energy productions [2,11,12]. Therefore, it is argued that addressing the emissions produced by all of the aforementioned sources should be considered when developing policies for a sustainable zero-carbon transport system. Otherwise, failure to reverse the current trends in the transport emission concentrations will only exacerbate the situation and lead to further adverse consequences in the future.

London's air quality is considered the worst among European cities [13], and the UK has been threatened with legal action and fines if it fails to reduce its air pollution emissions in some cities including the capital [14]. As a result, successive governments and policymakers have tried to implement new policies to tackle air pollution in London but, regrettably, the crisis still prevails [15]. In addition, recent studies have revealed that the vast majority of Londoners (82%) agreed that tackling air pollution must be a priority [16]. We therefore emphasise that effective new transport policies, tailored to the challenges that the capital faces, must be developed and introduced as a matter of urgency. This situation underpins the salience of this research paper.

In light of the current air pollution situation, the Mayor of London set a goal of achieving a zero-emission road transport by 2050 in the Mayor's Transport Strategy (MTS) [17], and the UK government is committed to lowering GHG emissions by 80% by 2050 through the 2008 Climate Change Act [11], as articulated in the Agenda 2030 for sustainable development [18]. It is argued that achieving this ambitious target can be attained by the decarbonisation of the transport sector [11]. Consequently, the UK government has announced its intention to ban the diesel and petrol vehicles by 2040 [19] in order to counteract air quality issues. The government justified the forthcoming ban by publishing reports [20] predicting that, by 2040, (1) high NO_x levels in the UK will be one of the biggest environmental risks to people's health in terms of diabetes, asthma, bronchitis, lung cancer, and heart disease; and (2) air pollution in the UK is expected to destroy 50% of the plant life as well as 40% of the wildlife habitats. However, there is a continuing debate about whether banning diesel vehicles can provide the answer to tackling air pollution. On the one hand, some German politicians as well as car companies, are not in favour of the ban and feel that it is "excessive" [21,22]. On the other hand, transport and environmental experts [7,15,23] realise the importance of the ban and have therefore made a strong case for policymakers to implement it as soon as possible. Recent research shows that banning diesel vehicles would have a significant impact on reducing the NO_x and $PM_{2.5}$ levels as well as improving the public health [7,15].

Studies examining the public opinion concerning the ban are scarce. Previous literature paid much attention to interventions such as the creation of low emission zones (LEZs) [24–26], but the authors concurred that LEZs do not bring about the desired improvements in the local air pollution. One recent study investigated the potential economic and health benefits of banning diesel vehicles in Dublin [7], while others have attempted to predict the attitudes of UK drivers towards switching to electric vehicles (EVs) after the ban is introduced [20]. However, to the best of our knowledge, there have not been any empirical investigations, specifically using London as a case study, to gauge public opinion regarding the forthcoming ban on diesel vehicles in 2040. The public opinion with respect to any putative ban has yet to be fully explored. Furthermore, it is argued that the reluctance to implement the ban in the UK is a contributing factor to thwarting the early uptake of EVs as well as weakening the EV policy [15,23]. For this reason, our research is significant, as it investigates whether the ban can be implemented earlier than 2040. The outcome of this investigation will provide clearer signals regarding the future of diesel vehicles, which in turn will strengthen the EV policy and uptake.

We argue that addressing London's air pollution is fundamentally linked to three factors, namely: (1) A strong political will, which includes strategies and action plans on the city, national, and international levels; (2) the cultural paradigm, e.g., city branding. This comprises the practices and marketing of the city, which has a direct influence on sustainability and people's behaviours [27,28]; and (3) public participation. This paper takes into consideration the third factor, i.e., public participation, paying particular attention to the case study of London using a bottom-up approach.

This paper aims to fill the gap in the literature by establishing the public opinion and attitudes towards the ban. It also aims to investigate the rate at which the Mayor should implement the ban, and to explore what factors may influence the public acceptance of this policy. Although the forthcoming ban is intended to apply to both diesel and petrol vehicles, the core focus of this paper is on the diesel vehicles since they produce more toxic emissions that require further attention.

2. Case Study, Data and Methods

2.1. Case Study Context

London is the capital city of the UK with an approximate population of 10 million and this figure is expected to increase dramatically in the near future [29,30]. This rapid growth in the population has led to greater car use and ownership. In 2018, the number of private licensed cars in London had reached about 2.7 million [31], causing serious problems in terms of air quality and health. For instance, a study conducted in London reported that the number of deaths caused by the NO_x and $PM_{2.5}$ as being about 9400 per annum [32]. London is a popular destination for tourists from all over the world, and its air quality is considered the worst among British cities. Therefore, it was selected as a case study since it has an impact on millions of people.

The capital also possesses several characteristics that make it a suitable case for the investigation of banning diesel vehicles. First, it is regarded as a world leading city in the transport innovation and cutting-edge technology, as well as hosting a number of top-class researchers [33]. These factors enable innovative solutions and effective intervention measures to be generated [34,35]. Second, it is equipped with suitable tools to facilitate a shift away from diesel vehicles to more sustainable alternatives, as well as encouraging active travel via its sophisticated transport system [36,37]. Third, it offers a rich dataset in terms of modelled emissions for different air pollutants that is easily available for research purposes [38].

2.2. Data and Methods

Data used in this study were collected via a face-to-face survey which was distributed to Londoners by the fieldwork conducted from June to August 2018. We utilised a random sampling method [39], similar to the approach used by Cao [40], and Cao and Hickman [41–43]. Prior to each interview, participants were asked if they were permanent residents in London and aged 17 or over. The targeted sample comprised those aged 17 or over since it is only license holders from this cohort who are legally permitted to drive. Only those who fulfilled these requirements were accepted to participate in this study and complete the survey. With a view to filling the knowledge gap in the literature regarding public attitudes towards the forthcoming ban on the internal combustion engine (ICE) vehicles (i.e., diesel and petrol), the survey sought to: (1) Collect insightful information about the Londoners' level of awareness about the harm caused by diesel vehicle emissions; (2) establish their views and attitudes regarding the forthcoming ban; and (3) discover whether they would be willing to shift to EVs. We argue that undertaking these investigations using a bottom-up approach is important because driving diesel vehicles can be cost-effective. As such, diesel vehicles use slightly less fuel than petrol ones, which offer cheaper running costs. Thus, diesel vehicles might be perceived as a more convenient option. The survey was piloted to make sure that all the questions were clearly worded without ambiguity and to avoid problems related to a low response rate. It was checked by a small number of colleagues and peers (n = 18) at the University of Westminster and University College London. As a result, 139 valid responses were collected. The survey consisted of eleven multiple-choice questions, while the twelfth was an open question designed to allow respondents to provide more detailed comments, which have been used for qualitative analysis in this research.

This section outlines the survey questions that were distributed to Londoners: (1) Since poor air quality affects the quality of life referring to the state of wellbeing in terms of health, comfort, and happiness [44], the first survey question sought to ascertain the Londoners' levels of satisfaction in respect to air quality. This question sought to discover to what extent they agreed with the following statement: "I am satisfied with the air quality in London". (2) The second question in the survey aimed at identifying their level of awareness about the harm caused by diesel vehicles since it is argued that the level of awareness influences people's attitudes towards environmentally-friendly products [45]. (3) The next survey question was consequently designed to find out how Londoners would feel about purchasing a diesel vehicle having been informed about some of the health problems that diesel vehicles cause. Respondents were asked: "Emissions from diesel vehicles can cause premature deaths and many diseases such as asthma and lung cancer. Would you, as a consequence of the statement above, still be willing to purchase a diesel vehicle?" (4) In response to the UK government's announcement that ICE vehicles will be banned from the beginning of 2040, the fourth survey question involved a hypothetical scenario whereby respondents were invited to envisage that the government wished to enforce a ban on diesel vehicles from the start of 2019. The purpose of this survey question was to determine how quickly the Mayor of London should introduce the policy. (5) The fifth question in the survey asked respondents about the type of vehicle that they owned (i.e., diesel, petrol, EV, etc.) so as to provide an insight into the proportion of car users in addition to the vehicle types. (6) The sixth survey question was essential to gain insights into how the diesel and petrol car drivers would act if the government incentivises EVs or bans ICE vehicles completely. Therefore, only respondents who represent the car-driving cohort were required to complete this question. Respondents were asked to what extent they agreed with the following statement: "I would replace my diesel/petrol vehicle with an electric one." (7) The purpose of the seventh survey question was to ascertain whether participants had become sensitised towards the environment after they had been exposed to information about the potential harm caused by diesel vehicles. Respondents were asked about the kind of vehicle they would purchase if they were going to purchase one today. Lastly, the eighth, ninth, tenth, and eleventh questions in the survey sought to obtain information about the participants' characteristics in terms of gender, age, highest educational attainment, and main transport mode for work [41,43,46–48], all of which are presented in Table 1 below. The final survey question was an open question that asked respondents to provide any suggestions and recommendations on how to improve London's air quality in relation to the diesel vehicle emissions.

Table 1. Sample characteristics.

		Frequency	Percentage
Gender	Male	76	55%
	Female	63	45%
Age	17–30 years old	73	52%
	Over 30 years old	66	48%
Educational attainment	Secondary school or below	1	1%
	Bachelor's degree	67	48%
	Master's degree	60	43%
	Doctoral degree	11	8%
Vehicle ownership	Yes	85	61%
	No	54	39%
Main transport mode for work	Car	20	14%
	Bus	25	18%
	Underground/Overground	75	54%
	Cycle	12	9%
	Walk	5	4%
	Others *	2	1%

* "Others" is comprised of working from home and/or not currently employed.

The present study utilises a non-parametric statistical testing method known as the Kruskal–Wallis H [49]. The test static, H, is calculated by the following formula:

$$H = \frac{12}{N(N-1)} \sum_{i=1}^{k} \frac{R_i^2}{n_i} - 3(N+1) \tag{1}$$

where:

- N: Total sample size
- K: Number of samples
- R_i: Sum of ranks in the ith sample
- n_i: Size of the ith sample.

The Kruskal–Wallis H Test (KWT) is a rank-based approach to the one-way ANOVA, which determines whether there are statistically significant differences between three or more groups for a dependent variable [50]. The KWT assumes that: (1) The sample population is drawn at random; (2) the observations of the samples are independent; and (3) the scale of measurement of the dependent variable should be at least ordinal [50]. This non-parametric statistical test has been used by many researchers [51–53], most commonly in behavioural sciences [54]. It is utilised in this study with the aim of gaining greater insight into the factors that may influence the public acceptance of the ban on diesel vehicles.

Perhaps a minor limitation of this method is that it indicates if there is a significant difference between groups but does not indicate which groups are different. Therefore, a qualitative analysis was conducted to overcome this limitation. In this paper, a mixture of quantitative and qualitative approaches is used [47]. The purpose of doing this is to minimise the disadvantages of each individual approach while maximising the benefits of both approaches combined [55].

3. Results and Discussion

3.1. Air Quality Satisfaction

The results show that the highest percentage of respondents (67%) said they "disagree" with the following statement: "I am satisfied with the air quality in London". By contrast, only 22% of respondents stated that they "agree" with the statement, and approximately 12% said they "neither agree nor disagree".

For 67% of the sample population to be dissatisfied with London's air quality calls into question the validity of current policies aimed at reducing the air pollution. This result also indicates that further and quicker actions should be taken by the Mayor of London and the responsible authorities [56] as well as car industries [37], perhaps in conjunction with community-led initiatives. Nevertheless, experts argue that such initiatives led by the community alone are unlikely to have an impact in the long term as they have to be supported by governmental interventions [37].

To highlight the importance of exploring respondents' satisfaction with the air quality, a study conducted by MacKerron and Mourato [57] to assess the relationship between the air quality in London and Londoners' level of life satisfaction, found that air pollution and life satisfaction are significantly correlated, so that, as air pollution increases, Londoners' life satisfaction diminishes. This may suggest that air pollution not only affects people's health, but also their perceptions of life satisfaction. Policymakers are aware of the hazard generated by air pollution and it is therefore incumbent on them to act promptly.

The last survey question was an open question, which sought the respondents' in-depth views on air pollution in London. Regarding air quality satisfaction, one participant stated:

"Air quality is good in London." (male, 24).

Analysing the aforementioned respondent's statement, current studies show that this is not true for London as a whole. In fact, estimates show that air pollution in London contributed to 4267 deaths in 2008 [58] and that this figure had increased to about 9400 in 2015, according to Walton et al. [32]. Since London has consistently breached the legal limits of the PM_{10} levels, the UK has been threatened with legal action [14]. Poor air quality in London is also a motivating factor that underpins this research since the majority of Londoners stated that tackling air pollution should be prioritised [16]. However, it could be argued that the respondent referred to above may be living in a less polluted area, which they may regard as being, in their judgement, relatively cleaner or at least sufficiently clean. Perhaps it would make more sense if the respondent had stated that air quality is good in their particular area rather than proposing that what discretely applies to one area applies to London as a whole. Respondents who stated that they "neither agree nor disagree" with the statement above might also be living in areas of London that have relatively low levels of air pollution.

3.2. Awareness Level of Diesel Vehicle Emissions

From the analysis in Figure 1, about 33% of the respondents considered themselves to be "moderately aware" of the harm caused by diesel vehicles, while a nearly equal number of respondents claimed they were "slightly aware" or "very aware" of the harm caused by diesel vehicles. Lastly, almost the same number of respondents said they were "not aware at all" as opposed to "extremely aware". This possibly indicates that the sample may have included a wide variety of respondents who come from different backgrounds. Only 9% of the respondents claimed to be "extremely aware", which suggests that they might belong to a group of experts, e.g., experienced transport planners and researchers with a doctoral or masters degree.

Additionally, 26% of the respondents were "very aware" of the harm caused by diesel vehicles. We might speculate that this percentage could include less experienced transport planners as well as researchers and students in related fields. Certainly, it is essential to determine the level of awareness among the population in order to understand how to proceed, and to plan for the next stages. For instance, if there is little awareness about the harm caused by diesel vehicles, then the next step to be taken is to put more efforts into prioritizing new policies aimed at increasing awareness. By contrast, if there is already a considerable level of awareness, then the next step would involve implementing new and more rigorous policies targeting air pollution such as the banning of diesel vehicles [2,7]. Indeed, those who are aware of the harm caused by diesel vehicles would already possess an appreciation of why the government might consider banning such vehicles.

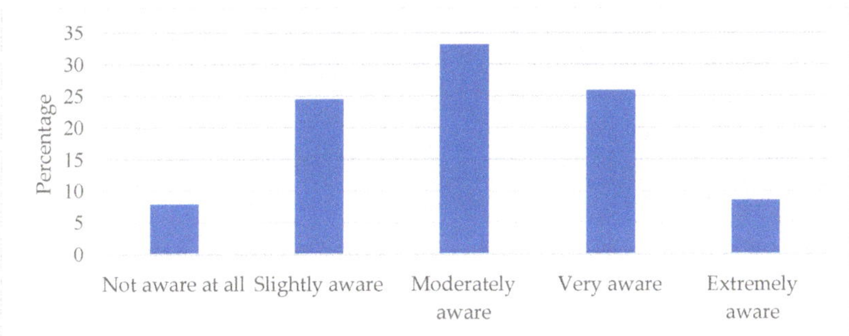

Figure 1. Awareness level of the harm caused by diesel vehicle emissions.

Levels of awareness could play a significant role in the shift away from diesel vehicles [56] and any move towards zero-emission vehicles, as stated in the MTS. This argument is also supported by the views of Londoners who completed the survey. In response to the last question, which invited respondents to offer their own recommendations for tackling air pollution in London, two respondents suggested the following:

"Have better knowledge on air pollution." (female, 23).

"To have a screen on the streets that displays real-time air pollution concentrations to make people more aware of the air pollution in the city." (female, 18).

Curtis and Headicar [59] found that raising the public awareness could play a substantial role in persuading people to switch from car usage to other forms of transport. Moreover, it is crucial to make people more aware of the fact that diesel vehicles emit toxic pollutants that are immensely harmful to health. Gaining their trust and confidence would undoubtedly help to implement new intervention measures to tackle air pollution effectively. Some scholars have argued [60] that educating the public about the possible harm caused by diesel vehicles would facilitate the more rapid implementation of the ban as well as facilitating a change in the current composition of vehicle fleets. It is further argued that people who possess awareness of environmental issues and concern about these tend to show positive attitudes towards EVs and other environmentally friendly products [45,61].

3.3. Information Provision

Interestingly, the analysis shows that slightly over half of those surveyed (51%) declared that they would not purchase a diesel vehicle after acknowledging the potential harm that they cause. However, a minority of the respondents (9%) said that they would still purchase a diesel vehicle. Furthermore, about 40% of the respondents claimed that they might buy a diesel vehicle, which implies that changes in policies and regulations related to LEZs and banning diesel vehicles should be more flexible. For example, these could include tightening the entry requirements for LEZs and extending LEZs to cover a wider area. Moreover, rather than banning diesel vehicles completely, a ban could be imposed on older vehicle models [15,56] as implemented in Hamburg [62], or diesel vehicles could be banned from entering certain roads that are severely affected by air pollution [63]. This might encourage those who said they "might" buy a diesel vehicle to compromise in the light of the aforementioned types of policy changes.

The KWT was used to ascertain if there was a relationship between the information provision about the harm caused by diesel vehicles and the likelihood of people supporting the ban. The results showed a significant relationship between those two factors with an H-value of 30.38, as can be seen in Table 2.

Table 2. Kruskal–Wallis H test results.

Category	H Value	
	Supporting Ban	**Influence on Kind of Vehicle to Buy**
Providing information about the harm caused by vehicle emissions	30.38 ***	11.33 *
Likelihood of switching to EVs	13.03 *	-

Notes: n = 139; * $p < 0.05$, ** $p < 0.01$, *** $p < 0.001$.

In the survey, prior to asking the participants whether they would purchase a diesel vehicle, they were given information about the potential harm caused by such vehicles. This led more than half of them to state that they would not buy a diesel vehicle, as opposed to only 9% who said they would still buy one. An important implication can be drawn from this, which is that the more awareness people have, the more likely they are to support the ban. This view is also supported by the testimony of Londoners in the survey:

"I was not aware of any damage that diesel vehicles can cause. But luckily with this survey, I am more aware now and will do what I can daily to save the society." (male, 31).

"I believe that what the government is doing is a good step forward. I have a diesel Mercedes CLS, and I really enjoy driving it, but now I am in the process of selling it for the sake of air quality. Thank you." (male, 25).

A key policy priority should, therefore, comprise a strategy to increase the public awareness about the negative health and environmental impacts caused by diesel vehicles emissions. This would facilitate the movement towards a zero-carbon transport system since it is argued that the decarbonisation of the road transport is strongly linked to people's travel behaviour [11].

3.4. Banning Diesel Vehicles

Perhaps, surprisingly, analysis of the survey results revealed that the vast majority of respondents would support the ban. About 65% of the sample population agreed with it, as opposed to 20% who would not be in favour. Nearly 15% of the sample population said they "neither agree nor disagree" with the ban, which suggests that they might be open to new policies and would not be considered problematic or non-compliant should such a policy come to be enforced.

A selection of the respondents are cited below expressing their views about the ban:

"I have only recently found out how bad diesel cars can be. It is a great initiative to ban them, but the government should give diesel car owners a huge incentive to compensate for the purchase of a new car that meets their requirements." (male, 42)

"Ban diesel cars, educate people to shift from cars to public transport, and increase diesel prices" (female, 36)

"Getting rid of diesel is just a start, we should work towards banning all private cars as soon as possible, including electrics which waste as much space." (female, 28)

"I would support more punitive measures including the expansion of the Congestion Charge Zone/London LEZ and even higher costs for diesel engines in Parking Control Zones. Also, a London based scrappage scheme which should incentivise diesel car users with active travel-based solutions, e.g., trading your diesel car for a new bike/public transport season ticket etc." (male, 37).

Of those surveyed, almost 13% advocated the ban when asked about recommendations for improving air quality in London in the open survey question. Therefore, support for a policy enforcing a ban is a conclusion that can legitimately be drawn from the data analysed in this research.

It is believed that banning diesel vehicles would create a positive impact on the environment by reducing the amount of air pollutants in the atmosphere. In Delhi, for example, slightly less than half of the diesel vehicles were replaced by compressed natural gas (CNG) alternatives [59]. Interestingly, in just one year, ambient air pollutants were reduced significantly. The case study reported reductions in sulphur dioxide (SO_2), carbon monoxide (CO), and NO_x by 22%, 10%, and 6%, respectively [59]. We can speculate about what the impact would have been if those diesel vehicles were replaced by zero emission EVs. Likewise, in the context of London, if a total ban on diesel vehicles is implemented, a positive impact would be seen in terms of reductions in pollutant concentrations, similar to, or possibly better than, those reported in the Delhi case study, since they would be replaced by EVs rather than CNG, which represent a more attractive alternative in terms of emissions produced.

Experts argue [7] that banning diesel vehicles would cause a substantial reduction in the NO_x and $PM_{2.5}$ emissions of 47% and 52%, respectively on the 2015 levels by the year 2030. Reductions in the NO_x and $PM_{2.5}$ levels would contribute to savings totalling 300 Disability Adjusted Life Years (DALYs). One DALY can be defined as the loss of the equivalent of one year of full health [64]. By the year 2024, diesel vehicles would be responsible for about 6300 DALYs. In the light of these findings, we can claim with confidence, that mitigation measures, i.e., banning diesel vehicles, should be considered as soon as possible. It is further argued that the ban would not only improve air quality and people's health but would also increase the uptake of EVs, especially if the appropriate infrastructure is in place [7].

Although some scholars [16] claim that there is little point in banning individual vehicles from certain roads because doing so would only improve air quality on those particular roads, but could consequently displace the pollution onto other roads, it can be argued that the ban first needs to be tested on specific roads for a certain period of time before implementing it across the entire city. People on low incomes who use their vehicles for their daily commute to work would be the group affected most by the ban since they may struggle to purchase newer models. If the ban is proven to achieve the desired outcomes in respect to improved air quality, then those people could be offered incentives to switch to more environmentally-friendly vehicles or, for instance, benefit from reduced public transport fares.

If policymakers also advocate implementing the ban but have concerns regarding the public acceptability, then the policy could be introduced gradually or on a trial basis. For instance, policymakers could initially enforce a ban on the most polluted roads by restricting the entry of diesel vehicles only. They could also enforce a ban in different areas for a limited amount of time, and then if this was found to result in significant reductions in pollutants levels by monitoring progress, the ban could be extended to more areas of London, thereby increasing the public confidence in respect to any proposed ban.

Overall, it is highly desirable to phase-out all old diesel vehicles since they cause higher levels of pollution than newer models [2,37,56]. In addition, new buses and taxis operating in London are more environmentally benign and thus should also be considered as an alternative to current diesel-fuelled vehicles [17,37].

3.5. Type of Vehicle Owned by Participants

The results of this survey question provide an insight into the proportion of car users in addition to the engine type of their vehicles. As can be seen from the analysis in Figure 2, at least 61% of the respondents were car users, while 39% were non-car users. It can be assumed that some respondents were also car users but did not own a car. The following questions were aimed at eliciting whether car drivers would be willing to switch from ICE vehicles to the use of EVs as the 2040 ban would apply to both petrol and diesel vehicles. If the number of car drivers in this survey had been very low, then the survey analysis would be inadequate because car drivers are the targeted group in respect to their attitudes towards the forthcoming ban and their willingness to shift to more environmentally-friendly alternatives. Therefore, as more than 61% of the respondents were car drivers, the analysis can be considered valid, although it may be desirable to increase the sample size in further research.

In 2016, there were 39% diesel vehicles, 60% petrol vehicles and 1.3% other vehicle types (EVs and hybrids) in Great Britain [3]. These statistics are consistent with, and to some extent similar to, the survey results in terms of the proportion of each vehicle type among the population, which suggests that the survey succeeded in reaching the targeted groups and being representative of the overall population.

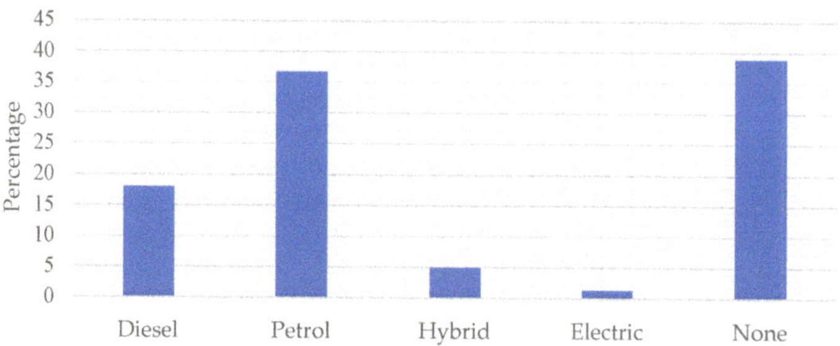

Figure 2. Type of vehicle owned.

3.6. Shifting to Electric Vehicles

Remarkably, over 68% of car drivers said they would agree to replace their current vehicles with EVs, as opposed to 20% who would refuse. Just under 12% of the respondents claimed to "neither agree nor disagree" with replacing their vehicles with EVs. It may be possible to persuade this group to switch to EVs by providing scrappage schemes or incentives to purchase them [37,56]. Therefore, one implication of this is that the government should provide a scrappage scheme for old vehicles to encourage both those who "disagree" and those who "neither agree nor disagree" to consider this option.

When participants were asked about their recommendations for tackling air pollution, nearly 12% responded that the adoption of EVs would provide a solution. Furthermore, about 8% of the participants said that incentives are necessary in order to persuade people to switch to EVs, while others complained about the high prices of EVs. It is therefore proposed that the government and Mayor of London should incentivise the purchase of EVs to increase their uptake rate. Below is a selection of the participants' opinions:

"London's air quality could be improved by the government giving incentives and benefits when people purchase an electric car." (male, 50).

"Diesel cars should be a relic of the past. We need to improve the performance of electric vehicles and lower their cost so that they are more widely adopted." (female, 39).

The KWT was employed to test the relationship between the willingness to switch to EVs and the likelihood of supporting the ban. The results revealed that an H value of 13.07 was obtained, as highlighted in Table 2, suggesting that a statistically significant relationship exists. As such, those who are willing to replace their diesel/petrol vehicles with electric ones are more likely to support the ban. Therefore, if the Mayor wishes to implement this policy soon, it is advised that the incentives for choosing EVs be put in place [37]. This programme could also be stimulated by the market. For instance, car manufacturers should be encouraged to be ambitious to produce more EVs and fewer ICE ones. After all, as discussed in the analysis, over 68% of car users agree on the need to shift to EVs.

As stated in the MTS [17], the goal is for all newly registered vehicles in London to be zero emissions so as to achieve the zero-emission road transport by 2050. Therefore, it is vital to discuss which factors would influence the acceptance of EVs in the capital to rapidly increase their adoption as a way to address air pollution.

Fundamentally, moving towards a more sustainable transport system through the uptake of EVs is critically linked to the level of public awareness and concern about the environmental impacts of ICE vehicles as well as EVs. For instance, Khaola et al. [65] investigated a possible correlation between people's concern for the environment and their attitudes towards green products (e.g., EVs) in general and found that there is a strong correlation between them. Additionally, Dogan and Ozmen [66] surveyed 752 respondents to investigate how the environmental concern would affect participants' attitudes towards buying EVs, and they discovered that participants with higher levels of environmental concern were more interested in EVs and more likely to purchase them.

Furthermore, it is argued that those who seek additional knowledge about EVs may be strongly influenced in terms of positive attitudes towards EVs. Many authors argue [20,67] that having a prior knowledge about EVs can make a huge difference regarding the decision-making process of buying EVs compared to those who have little or no knowledge.

Another factor that is believed to influence the adoption of EVs is incentives [68]. Studies show that providing incentives exerts a positive influence on purchasing EVs [69,70] given that the higher the incentive, the more likely people will purchase an EV. Incentives can be provided in different forms such as tax rebates and subsidies on purchase [71]. A tax rebate of £1000, for example, means that individuals will have a £1000 reduction in their tax when they purchase an EV. A study that was conducted in the United States showed that $1000 in tax rebate resulted in a 2.6% increase in EVs sales [71]. It is suggested that a similar method be used in London to encourage more people to purchase zero emission vehicles as an alternative to more polluting ones.

Although EVs are being introduced as an alternative to break fuel dependence and tackle other environmental problems [67], it is imperative to acknowledge that non-exhaust emissions should be also be taken into account when considering how best to address air pollution. There is a substantial volume of literature that has reached the conclusion that the volume of emissions from non-exhaust sources is much greater than emissions from vehicles exhausts [12,72,73]. Therefore, the goal of zero-emission road transport cited in the MTS is most likely to be achieved when there is almost no car use. Otherwise, it would be very challenging to realise this ambitious vision set out by the Mayor in the event that private cars, taxis, and buses continue to operate on London's roads. However, the phasing-out of ICE vehicles would significantly improve the NO_x reductions [7] since EVs do not emit those kinds of pollutants.

3.7. Influence on which Kind of Vehicle to Purchase

The findings from Figure 3 highlight the importance of providing information (i.e., raising awareness) about the harm caused by diesel vehicles, as the participants showed positive attitudes towards low emission vehicles such as hybrids and EVs. Interestingly, about 60% of those surveyed said that they would purchase a low-emission vehicle including hybrids and EVs. Furthermore, about 4% of the respondents stated that they would not purchase any vehicle because they prefer to use other types of public transport such as trains, buses, and cycling. In addition, the most remarkable observation to emerge from comparing Figures 2 and 3 is that the percentages of diesel and petrol vehicles dropped by more than half, from 18% to 8% and from 37% to 17%, respectively.

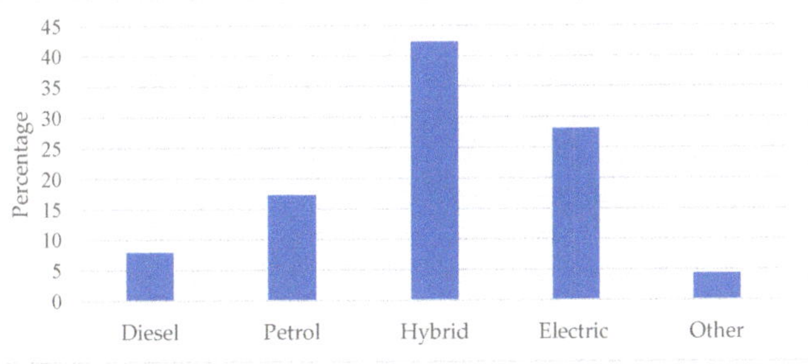

Figure 3. Which kind of vehicle to purchase in the future.

The KWT was also employed here to ascertain the effect of providing information on people's decision-making when buying a new vehicle. The test results presented in Table 1 reveal that providing people with relevant information about the potential harm caused by emissions from diesel vehicles will influence their decision-making process and, ultimately, lead them to make more environmentally friendly choices. It might be argued that those who are considering purchasing a more environmentally friendly vehicle are more likely to agree with the ban on diesel vehicles. Therefore, it is essential that the motor industries get involved [37] with this movement towards a sustainable zero-carbon road transport system by providing more environmentally friendly vehicles to help more people to make the shift away from ICE vehicles. This view was also expressed by one survey respondent:

"I think the car industry needs to be the leader. Technology can be used to assist in sustainable & cost-effective vehicles and transport modes." (female, 29).

Below is a selection of the participants' views about, and recommendations for, improving the local air quality in London:

"Air quality could be improved by creating more green spaces and trees as well as encouraging people to use public transport. Electric vehicles are the future, but at the moment they do not seem to be viable on a large scale due to their high cost." (male, 25).

"Encourage people to use environmentally friendly vehicles by reducing their price." (male, 19).

Although the findings in Figure 3 indicate positive attitudes towards different types of EVs, many researchers [74–76] believe that the high cost of purchase represents a barrier to the adoption of EVs, which concurs with the views cited above. Additionally, several authors [67,68] have identified other potential barriers to the adoption of EVs, such as their limited range, as well as style, size, performance and safety.

Nonetheless, some of these barriers could potentially be overcome by driver training, for example to address the problem of EV's limited mileage range [67]. Moreover, it is important to understand the specific needs and requirements of Londoners in relation to a vehicle. This could help to develop effective policies and stepping-stone guidelines for car manufacturers so as to overcome some of the aforementioned barriers. Two key factors could foster the acceptance of EVs, namely providing incentives in different forms, and raising consciousness about and concern for the environment through awareness campaigns. Acceptance can also be increased by advertising and marketing, which might include word-of-mouth recommendation [77]. Additionally, Globisch et al. [78] stressed the importance of providing EV users with the necessary information, support and assistance to overcome technical failures, as a way to foster the EV deployment. They also found that focusing on the vehicle design (perceived enjoyment) of EVs can significantly influence the choice of the car purchased.

Intervention by policymakers could also make a difference [79]. For instance, a new policy could be introduced stipulating that, for every sports car sold by manufacturers, five EVs should be sold. This would even-out the total amount of emissions produced by a particular car manufacturer since sports cars emit more pollutants than EVs [80]. It would also potentially encourage car manufacturers to set ambitious goals in terms of increasing the EV sales using different methods such as awareness-raising campaigns, and more efficient marketing tools, as well as providing customised assistance to their users.

4. Conclusions and Policy Implications

Using London as a case study, this paper explored the public opinion and respondents' views regarding the forthcoming ban on diesel vehicles, examined whether the ban could be implemented earlier than 2040, and set out to discover Londoners' attitudes towards switching to EVs.

The analysis from the KWT revealed that being concerned about the environment, together with an appropriate level of awareness surrounding its vulnerability to certain hazards, could play a huge role in supporting the ban. In order to facilitate the implementation of the ban, it is highly recommended that Londoners' level of awareness about the harm caused by diesel vehicles needs to be increased, which can be achieved by awareness-raising campaigns. Findings from the survey also revealed that the majority of participants "agree" with the ban, suggesting that the policy can be implemented well before 2040. These results have been very encouraging, and the discussion proposed several courses of action for the policy of banning diesel vehicles.

This work has shown substantial evidence that the majority of car drivers—slightly less than 70%—would agree to switch from driving ICE vehicles to EVs. However, some participants raised concerns about the high price of EVs whereas others proposed that the government should provide incentives for people to switch to EVs, which suggests that Londoners may not be willing to pay more for a private transport. A further important implication is that Londoners must be provided with incentives tailored to their needs, as well as ways of increasing their awareness of and concern for the environment as a solution to facilitate the acceptance of EVs. Another factor likely to enhance the EV policy and uptake rate is the banning of diesel vehicles, if the appropriate infrastructure is provided. This may include creating sufficient charging stations and extending the installation of charging points to cover more residential areas as well as carparks [20,67,81].

Our findings will be useful for the Mayor of London and policymakers to push the Agenda 2030 forward so as to achieve the goals of sustainable cities and communities as well as climate action [82]. This paper has led us to conclude that the ban should be enforced as soon as possible since the majority of Londoners appear to support the initiative, according to the sample in this research. We would argue that even if CO_2 emissions were to be controlled and reduced, the overall climate change effect would still exist because of the time scales involved, meaning that it takes a long time to reverse the severe impacts of poor air quality [18]. In order to achieve the desired outcomes for air quality in the long-term, critical decisions must be taken and "aggressive" policies implemented now to determine the quality of London's air in the future.

Author Contributions: M.S. primarily ran the modelling, produced the figures and tables, designed the questionnaire, collected the surveys, and co-wrote the manuscript. M.C. designed the questionnaire, further interpreted the modelling results, commented on the interview results, and edited the manuscript. Y.Z., C.P., X.G., and Y.L. commented on and edited the manuscript.

Funding: This research received no external funding.

Acknowledgments: The authors would like to thank the two special issue editors and the three anonymous reviewers for their suggestions regarding the improvement of the initial draft. This research did not receive any specific grant from funding agencies in the public, commercial, or not-for-profit sectors.

Conflicts of Interest: The authors declare no conflict of interest.

Abbreviations and Acronyms

CO_2	Carbon Dioxide
NO_x	Nitrogen Oxides
PM	Particulate Matter
GHG	Greenhouse Gas
MTS	Mayor's Transport Strategy
LEZs	Low Emission Zones
EVs	Electric Vehicles
ICE	Internal Combustion Engine
N	Total Sample Size
K	Number of Samples
R_i	Sum of Ranks in the ith Sample
n_i	Size of the ith Sample
KWT	Kruskal–Wallis Test
CNG	Compressed Natural Gas
SO_2	Sulphur Dioxide
CO	Carbon Monoxide
DALYs	Disability Adjusted Life Years

References

1. European Commission. Commission and ACEA Agree on CO_2 Emissions from Cars. Available online: http://europa.eu/rapid/press-release_IP-98-734_en.htm?locale=en (accessed on 16 July 2019).
2. Laybourn-Langton, L.; Quilter-Pinner, H.; Ho, H. *Lethal and Illegal: London's Air Pollution Crisis*; Institute for Public Policy Research: London, UK, 2016; Available online: https://www.ippr.org/research/publications/lethal-and-illegal-londons-air-pollution-crisis (accessed on 26 July 2019).

3. Department for Transport. *Transport Statistics Great Britain 2017*; Department for Transport: London, UK, 2017. Available online: https://www.gov.uk/government/statistics/transport-statistics-great-britain-2017 (accessed on 16 July 2019).

4. Turcksin, L.; Mairesse, O.; Macharis, C.; Mierlo, J.V. Encouraging environmentally friendlier cars via fiscal measures: General methodology and application to Belgium. *Energies* **2013**, *6*, 471–491. [CrossRef]

5. Sullivan, J.L.; Baker, R.E.; Boyer, B.A.; Hammerle, R.H.; Kenney, T.E.; Muniz, L.; Wallington, T.J. CO_2 emission benefit of diesel (versus Gasoline) powered vehicles. *Environ. Sci. Technol.* **2004**, *38*, 3217–3223. [CrossRef] [PubMed]

6. Helmers, E.; Leitão, J.; Tietge, U.; Butler, T. CO_2-equivalent emissions from European passenger vehicles in the years 1995–2015 based on real-world use: Assessing the climate benefit of the European "diesel boom". *Atmos. Environ.* **2019**, *198*, 122–132. [CrossRef]

7. Dey, S.; Caulfield, B.; Ghosh, B. Potential health and economic benefits of banning diesel traffic in Dublin, Ireland. *J. Transp. Health* **2018**, *10*, 156–166. [CrossRef]

8. Andrés, L.; Padilla, E. Driving factors of GHG emissions in the EU transport activity. *Transp. Policy* **2018**, *61*, 60–74. [CrossRef]

9. Seo, J.; Park, J.; Oh, Y.; Park, S. Estimation of total transport CO_2 emissions generated by medium-and heavy-duty vehicles (MHDVs) in a sector of Korea. *Energies* **2016**, *9*, 638. [CrossRef]

10. Santos, G. Road transport and CO_2 emissions: What are the challenges? *Transp. Policy* **2017**, *59*, 71–74. [CrossRef]

11. Vallack, H.; Haq, G.; Whitelegg, J.; Cambridge, H. Policy pathways towards achieving a zero-carbon transport sector in the UK in 2050. *World Transp. Policy Pract.* **2014**, *20*, 28–42.

12. Keuken, M.P.; Moerman, M.; Voogt, M.; Blom, M.; Weijers, E.P.; Röckmann, T.; Dusek, U. Source contributions to PM2.5 and PM10 at an urban background and a street location. *Atmos. Environ.* **2013**, *71*, 26–35. [CrossRef]

13. Greater London Authority. *Comparison of Air Quality in London with a Number of World and European Cities*; Greater London Authority: London, UK, 2014. Available online: https://www.london.gov.uk/sites/default/files/comparison_of_air_quality_in_world_cities_study_final.pdf (accessed on 16 July 2019).

14. European Commission. *Air Quality: Commission Sends Final Warning to the UK Over Levels of Fine Particle Pollution*; European Commission: Brussels, Belgium, 2010. Available online: https://europa.eu/rapid/press-release_IP-10-687_en.htm (accessed on 15 July 2019).

15. Brand, C. Beyond 'Dieselgate': Implications of unaccounted and future air pollutant emissions and energy use for cars in the United Kingdom. *Energy Policy* **2016**, *97*, 1–12. [CrossRef]

16. London Councils. 2018 Air Quality Polling. Available online: https://www.londoncouncils.gov.uk/our-key-themes/environment/air-quality-london/air-quality-public-polling/2018-air-quality-polling (accessed on 26 August 2019).

17. Greater London Authority. *The Mayor's Transport Strategy*; Greater London Authority: London, UK, 2018. Available online: https://www.london.gov.uk/what-we-do/transport/our-vision-transport/mayors-transport-strategy-2018 (accessed on 16 July 2019).

18. The UK Government's Approach to Delivering the Global Goals for Sustainable Development—At Home and Around the World. Available online: https://assets.publishing.service.gov.uk/government/uploads/system/uploads/attachment_data/file/603500/Agenda-2030-Report4.pdf (accessed on 12 August 2019).

19. UK Plan for Tackling Roadside Nitrogen Dioxide Concentrations: An Overview. Available online: https://assets.publishing.service.gov.uk/government/uploads/system/uploads/attachment_data/file/633269/air-quality-plan-overview.pdf (accessed on 16 July 2019).

20. Bennett, R.; Vijaygopal, R. An assessment of UK drivers' attitudes regarding the forthcoming ban on the sale of petrol and diesel vehicles. *Transp. Res. Part D Transp. Environ.* **2018**, *62*, 330–344. [CrossRef]

21. Meyer, D. Germany's First Diesel Vehicle Ban Has Started in Hamburg. Available online: https://fortune.com/2018/05/31/germany-hamburg-diesel-ban/ (accessed on 16 July 2019).

22. Möhner, M. Driving ban for diesel-powered vehicles in major cities: An appropriate penalty for exceeding the limit value for nitrogen dioxide? *Int. Arch. Occup Environ. Health* **2018**, *91*, 373–376. [CrossRef]

23. O'Neill, E.; Moore, D.; Kelleher, L.; Brereton, F. Barriers to electric vehicle uptake in Ireland: Perspectives of car-dealers and policy-makers. *Case Stud. Transp. Policy* **2019**, *7*, 118–127. [CrossRef]

24. Ellison, R.B.; Greaves, S.P.; Hensher, D.A. Five years of London's low emission zone: Effects on vehicle fleet composition and air quality. *Transp. Res. Part D Transp. Environ.* **2013**, *23*, 25–33. [CrossRef]

25. Ferreira, F.; Gomes, P.; Tente, H.; Carvalho, A.C.; Pereira, P.; Monjardino, J. Air quality improvements following implementation of Lisbon's Low Emission Zone. *Atmos. Environ.* **2015**, *122*, 373–381. [CrossRef]

26. Fensterer, V.; Küchenhoff, H.; Maier, V.; Wichmann, H.-E.; Breitner, S.; Peters, A.; Gu, J.; Cyrys, J. Evaluation of the Impact of Low Emission Zone and Heavy Traffic Ban in Munich (Germany) on the Reduction of PM10 in Ambient Air. *Int. J. Environ. Res. Public Health* **2014**, *11*, 5094–5112. [CrossRef] [PubMed]

27. Dastgerdi, A.S.; De Luca, G. Strengthening the city's reputation in the age of cities: An insight in the city branding theory. *City Territ. Archit.* **2019**, *6*, 2. [CrossRef]

28. De Jong, M.; Hoppe, T.; Noori, N. City Branding, Sustainable Urban Development and the Rentier State. How do Qatar, Abu Dhabi and Dubai present Themselves in the Age of Post Oil and Global Warming? *Energies* **2019**, *12*, 1657. [CrossRef]

29. Greater London Authority. A Growing Population. Available online: https://www.london.gov.uk//what-we-do/planning/london-plan/current-london-plan/london-plan-chapter-one-context-and-strategy-0 (accessed on 16 July 2019).

30. Font, A.; Guiseppin, L.; Blangiardo, M.; Ghersi, V.; Fuller, G.W. A tale of two cities: Is air pollution improving in Paris and London? *Environ. Pollut.* **2019**, *249*, 1–12. [CrossRef]

31. Statista. Licensed Cars in London, England 1995–2018. Available online: https://www.statista.com/statistics/314980/licensed-cars-in-london-england-united-kingdom/ (accessed on 27 August 2019).

32. Walton, H.; Dajnak, D.; Beevers, S.; Williams, M.; Watkiss, P.; Hunt, A. Understanding the Health Impacts of Air Pollution in London. 2015. Available online: https://www.london.gov.uk/WHAT-WE-DO/environment/environment-publications/understanding-health-impacts-air-pollution-london (accessed on 15 July 2019).

33. Gordon, I.R.; McCann, P. Innovation, agglomeration, and regional development. *J. Econ. Geogr.* **2005**, *5*, 523–543. [CrossRef]

34. Cao, M.; Hickman, R. Car dependence and housing affordability: An emerging social deprivation issue in London? *Urban Stud.* **2018**, *55*, 2088–2105. [CrossRef]

35. Davies, A.; MacAulay, S.; DeBarro, T.; Thurston, M. Making Innovation Happen in a Megaproject: London's Crossrail Suburban Railway System. *Proj. Manag. J.* **2014**, *45*, 25–37. [CrossRef]

36. Mindell, J.S.; Cohen, J.M.; Watkins, S.; Tyler, N. Synergies between low-carbon and healthy transport policies. *Proc. Inst. Civ. Eng. Transp.* **2011**, *164*, 127–139. [CrossRef]

37. Quarmby, S.; Santos, G.; Mathias, M. Air Quality Strategies and Technologies: A Rapid Review of the International Evidence. *Sustainability* **2019**, *11*, 2757. [CrossRef]

38. Greater London Authority. London Atmospheric Emissions Inventory (LAEI) 2013—London Datastore. Available online: https://data.london.gov.uk/dataset/london-atmospheric-emissions-inventory-2013 (accessed on 25 July 2019).

39. Valliant, R.; Dever, J.A.; Kreuter, F. Basic steps in weighting. In *Practical Tools for Designing and Weighting Survey Samples. Statistics for Social and Behavioral Sciences*; Springer: London, UK, 2013.

40. Cao, M. Exploring the Relation between Transport and Social Equity: Empirical Evidence from London and Beijing. Ph.D. Thesis, University College London, London, UK, 2019.

41. Cao, M.; Hickman, R. Urban transport and social inequities in neighbourhoods near underground stations in Greater London. *Transp. Plan. Technol.* **2019**, *42*, 419–441. [CrossRef]

42. Cao, M.; Robin, H. Transport, social equity and capabilities in East Beijing. In *Handbook on Transport and Urban Transformation in Contemporary China*; Chen, C., Pan, H., Shen, Q., Wang, J., Eds.; Edward Elgar: London, UK, 2019.

43. Cao, M.; Hickman, R. Understanding travel and differential capabilities and functionings in Beijing. *Transp. Policy* **2019**. [CrossRef]

44. Capata, R. Urban and extra-urban hybrid vehicles: A technological review. *Energies* **2018**, *11*, 2924. [CrossRef]

45. Bamberg, S. How does environmental concern influence specific environmentally related behaviors? A new answer to an old question. *J. Environ. Psychol.* **2003**, *23*, 21–32. [CrossRef]

46. Zhao, P.; Li, S.; Li, P.; Liu, J.; Long, K. How does air pollution influence cycling behaviour? Evidence from Beijing. *Transp. Res. Part D Transp. Environ.* **2018**, *63*, 826–838. [CrossRef]

47. Zhang, M.; He, S.; Zhao, P. Revisiting inequalities in the commuting burden: Institutional constraints and job-housing relationships in Beijing. *J. Transp. Geogr.* **2018**, *71*, 58–71. [CrossRef]

48. Hickman, R.; Cao, M.; Mella Lira, B.; Fillone, A.; Bienvenido Biona, J. Understanding capabilities, functionings and travel in high and low income neighbourhoods in Manila. *Soc. Incl.* **2017**, *5*, 161–174. [CrossRef]

49. Dalgaard, P. Analysis of variance and the Kruskal—Wallis Test. In *Introductory Statistics with R*; Springer: New York, NY, USA, 2008; pp. 127–143.

50. Leon, A.C. Descriptive and inferential statistics. In *Comprehensive Clinical Psychology*; Bellack, A.S., Hersen, M., Eds.; Pergamon: Oxford, UK, 1998; pp. 243–285.

51. Martins, F.; Felgueiras, C.; Smitkova, M.; Caetano, N. Analysis of fossil fuel energy consumption and environmental impacts in European Countries. *Energies* **2019**, *12*, 964. [CrossRef]

52. Vergura, S. Hypothesis tests-based analysis for anomaly detection in photovoltaic systems in the absence of environmental parameters. *Energies* **2018**, *11*, 485. [CrossRef]

53. Zhang, L.; Mu, Z.; Gao, X. Coupling analysis and performance study of commercial 18650 lithium-ion batteries under conditions of temperature and vibration. *Energies* **2018**, *11*, 2856. [CrossRef]

54. Siegel, S., Jr.; Castellan, N.J. *Nonparametric Statistics for The Behavioral Sciences*, 2nd ed.; McGraw-Hill: Boston, MA, USA, 1988; ISBN 978-0-07-057357-4.

55. Fellows, R.; Liu, A. *Research Methods for Construction*, 4th ed.; Wiley: Chichester, UK; Blackwell: Malden, MA, USA, 2015; ISBN 978-1-118-91572-1.

56. Russell-Jones, R. Air pollution in the UK: Better ways to solve the problem. *BMJ* **2017**, *357*, j2713. [CrossRef] [PubMed]

57. MacKerron, G.; Mourato, S. Life satisfaction and air quality in London. *Ecol. Econ.* **2009**, *68*, 1441–1453. [CrossRef]

58. Air Quality in Hackney: A Guide for Public Health Professionals. Available online: https://www.london.gov.uk/sites/default/files/air_quality_for_public_health_professionals_-_lb_hackney.pdf (accessed on 15 July 2019).

59. Curtis, C.; Headicar, P. Targeting travel awareness campaigns: Which individuals are more likely to switch from car to other transport for the journey to work? *Transp. Policy* **1997**, *4*, 57–65. [CrossRef]

60. Goyal, P. Present scenario of air quality in Delhi: A case study of CNG implementation. *Atmos. Environ.* **2003**, *37*, 5423–5431. [CrossRef]

61. Vidhi, R.; Shrivastava, P. A review of electric vehicle lifecycle emissions and policy recommendations to increase EV penetration in India. *Energies* **2018**, *11*, 483. [CrossRef]

62. Tietge, U. Cities Driving Diesel Out of the European Car Market. Available online: https://www.fiafoundation.org/media/597008/uwe-tietge.pdf (accessed on 27 July 2019).

63. German Cities Ban Older Diesel Cars. Available online. https://www.reuters.com/article/us-germany-emissions-factbox/factbox-german-cities-ban-older-diesel-cars-idUSKCN1NK28L (accessed on 27 July 2019).

64. World Health Organization. Disability-Adjusted Life Years (DALYs). Available online: http://www.who.int/gho/mortality_burden_disease/daly_rates/text/en/ (accessed on 16 July 2019).

65. Khaola, P.P.; Potiane, B.; Mokhethi, M. Environmental concern, attitude towards green products and green purchase intentions of consumers in Lesotho. *Ethiop. J. Environ. Stud. Manag.* **2014**, *7*, 361–370. [CrossRef]

66. Dogan, V.; Ozmen, M. Belief in environmentalism and independent/interdependent self-construal as factors predicting interest in and intention to purchase hybrid electric vehicles. *Curr. Psychol.* **2017**. [CrossRef]

67. Rezvani, Z.; Jansson, J.; Bodin, J. Advances in consumer electric vehicle adoption research: A review and research agenda. *Transp. Res. Part D Transp. Environ.* **2015**, *34*, 122–136. [CrossRef]

68. Kwon, Y.; Son, S.; Jang, K. Evaluation of incentive policies for electric vehicles: An experimental study on Jeju Island. *Transp. Res. Part A Policy Pract.* **2018**, *116*, 404–412. [CrossRef]

69. Krupa, J.S.; Rizzo, D.M.; Eppstein, M.J.; Brad Lanute, D.; Gaalema, D.E.; Lakkaraju, K.; Warrender, C.E. Analysis of a consumer survey on plug-in hybrid electric vehicles. *Transp. Res. Part A Policy Pract.* **2014**, *64*, 14–31. [CrossRef]

70. Zhang, Y.; Yu, Y.; Zou, B. Analyzing public awareness and acceptance of alternative fuel vehicles in China: The case of EV. *Energy Policy* **2011**, *39*, 7015–7024. [CrossRef]

71. Jenn, A.; Springel, K.; Gopal, A.R. Effectiveness of electric vehicle incentives in the United States. *Energy Policy* **2018**, *119*, 349–356. [CrossRef]

72. Ketzel, M.; Omstedt, G.; Johansson, C.; Düring, I.; Pohjola, M.; Oettl, D.; Gidhagen, L.; Wåhlin, P.; Lohmeyer, A.; Haakana, M.; et al. Estimation and validation of PM2.5/PM10 exhaust and non-exhaust emission factors for practical street pollution modelling. *Atmos. Environ.* **2007**, *41*, 9370–9385. [CrossRef]

73. Howard, R. Up in the Air: How to Solve London's Air Quality Crisis—Part 1. 2015. Available online: https://Policyexchange.Org.Uk/Publication/Up-In-The-Air-How-To-Solve-Londons-Air-Quality-Crisis-Part-1/ (accessed on 15 July 2019).
74. Graham-Rowe, E.; Gardner, B.; Abraham, C.; Skippon, S.; Dittmar, H.; Hutchins, R.; Stannard, J. Mainstream consumers driving plug-in battery-electric and plug-in hybrid electric cars: A qualitative analysis of responses and evaluations. *Transp. Res. Part A Policy Pract.* **2012**, *46*, 140–153. [CrossRef]
75. Jensen, A.F.; Cherchi, E.; Mabit, S.L. On the stability of preferences and attitudes before and after experiencing an electric vehicle. *Transp. Res. Part D Transp. Environ.* **2013**, *25*, 24–32. [CrossRef]
76. Lieven, T.; Mühlmeier, S.; Henkel, S.; Waller, J.F. Who will buy electric cars? An empirical study in Germany. *Transp. Res. Part D Transp. Environ.* **2011**, *16*, 236–243. [CrossRef]
77. Wikström, M.; Hansson, L.; Alvfors, P. Investigating barriers for plug-in electric vehicle deployment in fleets. *Transp. Res. Part D Transp. Environ.* **2016**, *49*, 59–67. [CrossRef]
78. Globisch, J.; Dütschke, E.; Schleich, J. Acceptance of electric passenger cars in commercial fleets. *Transp. Res. Part A Policy Pract.* **2018**, *116*, 122–129. [CrossRef]
79. Cao, M.; Chen, C.-L.; Hickman, R. Transport emissions in Beijing: A scenario planning approach. *Proc. Inst. Civ. Eng. Transp.* **2017**, *170*, 65–75. [CrossRef]
80. Hooftman, N.; Messagie, M.; Van Mierlo, J.; Coosemans, T. A review of the European passenger car regulations—Real driving emissions vs local air quality. *Renew. Sustain. Energy Rev.* **2018**, *86*, 1–21. [CrossRef]
81. Egbue, O.; Long, S. Barriers to widespread adoption of electric vehicles: An analysis of consumer attitudes and perceptions. *Energy Policy* **2012**, *48*, 717–729. [CrossRef]
82. Contribution of Working Group I to the Fourth Assessment Report of the Intergovernmental Panel on Climate Change. Available online: https://www.ipcc.ch/site/assets/uploads/2018/02/ar4_syr_full_report.pdf (accessed on 12 August 2019).

 *energies*

Article

Uncovering Household Carbon Footprint Drivers in an Aging, Shrinking Society

Yuzhuo Huang [1], Yosuke Shigetomi [1,*], Andrew Chapman [2] and Ken'ichi Matsumoto [1,*]

[1] Graduate School of Fisheries and Environmental Sciences, Nagasaki University, 1-14 Bunkyo-machi,
    Nagasaki 852-8521, Japan; bb53618016@ms.nagasaki-u.ac.jp
[2] International Institute for Carbon-Neutral Energy Research, Kyushu University, 744 Motooka, Nishi-ku,
    Fukuoka 819-0395, Japan; chapman@i2cner.kyushu-u.ac.jp
*   Correspondence: y-shigetomi@nagasaki-u.ac.jp (Y.S.); kenichimatsu@nagasaki-u.ac.jp (K.M.);
    Tel.: +81-95-819-2785 (Y.S.); Tel.: +81-95-819-2735 (K.M.)

Received: 31 July 2019; Accepted: 28 September 2019; Published: 30 September 2019

Abstract: In order to meet climate change mitigation goals, nations such as Japan need to consider strategies to reduce the impact that lifestyles have on overall emission levels. This study analyzes carbon footprints from household consumption (i.e., lifestyles) using index and structural decomposition analysis for the time period from 1990 to 2005. The analysis identified that households in their 40s and 50s had the highest levels of both direct and indirect CO_2 emissions, with decomposition identifying consumption patterns as the driving force behind these emissions and advances in CO_2 reduction technology having a reducing effect on lifestyle emissions. An additional challenge addressed by this study is the aging, shrinking population phenomenon in Japan. The increase in the number of few-member and elderly households places upward pressure on emissions as the aging population and declining national birth rate continues. The analysis results offer two mitigatory policy suggestions: the focusing of carbon reduction policies on older and smaller households, and the education of consumers toward low-carbon consumption habits. As the aging, shrinking population phenomenon is not unique to Japan, the findings of this research have broad applications globally where these demographic shifts are being experienced.

Keywords: CO_2 emissions; carbon footprint; household consumption; index decomposition analysis; structural decomposition analysis; aging society; Japan

---

1. Introduction

To meet the global climate change mitigation target of limiting global warming to just 1.5 °C [1], an extensive effort toward reducing carbon emissions throughout our lifestyles is critical. Leading this effort is particularly essential for developed nations, which are more likely to emit larger amounts of carbon dioxide (CO_2) and other greenhouse gases (GHG) and who have access to cleaner technologies than developing nations. Consumption-based accounting [2,3] quantifies not only direct or territorial emissions due to fuel combustion but also indirect emissions generated through the supply chains of goods and services, allowing for the consideration of broader abatement options from both the demand and supply sides [4]. This is known as a "cradle-to-gate" (i.e., from raw material extraction to final consumption) assessment. From this point of view, our lifestyle (household consumption) has been highlighted as playing a dominant role in cradle-to-gate GHG emissions, as measured through national carbon footprints (CF) [5–7]. Globally, almost 65% of CF are induced by household consumption [8]. In order to achieve a low-carbon future, drastic changes are needed not only in industrial supply structures but also in our daily lifestyle structures.

Japan, selected as the target nation for this study, is among the most industrialized GHG-emitting nations [9]. In line with Japan's Paris Agreement goals, GHG emissions need to be reduced by 26% by

2030 compared with 2013 levels. To achieve this, it is necessary to implement effective measures with respect to technological innovation and envisaged demographic shifts [10,11]. Japan's CF structure has been analyzed, identifying that household consumption is dominant, similar to other developed nations [12]. Further, previous research has analyzed the household CF in Japan, identifying key consumption and behavior domains for effective reduction [13–18]. However, this precedential research did not examine time trends of household CF or the crucial supply and demand drivers.

This study sheds light on structural changes in Japanese household CF to elucidate levers which enable the reduction of CF and enable an understanding of how supply and demand drivers underpin these CF. In addition, this research focuses on demographic trends as a demand driver of changes in CF due to an aging, shrinking population—a serious issue faced in Japan and other nations, which may affect consumption and resultant CF values [13,18,19]. For instance, Shigetomi et al. [11] evaluated the impact of demographic trends on energy-related CO_2 emissions from residents during 1990–2015 using decomposition analysis, indicating that as the number of single households increased, especially older households, a commensurate increase in emissions was observed.

In order to analyze the underpinning factors of CF, this study adopts index decomposition analysis (IDA) [20] and structural decomposition analysis (SDA) [21] on a time-series data set of both direct and indirect CO_2 emissions induced by Japanese households. To date, IDA has been applied mainly to examine the driving forces of energy consumption and its relation to CO_2 emissions by sector [22,23]. SDA has clarified the determinants of lifecycle emissions within CF, utilizing input–output table data [24,25]. These two approaches have been utilized to analyze key determinants of home energy-related CO_2 emissions [26–28] and household CF [29–31], respectively, regarding the carbon emission intensity, supply chain structure, consumption volume and composition, and population.

Studies relevant to Japan have assessed energy-related CO_2 emissions at the national level [32], specific to services [33], transport [34], manufacturing [35], and residential sectors [11], identifying the key drivers for reducing emissions by sector. For instance, related to household consumption, Shigetomi et al. [11] examined the trends in residential energy-related CO_2 emissions across 47 regions in Japan during 1990–2015 using the regional energy balance tables, demographic census and consumer expenditure survey, demonstrating the importance of prioritizing national and local policy interventions of the supply and demand sides in light of the differences in the key drivers in each region. They do not consider, however, the upstream (indirect) CO_2 emissions created by the overall household consumption via the supply chain. To the best of our knowledge, this study highlights the structural changes in Japanese household CF detailing both direct and indirect energy-related CO_2 emissions for the first time. This novel research uncovers insights for mitigating household CF with respect to both supply and demand factors and demographics, notably an aging, shrinking society.

The research is organized as follows: Section 2 describes the methods and data employed to calculate household CF and to estimate the contribution of drivers toward structural changes. Section 3 presents the results and discussion. Finally, Section 4 concludes with detailed policy implications and a future outlook based on the obtained results.

2. Materials and Methods

Here, the materials and methodology are defined, including the quantification and decomposition approaches, data utilized, and the methodological and data limitations.

2.1. Quantification of Carbon Footprint by Household Consumption

Household CF is defined as the sum of direct carbon emissions induced by driving a passenger motor car, cooking and household heating (D), and indirect (embodied) carbon emissions generated through the supply chain due to household consumption (S). D is calculated using Equation (1).

$$D = \sum_k \sum_b e_k^{dir} f_{kb}, \tag{1}$$

where e_k^{dir} represents the direct CO_2 emissions per consumption expenditure for energy item k. f_{kb} denotes the household's final consumption by attribute b for energy item k. Next, S is quantified by Equations (2) and (3).

$$S = \sum_i \sum_b e_i^{ind} f_{ib},\tag{2}$$

$$e_i^{ind} = \sum_j q_i L_{ij},\tag{3}$$

where e_i^{ind} represents the upstream CO_2 emissions per consumption expenditure (embodied CO_2 emission intensity) for commodity $i \ni k$. j denotes the commodity sector. It is estimated by using q_i and $L_{ij} = \left(I - A_{ij}\right)^{-1}$ which denote the vector containing the direct CO_2 emissions per unit production output for commodity i and upstream requirements per unit production, respectively. L_{ij} is an element of the Leontief inverse matrix [36] obtained from the input–output table.

2.2. Index Decomposition Analysis and Structural Decomposition Analysis

To comprehend the contribution of various indicators to the changes in household CF by using IDA and SDA, we decomposed both the direct CO_2 emissions derived from home energy and the indirect CO_2 emissions generated through the supply chain of goods and services (commodities) purchased by households, as shown in Equations (4) and (5).

$$\begin{aligned}
D &= \sum_k \sum_b e_k^{dir} \frac{f_{kb}}{f_b} \frac{f_b}{p_b} \frac{p_b}{H_b} \frac{H_b}{H} H \\
&= \sum_k \sum_b e_k^{dir} y_{kb} w_b s_b d_b H
\end{aligned}\tag{4}$$

$$\begin{aligned}
S &= \sum_i \sum_b q_i L_{ij} \frac{f_{ib}}{f_b'} \frac{f_b'}{p_b} \frac{p_b}{H_b} \frac{H_b}{H} H \\
&= \sum_i \sum_b q_i L_{ij} y_{ib} w_b' s_b d_b H
\end{aligned}\tag{5}$$

where H and p represent the total number of households and population, respectively. Both y_{kb} and y_{ib} refer to consumption patterns (e.g., medical services are more heavily consumed by elderly households than younger households). w_b and w_b' represent the average per-capita consumption volume for energy items and that for all commodities, respectively. s_b represents the average number of members in each household (i.e., family size). d_b describes the distribution of households (i.e., the proportion of younger households to total households). Thus, Equations (4) and (5) are based on Equations (1) and (2), with household final consumption decomposed into the five factors in line with consumption pattern, consumption volume, family size, household distribution, and number of households. Overall, six drivers are considered for direct CO_2 emissions, and seven drivers for indirect CO_2 emissions.

When D and S shift from year t to year $t + 1$, there are no unique solutions for how the decomposition should be solved. To quantify the contributions of each factor, this study used the Shapley–Sun decomposition approach (S-S method) [37] for D, and the Dietzenbacher and Los decomposition approach (D-L method) [38] for S, cognizant of identical decomposition without any residues and the commonality of results [24]. For example, the total difference of Equation (4) can be represented by Equation (6).

$$\begin{aligned}
\Delta D = &\sum_k \sum_b \Delta e_k^{dir} y_{kb} w_b s_b d_b H + e_k^{dir} \Delta y_{kb} w_b s_b d_b H + e_k^{dir} y_{kb} \Delta w_b s_b d_b H \\
&+ e_k^{dir} y_{kb} w_b \Delta s_b d_b H + e_k^{dir} y_{kb} w_b s_b \Delta d_b H + e_k^{dir} y_{kb} w_b s_b d_b \Delta H
\end{aligned}\tag{6}$$

where Δ indicates the difference operator. Equation (6) converts six multiplicative terms in the first term of Equation (4) into six additive terms. Each additive term in Equation (6) denotes the contribution to changes in D induced by a targeted factor while all other factors are constant. For instance, the first term in Equation (6) refers to the effect on direct CO_2 emissions of changes in direct emission intensity, while consumption patterns, consumption volume, family size, household distribution, and total number of households are constant between t and $t + 1$. Each of the contributions were estimated by taking the average of the $6! = 720$ decomposition equations possible [37]. Here, the effects on direct CO_2 emissions that are related to the first, second, third, fourth, fifth, and sixth terms in Equation (6) are referred to as the intensity effect (direct), consumption pattern effect, consumption volume effect, family size effect, household distribution effect, and household number effect, respectively.

In a similar manner, the total difference of Equation (5) can be demonstrated by Equation (7).

$$\Delta S = \sum_i \sum_b \Delta q_i L_{ij} y_{ib} w_b s_b d_b H + q_i \Delta L_{ij} y_{ib} w_b s_b d_b H + q_i L_{ij} \Delta y_{ib} w_b s_b d_b H$$
$$+ q_i L_{ij} y_{ib} \Delta w_b s_b d_b H + q_i L_{ij} y_{ib} w_b \Delta s_b d_b H + q_i L_{ij} y_{ib} w_b s_b \Delta d_b H \qquad (7)$$
$$+ q_i L_{ij} y_{ib} w_b s_b d_b \Delta H$$

Finally, each of the contributions were estimated by taking the average of the $7! = 5040$ decomposition equations possible [38]. Here, the effects on indirect CO_2 emissions that are related to the first, second, third, fourth, fifth, sixth, and seventh terms are referred to as the intensity effect (indirect), supply chain effect, consumption pattern effect, consumption volume effect, family size effect, household distribution effect, and household number effect, respectively.

2.3. Data

This study uses the time-series Japan input–output tables (TJIO) consisting of the economic transaction (L_{ij}) and household final demand structures ($\sum_b f_{ib}$) for 1990, 1995, 2000, and 2005 based on the 2005 price with 397 common commodities. Hence, the data are comparable among periods. Further, we disaggregated the commodities within petroleum products into six detailed commodities including gasoline, light oil, kerosene, liquefied petroleum gas (LPG), and other petroleum products (i.e., lubricants) by using the Comprehensive Energy Statistics [39]. To identify the relationship between household CF and demographic trends, the final household demands from the TJIO were divided into consumption expenditures for six age groups of the highest income earner in the household ($29\leq$, 30–39, 40–49, 50–59, 60–69, $\geq$70) for each year using the national survey of family income and expenditure (NSFIE) [10]. Then, the consumption share by attribute, summation of each final consumption type by commodity and the difference between the producer-based price and consumer-based price were considered as detailed in a previous study [14]. e_i^{ind} was obtained based on Equation (3). q_i was calculated by dividing sectoral CO_2 emissions obtained from the 3EID [41] by total output utilizing the TJIO. b denotes the age of the highest income earner (1: $\leq$29, 2: 30–39, 3: 40–49, 4: 50–59, 5: 60–69, 6: $\geq$70).

2.4. Limitations

The approach used in this study has several limitations with regard to the data and methodology employed. First, to conduct SDA, it is necessary to prepare time-series data on both household consumption expenditures and environmental burden intensities that are consistent with the IO table and to deflate according to the base year price information. In this regard, this study used the TJIO covering the periods 1990–1995–2000–2005. We recognize that the latest year analyzed in this study is more than 10 years in the past; however, it is currently impossible to prepare more recent deflated data for consumption expenditure and embodied CO_2 emission intensity. The latest domestic IO table describes the economic transactions accounting for household consumption expenditures in 2011, and the embodied CO_2 emission intensity values for 2011 have already been published. However, it is relatively difficult to deflate consumption expenditures and intensities due to the disconnection between commodity sectors and their definitions.

3. Results and Discussion

The results and discussion are divided into the three sections of overall trends in emissions, changes within sectors and age groups, and the decomposition of the driving forces underpinning emissions.

3.1. Overall Trends of Total Direct and Indirect CO_2 Emissions

CO_2 emissions induced by Japanese household consumption from 1990 to 2005 are shown in Table 1. Both direct and indirect CO_2 emissions showed a significant growth trend during the analyzed period. In terms of overall emissions, indirect CO_2 emissions levels remain nearly four times those of direct CO_2 emissions. The annual average increase of direct CO_2 emissions is relatively small, and emissions in 2005 are slightly lower than those in the year 2000. However, indirect CO_2 emissions continuously increased, with an annual average increase of 6.6 Mt in 1990–2005. In terms of the growth rate of emissions, that of direct emissions was higher than that of indirect emissions during the studied period (see Table 1). These characteristics will be elaborated by observing the trends in direct and indirect CO_2 emissions by sector, as presented in the following section.

Table 1. Compositions of Japanese household carbon footprints (CF) in 1990–2005

	Year				Average Annualized Increase	Growth Rate (between 1990 and 2005)
	1990	1995	2000	2005		
Direct CO_2 emissions (Mt-CO_2)	103.6	124.2	143.6	142.5	2.6	37.5%
Indirect CO_2 emissions (Mt-CO_2)	473.5	542.6	537.9	572.3	6.6	20.9%
Total CO_2 emissions (Mt-CO_2)	577.2	666.9	681.5	714.8	9.2	23.8%

3.2. Changes in Direct and Indirect CO_2 Emissions in Different Sectors and Age Groups

The total direct and indirect CO_2 emissions are disaggregated into sectors and household age groups to evaluate their impact on CO_2 emissions. Fourteen sectors are considered: "food and non-alcoholic beverages," "alcoholic beverages and tobacco," "clothing and footwear," "housing," "furnishings," "medicals," "private vehicles," "public transport," "information and communication," "recreation and culture," "education," "restaurants and hotels," "consumable goods," and "margins, religions and other services." These sectors are determined based on the "Classification of Individual Consumption by Purpose."

Figure 1 presents the trends in total direct CO_2 emissions by sector and average direct CO_2 emissions per capita across age groups. To observe the impact of demographic factors on carbon emissions in an aging society in greater detail, we examined the per capita CO_2 emissions as well as the per household emissions across different age groups. Here, we focus on the per capita emissions (Figure 1b), because the per household emissions are affected by the average family size and composition of households attributed to each age group. The results of per household direct and indirect emissions are detailed in Appendix A.

Private vehicles and housing are the only two sectors that generate direct CO_2 emissions from households, as fossil fuels are used for these activities. For instance, the private vehicles sector includes gasoline and light oil, while the housing sector includes kerosene, LPG, coal products, and city gas. As shown in Figure 1a, the private vehicles sector accounts for a large proportion of direct CO_2 emissions throughout the analyzed period, impacting the high growth rate of direct emissions, as referred to above. From 1990 to 2005, Japanese car ownership rose from 57.99 million to 78.28 million vehicles—an increase of approximately 35% [42]. With this increase in car ownership, the growth rate of gasoline consumption was much larger than for other fuels. Consequently, the direct emissions for the private vehicles sector increased significantly compared to those for the housing sector.

Considering different age groups, direct CO_2 emissions (per capita) from the private vehicles sector are concentrated within two age groups—40s and 50s—as shown in Figure 1b. For those in their 40s, this may be due to work and family needs. With the increase of household savings in these age groups, many families tend to own their own cars and use them frequently. The 50s age group has the highest direct CO_2 emissions across all age groups, possibly because the annual income in the 50s is higher than that of other age groups [40]. In addition, with the increase in family members, the household size has also expanded to a certain extent when compared to others, potentially expanding the demand for private vehicles, particularly large-sized cars. As for the direct CO_2 emissions from the housing sector, these are also concentrated in the 40s and 50s age groups. This may be because more people in their 40s and 50s are married and living with their children, and they tend to live in relatively large, energy-consuming houses. Furthermore, the changes in direct CO_2 emissions in different age groups also showed certain peculiarities, such as CO_2 emissions gradually increasing for householders in their 20s, reaching their peak in the 50s and then subsequently declining.

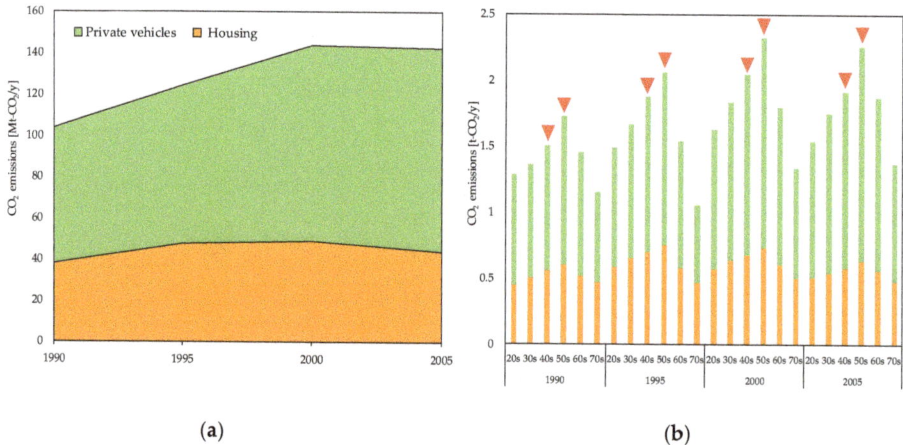

Figure 1. Sectoral composition of direct CO_2 emissions from 1990–2005. (**a**) Total (Mt-CO_2/y) and (**b**) per capita by age group (t CO_2/y). Inverted triangles denote the noteworthy household age groups as detailed in the main text.

The change in indirect CO_2 emissions in different age groups is similar to that of direct CO_2 emissions (see Figures 1b and 2b). Considering the order of growth in sectoral indirect emissions, we selected food and non-alcoholic beverages, housing, and public transport sectors for discussion here, as shown in Figure 2a. First, the indirect CO_2 emissions generated by the housing sector were mainly concentrated in the three age groups of the 40s, 50s, and 60s, as demonstrated by Figure 2b. The reason behind the similarity between direct and indirect CO_2 emissions of those in their 40s and 50s is that residents modify their houses to meet the needs of family life and child-rearing. For the 60s age group, a consideration of the living environment and living conditions for old-age life could increase the cost of housing and lead to the production of more indirect CO_2 emissions.

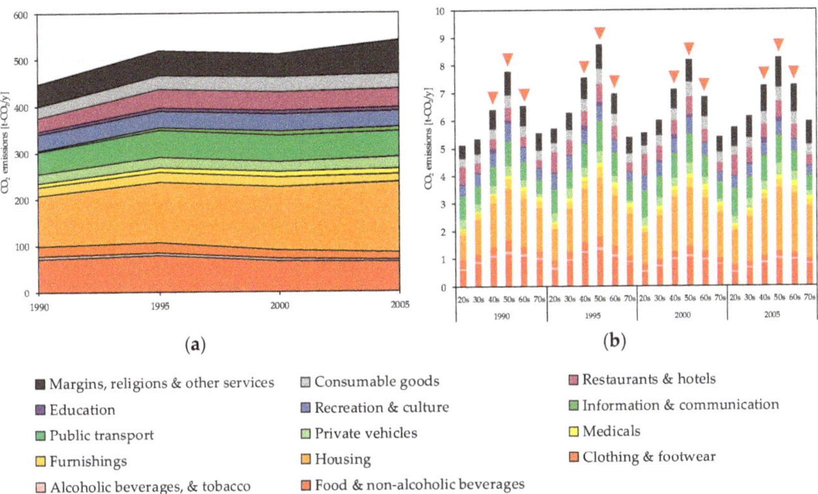

Figure 2. Sectoral composition of indirect CO_2 emissions from 1990–2005. (**a**) Total (Mt-CO_2/y) and (**b**) per capita by age group (t-CO_2/y). Inverted triangles denote the noteworthy household age groups as detailed in the main text.

Indirect CO_2 emissions from the food and non-alcoholic beverages sector were also concentrated in the 40s, 50s and 60s age groups. In the 40s and 50s age groups, the expansion of household size may lead to an increase in food consumption, which could increase indirect CO_2 emissions. As for the 60s age group, elderly people tend to spend more money on high-quality, expensive food compared to the other age groups. In addition, although the household size is smaller than the 40s and 50s age groups, it is still larger than the age groups of the 20s, 30s, and 70s. These factors combined make the 60s age group the third-largest indirect CO_2 emitter on a per capita basis.

Indirect CO_2 emissions from the public transport sector were concentrated in the 20s and 50s age groups. Compared with other age groups, the proportion of private vehicle possession in the 20s age group is relatively low, leading to an increased use of public transport, which could increase indirect CO_2 emissions from this sector. People in the households with in the 50s age group produced the highest CO_2 emissions from both private vehicles and public transport sectors. Generally, people in the 50s age group have the highest annual income, and with the expansion of household size, families are more likely to use private vehicles alongside the children of these families using public transport for attending school.

While the overall CO_2 emissions of the public transport sector are increasing, the growth rate is not as fast as that of the private vehicles sector. In addition, there are different trends in indirect emissions from the private vehicles and public transport sectors by age group. In 2005, indirect emissions for the private vehicles sector for those in their 20s decreased by 21%, while those in other age groups, particularly the 30s and 60s, increased by 13–35% compared to 1990. The reasoning for such a decline in the 20s age group is that they were more likely to purchase smaller, less expensive vehicles such as lightweight automobiles (known as kei-cars in Japan) due to financial aspects. On the other hand, because age groups from the 30s to 50s are more likely to be involved in child-raising activities, family life has led to an increased demand for private transport. Furthermore, due to the demographic shift related to an aging, shrinking population leading to a postponement of childbearing age, those in their 60s also have experienced an increase in their use of private vehicles.

For public transport, growth trends were evident across all age groups. Among them, the growth is particularly significant for those in their 20s and 40s. In the 20s age group, the reduction in the use of private vehicles led to an increase in public transport demand. The use of public transport in the 40s age group also increased, which is likely due to the expansion of travel needs for work and school as well as an increase in the use of private vehicles, although at a lower rate than for the 30s, 50s, and 60s.

On average, annual income increases from the 20s to the 50s, subsequently declining and leading to an increase in consumption and savings, creating upward pressure on CO_2 emissions [40]. Meanwhile, because of the decline in income after the 50s, indirect CO_2 emissions also tend to decline. Further, the household size expands from the 20s to the 50s, while post 50s, the household begins to shrink as children become independent and establish their own households, which is likely linked to the peak in CO_2 emissions observed for the 50s age group.

3.3. Decomposition Results

3.3.1. Driving Forces of Total Direct and Indirect CO_2 Emissions

Figure 3 shows the results of examining the factors affecting direct CO_2 emissions during 1990–2005 using the IDA.

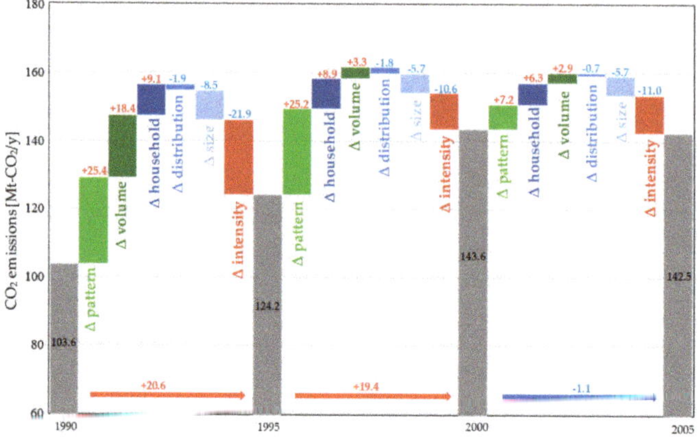

Figure 3. Driving forces of direct household CO_2 emissions in Japan during 1990-2005. Δintensity is the intensity effect, Δpattern is the consumption pattern effect, Δvolume is the consumption volume effect, Δsize is the household size effect, Δdistribution is the household distribution effect, and Δhousehold is the household number effect.

Among factors, the intensity effect was the main driver in reducing direct CO_2 emissions, while the consumption pattern was the main driver which increased emissions. The negative impact of the intensity effect on direct CO_2 emissions has progressed over time, indicating that Japan has made substantial progress in the carbon reduction technology used in daily life since 1990.

Figure 4 describes the direct CO_2 emissions by household energy item. The consumption of gasoline increased significantly prior to 2000, which is one of the reasons for the growth of direct CO_2 emissions driven by the consumption pattern during this period. After 2000, along with the slowdown of the growth of gasoline consumption, the positive impact of the consumption pattern has also weakened. In addition, the total number of households in Japan increased from 40.67 million to 49.06 million between 1990 and 2005 [11], promoting a positive impact of the number of households on direct CO_2 emissions. The positive effect of consumption volume on direct CO_2 emissions weakened consistently between 1990 and 2005. This is probably related to the increase of small-scale households which could reduce their energy consumption to a certain extent. Moreover, the negative impact of household size and household distribution on direct CO_2 emissions is gradually increasing, which is likely due to the influence of recent demographic trends such as an increase in one-person households and a reduction in household size because of an aging society with fewer children.

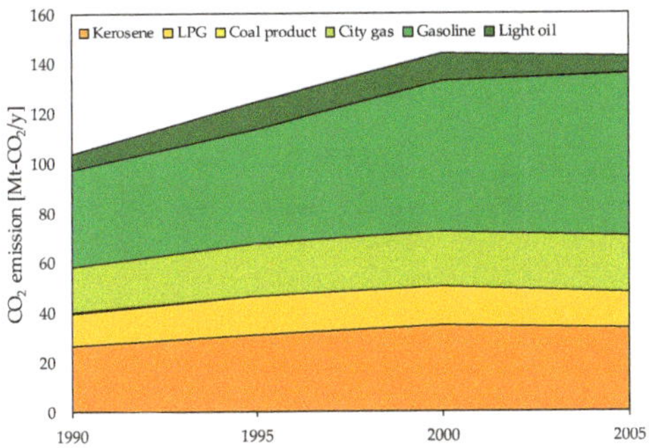

Figure 4. Trend in the direct CO$_2$ emissions by energy item.

Considering the effects of household size, the number of households and household distribution in 1995, 2000 and 2005, we find that the sum of these effects gradually decreased direct CO$_2$ emissions. This suggests that if Japan maintains an aging society with a low birth rate, direct CO$_2$ emissions generated by the household sector will gradually decrease.

As for indirect CO$_2$ emissions, both the total emissions and change in emission levels were higher than for direct CO$_2$ emissions, and the sectoral composition of indirect CO$_2$ emissions is more diverse. Further, the change in driving forces of indirect CO$_2$ emissions are relatively complex. Figure 5 shows the SDA result for indirect CO$_2$ emissions in 1990–2005. Indirect CO$_2$ emissions grew rapidly from 1990 to 1995 with growth slowing down during 1995–2000 and slightly accelerating thereafter. This may be due to the post-bubble economy in which Japan adopted government intervention policies to stimulate the recovery of the economy and increase residential consumption, resulting in an increase in indirect CO$_2$ emissions between 1990 and 1995. With the change in policy direction from economic stimulus to economic constraint (including raising the consumption tax and increasing medical expenses in 1997) [43,44], the growth rate of indirect CO$_2$ emissions slowed down in 1995–2000. After 2000, due to the effect of the internet bubble in the United States [45], Japan was forced to introduce looser monetary policies to stimulate economic development which increased its indirect emissions to some extent. With the increase in the number of households, the positive impact of the number of households on indirect CO$_2$ emissions has gradually increased, becoming one of the main factors promoting indirect CO$_2$ emissions post-2000. The positive impact of the supply chain structure on indirect CO$_2$ emissions increased slightly before 2000, shifting to a negative effect after 2000. This may be caused by the transformation and maturing of enterprise, eliminating excess employment, equipment and debt through a severe restructuring process from the late 1990s to the early 2000s. From the 1990s, along with an emphasis on environmental protection and energy saving (e.g., the Kyoto Protocol adopted in 1997), the impact of the supply chain structure toward CO$_2$ emissions became negative, which reflects the great development of low-carbon technology in the whole supply chain. The negative impact of the intensity effect on indirect CO$_2$ emissions increased from 1990 to 2000 and weakened thereafter. After the bubble economy, economic recovery may be an important reason for the change in intensity effect. The impact of the consumption pattern on indirect CO$_2$ emissions changed from negative to positive from 1995 to 2000, turning to negative once more after 2000, although this impact was relatively small. Because of the economic stimulus policy post-bubble economy, the choices of consumers tended to be toward high-quality and environment-friendly consumption, causing the consumption pattern to inhibit indirect CO$_2$ emissions. However, during the period of economic constraint policies, consumers tended to choose goods with a high performance and low price, reducing the environmental awareness of consumption, resulting in a positive consumption pattern impact.

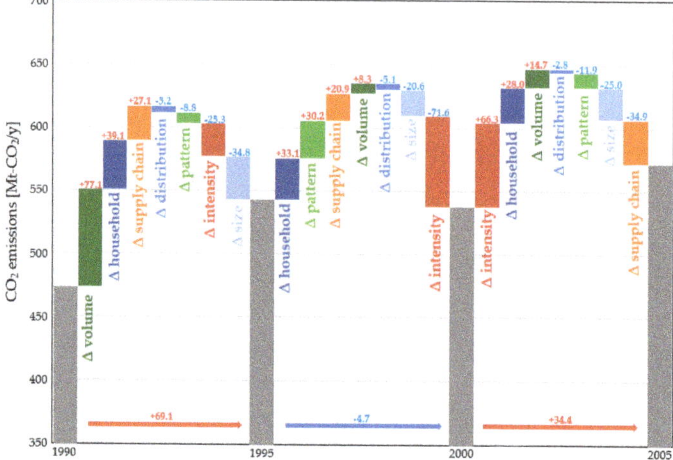

Figure 5. Driving forces of indirect CO_2 emissions during 1990–2005. Δintensity is the intensity effect, Δsupply chain is the supply chain structure effect, Δpattern is the consumption pattern effect, Δvolume is the consumption volume effect, Δsize is the household size effect, Δdistribution is the household distribution effect, and Δhousehold is the household number effect.

Summarizing the effect of household size, the number of households and household distribution in 1995, 2000 and 2005 and observing the changes, it was identified that the effect is gradually changing toward the positive and increasing. This shows that if Japan maintains an aging society with a low birth rate, the indirect CO_2 emissions generated by households will continue to grow.

3.3.2. Driving Forces of Indirect CO_2 Emissions of Key Sectors

By observing the changes in indirect CO_2 emissions in different sectors, this study identified four sectors with significant growth (amount) in indirect CO_2 emissions from 1990–2005. These are the housing, medical, private vehicle and public transport sectors. We emphasize discussion about the decomposition results for these four sectors, as shown in Figure 6, because we consider it essential with regard to the relationship between an aging, shrinking population and increasing household CO_2; those for the other 10 sectors and additional insights for the two most remarkable sectors showing the highest and lowest growth rate (i.e., the information and communication and clothes and footwear sectors, respectively) are detailed in Appendix A.

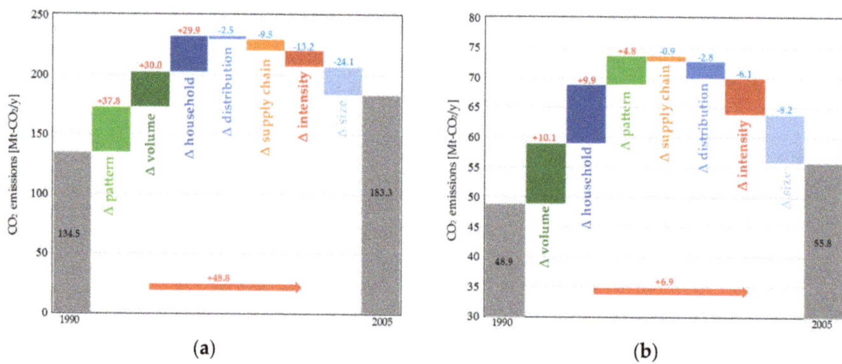

(a)   (b)

Figure 6. *Cont.*

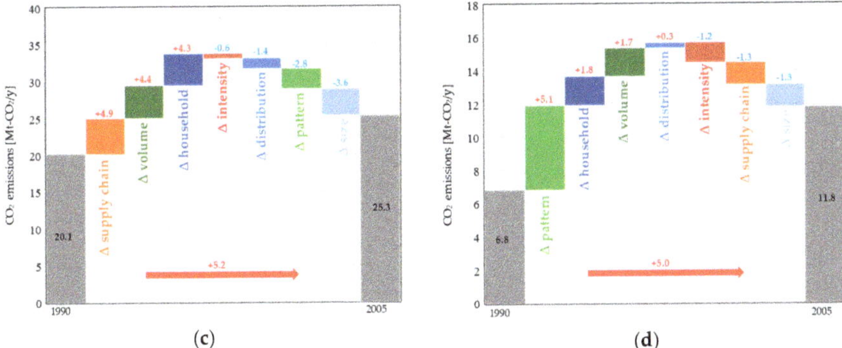

Figure 6. Structural decomposition analysis (SDA) results for indirect CO_2 emissions between 1990 and 2005 for four selected sectors. Δintensity is the intensity effect, Δsupply chain is the supply chain structure effect, Δpattern is the consumption pattern effect, Δvolume is the consumption volume effect, Δsize is the household size effect, Δdistribution is the household distribution effect, and Δhousehold is the household number effect. (**a**) Housing sector, (**b**) public transport, (**c**) private vehicles, and (**d**) medical sector.

First, the housing sector (Figure 6a), which produced the largest indirect CO_2 emissions at any time between 1990 and 2005, increased consistently between 1990 and 2005. The technology and size effects are the main drivers which reduce indirect CO_2 emissions. The rapid development of energy-related technologies has greatly reduced the indirect CO_2 emissions from housing. At the same time, with the growth of single-member families, the proportion of small-scale households has gradually increased [46]. Compared with average-sized households, small-scale households utilize a lower number of consumables and appliances (e.g., air-conditioners, etc.), leading to a lower level of indirect CO_2 emissions. The number of households, consumption volume, and consumption pattern were the main drivers of CO_2 emissions growth. From 1990 to 2005, the number of households in Japan expanded. At the same time, the increase in single-member households has increased the demand for housing, further expanding consumption volumes. Therefore, the positive impact of the number of households and consumption volume on indirect CO_2 emissions increased. As for the consumption pattern, residents seeking a better quality of life tend to invest in quality of life outcomes, leading to an increase in indirect CO_2 emissions.

The public transport sector is the second-largest indirect CO_2 emitter among the four sectors. As shown in Figure 6b, changes in consumption volume and the number of households were the main drivers which promoted the growth of indirect CO_2 emissions in this sector. The number of households in Japan continued to grow from 1990 to 2005, increasing the positive effect, while, due to economic situation changes, residents might choose public transport in order to reduce living costs. An increase in environmental awareness has also prompted people to use more public transport. In addition, indirect CO_2 emissions have been restricted to a large extent by the household size and household distribution effects. The reason for this may be that small-scale households may prefer to travel by lower-cost public transport when compared to private transport, with the same trend shown by the elderly. Moreover, the development of low-carbon technologies has also greatly reduced the CO_2 emissions associated with the public transport sector.

The private vehicle sector is one of the main sources of direct CO_2 emissions from households (see Figure 1), while this sector also has a strong impact on indirect CO_2 emissions. With the increasing number of households in Japan [46], the demand for private cars is also expanding [42], causing the number of households and consumption volume effects to become important drivers in promoting indirect CO_2 emissions in this sector, as shown in Figure 6c. The supply chain effect has increased indirect CO_2 emissions, which is likely due to economic globalization, causing the production and manufacture of automobiles to be regionally diversified, whereby technological differences in production and transport between regions may place upward pressure on indirect CO_2 emissions. The consumption pattern has become the main driver restricting indirect CO_2 emissions, perhaps because people have become more likely to use public transport due to the abovementioned reasons, although many households still have a need for a private vehicle. With increasing awareness of environmental protection, households are willing to consider the purchase of environmentally friendly automobile models. Meanwhile, with the development of automobile manufacturing technology, consumers are more willing to buy fuel-efficient vehicles. Household size and household distribution effects had a stable, inhibitory effect on indirect CO_2 emissions, mainly because an aging society and the increase in households with less members reduced the consumption of private vehicles.

Although the medical sector has the lowest indirect CO_2 emissions among the four sectors analyzed, it has the highest growth rate, at 73% (Figure 6d). Considering the current situation of Japan's aging society, the medical sector has a great impact on the lives of elderly people. Therefore, it is of great practical significance to analyze the influencing factors of indirect CO_2 emissions in the medical sector. Consumption patterns and consumption volume have strongly contributed to the increase in indirect CO_2 emissions as elderly households increase. In the context of an aging society, Japan's elderly population continues to grow, leading to the continued expansion of national medical expenditure. Furthermore, the proportion of imported drugs has been gradually increasing [42], which is also the main reason for the decreasing impact of supply chain structure on indirect CO_2 emissions. Though the impact of technology on indirect CO_2 emissions changed during 1990–2005, it has always been negative. This is probably due to the continuous development of medical manufacturing technologies which are more environmentally conscious. Finally, the household size also plays an important role in inhibiting indirect CO_2 emissions. The decrease in household size (i.e., a relative increase in the share of small households in the total households) has greatly reduced household consumption of medicine and consequently reduced indirect CO_2 emissions to a certain extent.

4. Conclusions

This study identified the driving forces of direct and indirect household CO_2 emissions (i.e., household CF) both overall and at the sectoral level using IDA and SDA. In the analysis, emissions from different age groups were also considered. The main findings from our analysis are as follows:

1. Among household-related CF during 1990–2005, indirect CO_2 emissions kept increasing from 1990, although direct CO_2 emissions slowed down between 2000 and 2005.
2. Per capita CO_2 emissions of direct and indirect emissions by household age group showed similar distributions during the studied period. Emissions begin to rise from the 20s and decline after peaking in the 50s. In addition, the level of both direct and indirect emissions per capita did not change radically during the analyzed 15-year period.
3. The two decomposition analyses for direct and indirect CO_2 emissions showed that the effects of changes in household size due to the trend away from nuclear family structures and production technology progress restrained indirect CO_2 emissions to a large extent. On the other hand, the results also showed that if Japan continues to follow current consumption and demographic trajectories (i.e., the aging society becoming an 'aged' society), both of those emissions will increase regarding the contributions from related drivers.
4. Decomposition analysis for the sectoral CF showed that the main factors leading to the increase in indirect CO_2 emissions were the increase in the number of households and consumption volume. Regarding aging and a reduced birth rate, the increase in small-scale households has increased the overall number of households in Japan due to changing family structures, which has expanded household consumption volume per capita—an important driving force behind increasing household CF.

Based on these findings, the following conclusions are offered to contribute toward achieving Japan's future carbon reduction target. It is necessary to comprehensively consider the household CF, and, considering the impact of an aging society on Japan's overall CO_2 emissions, more practical measures should be taken to reduce CO_2 emissions by sector. Specifically, the following policy implications are proposed based on our findings.

First, carbon-reduction policies in an aging society should consider demographic influences. Under an aging society with an increase in the proportion of elderly people, the proportion of young people will gradually decrease. Therefore, household-related sectors need to pay more attention to the CO_2 emissions from consumption by elderly people. In addition, from the perspective of factors affecting CO_2 emissions within sectors, the increase in small-sized households greatly increase CO_2 emissions. Thus, for further CO_2 emission reductions, enterprises in household-related sectors should adjust their business orientation to meet the needs of small-scale households.

Second, consumers should be encouraged to consume responsibly; i.e., by choosing environmentally friendly products or restricting the irresponsible consumption of goods and services (e.g., less food waste). As a first step toward doing so, considering the fact that the indirect CO_2 emissions of residents' lives are higher than their direct CO_2 emissions, it is imperative that householders are made aware of this fact. This educational aspect will be essential in order to reduce indirect CO_2 emissions as well as promoting low-carbon production for industry. The government should expand financial support for enterprises to develop low-carbon technologies to alleviate financial pressures and stimulate enterprise to invest in related research and development. From the consumer side, the key is to develop green-consumption awareness and to encourage the purchase of environmental-friendly commodities. For example, for household products, which account for a large proportion of households' indirect CO_2 emissions, financial support should be provided when purchasing environmentally friendly products. Moreover, a healthy lifestyle is of great significance for residents to reduce potential CO_2 emissions; for instance, healthy living habits can keep the body in good condition, which is likely to reduce certain medical expenses and decrease the indirect CO_2 emissions from the medical sector.

In addition to policy implications from the domestic perspective, there is also an important implication from the international perspective. Although the aging society is a serious issue in Japan, this issue is also emerging in developed countries. Even for developing countries, China is also beginning to experience an aging society. Therefore, achieving the goal of CO_2 mitigation under the constraints of an aging society has important practical significance for many countries. In order to reduce per household emissions, policy encouraging multi-generational households may be appropriate. A complementary policy for per capita emission amelioration may be to encourage the sharing economy, whereby individual consumption is offset by shared facilities and conveniences (i.e., car sharing, etc.). As illustrated above, such a demographic shift would gradually change toward the positive and increase indirect CO_2 emissions to a level which is higher than direct emissions, which is associated with an increase in the number of households rather than population. Thus, it should be noted that efforts for recovering the population and labor force through policy measures will eventually lead to an increase in overall CO_2 emissions. On the basis of clarifying influential factors toward the CF, particularly those for sectors which presented an increasing trend such as food, public transport, private transport, and the medical sector, the implementation of carbon reduction schemes are crucial. Furthermore, the government needs to formulate fit-for-purpose carbon reduction policies based on the characteristics of CF for different age groups. Finally, because the aging society issue occurs over a very long time-span, CO_2 emission reduction policies should be continuously adjusted according to the evolution of demographic changes.

Author Contributions: Conceptualization, Y.S. and K.M.; methodology, Y.S.; software, Y.S.; validation, K.M; formal analysis, Y.S.; investigation, Y.H.; resources, Y.S.; data curation, Y.S.; writing—original draft preparation, Y.H. and Y.S.; writing—review and editing, A.C. and K.M.; visualization, Y.H. and Y.S.; supervision, K.M.; project administration, Y.S.; funding acquisition, Y.S. and K.M.

Funding: This research was supported by JSPS KAKENHI Grant Numbers JP18K11754, JP18K11800, and JP18K18231; and the Integrated Research Program for Advancing Climate Models (TOUGOU program) of MEXT. These organizations did not have any involvement or influence in the implementation of this study.

Conflicts of Interest: The authors declare no conflict of interest. The funding sponsors had no role in the design of the study; in the collection, analyses, or interpretation of data; in the writing of the manuscript; or in the decision to publish the results.

Appendix A

Figure A1 depicts direct and indirect CO_2 emissions per household during the period 1990–2005. Among age groups, both direct and indirect CO_2 emissions per household for those in their 50s were estimated to be the largest, followed by those in their 40s during all periods. This is mainly because their average household income and family size were larger than for other households (e.g., in 2005, the average household income and family size in their 50s and 40s are 7.83 and 7.25 M-JPY/y and 2.87 and 3.17 people/household, respectively). While indirect CO_2 emissions for those in their 60s were the third largest for the 15 years investigated, their direct CO_2 emissions were fourth—smaller than for those in their 30s from 1990 to 2000. In 2005, emissions in the 60s group increased to become larger than for the 30s. Comparing the per household results to the per capita results, as presented in Figures 1b and 2b, the largest emissions were seen from those in their 50s. However, the orders of magnitude of indirect CO_2 emissions for those in their 40s and 60s differs between per household and per capita results, as seen in the emissions levels for 1990 and 2005. This is also observed for direct CO_2 emissions for households aged between their 20s and 70s. These differences between per household and per capita are most affected by the average family size and composition of households (as described in the body of this study).

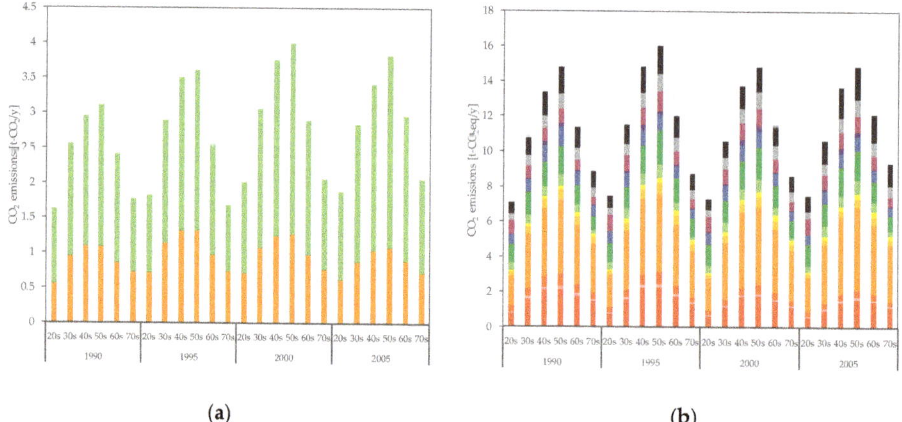

Figure A1. Sectoral composition of per household CO_2 emissions by age group (t-CO_2/y) from 1990 to 2005. (**a**) Direct and (**b**) indirect.

Figure A2 depicts a sectoral-level decomposition analysis for indirect CO_2 emissions, except for the sectors discussed in Section 3.3.2. Note that the highest and lowest growth rates of indirect emissions were shown in the information and communication and clothing and footwear sectors, respectively. The growth rate in the information and communication sector was 267%, while that in the clothing and footwear sector was −31% during the studied period. The reasons behind these changes can be explained as follows. In the information and communication sector, the pattern effect was the major factor underpinning an increase in emissions (Figure A2e). From the end of the 20th century, the world, including Japan, began to enter into the information technology age, and computers and the internet began to spread. At the same time, mobile phones also became more common. This means that information and communication technology penetrated into daily life, and the consumption of such technologies has rapidly increased, having a commensurate effect on indirect emissions.

In the clothing and footwear sector, the pattern effect was also the main factor responsible for reducing indirect emissions (Figure A2c). During the studied period, the outsourcing of production in this sector to developing Asian countries reduced production costs (as a result, the supply chain effect became positive during this period). At the same time, fast fashion became popular (Uniqlo, etc.). These complementary phenomena resulted in a lower consumption of apparel and led to a reduction in indirect emissions.

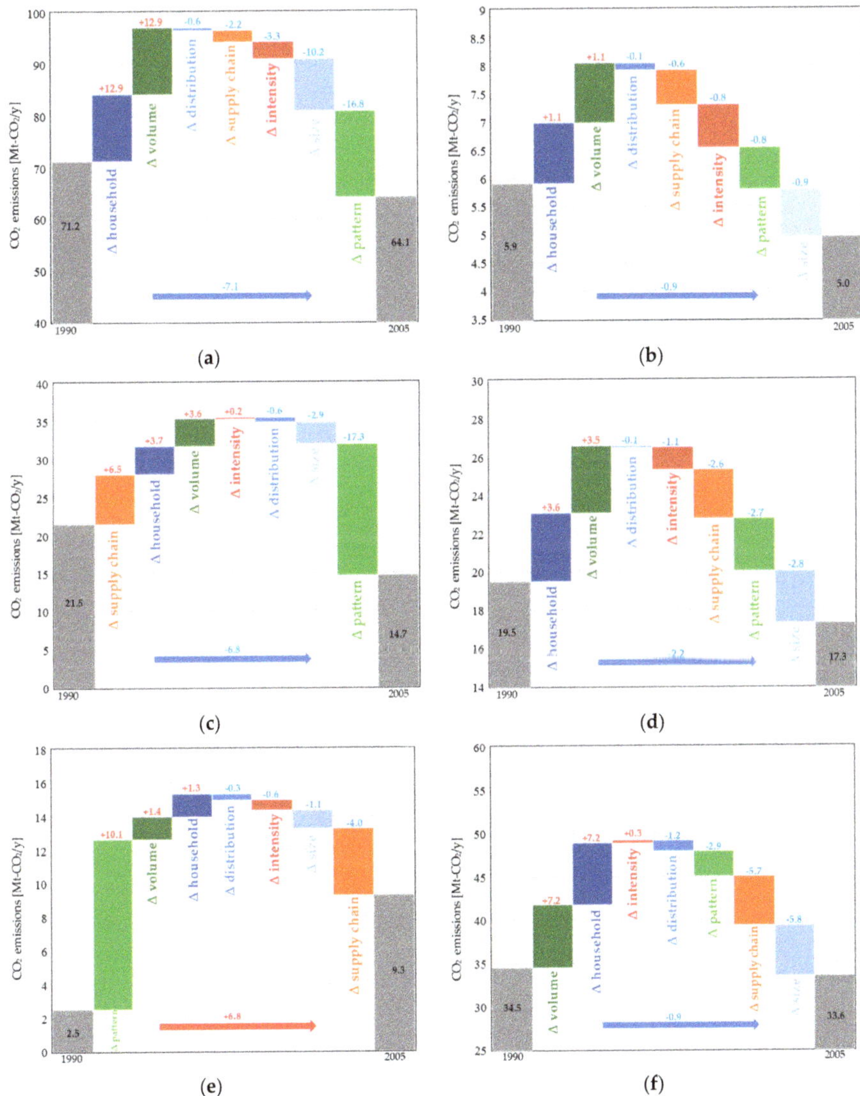

Figure A2. *Cont.*

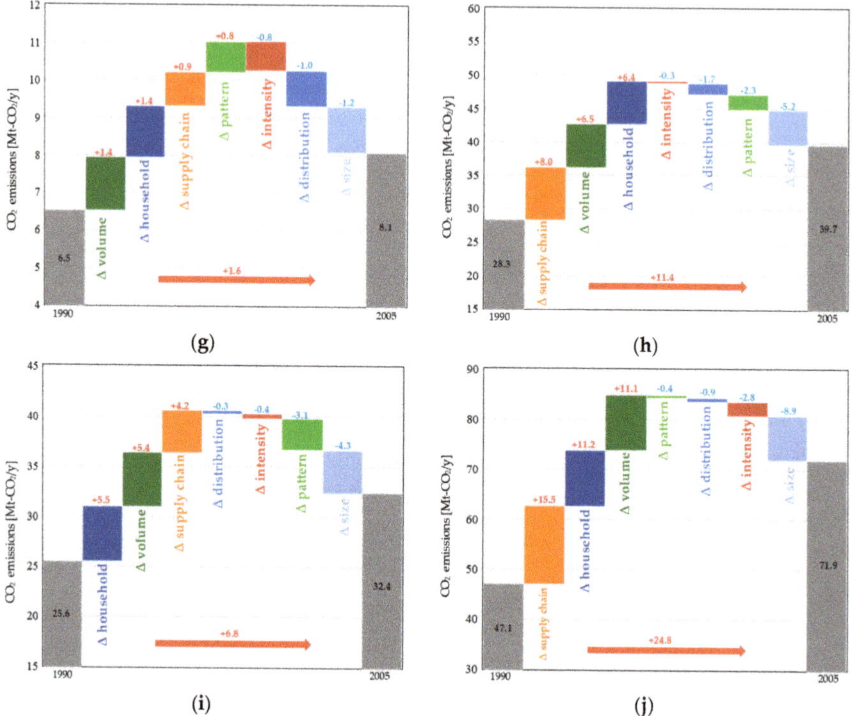

Figure A2. SDA results for indirect CO_2 emissions between 1990 and 2005 for ten sectors not shown in Section 3.3.2. Δintensity is intensity effect, Δsupply chain is the supply chain structure effect, Δpattern is the consumption pattern effect, Δvolume is the consumption volume effect, Δsize is the household size effect, Δdistribution is the household distribution effect, and Δhousehold is the household number effect. (**a**) Food and non-alcoholic beverages, (**b**) alcoholic beverages and tobacco, (**c**) clothing and footwear, (**d**) furnishings, (**e**) information and communication, (**f**) recreation and culture, (**g**) education, (**h**) restaurants and hotels, (**i**) consumable goods, and (**j**) margins, religions and other services.

References

1. IPCC Global Warming of 1.5 °C. Summary for Policymakers. Available online: https://www.ipcc.ch/sr15/ (accessed on 20 July 2019).
2. Peters, G.P. From production-based to consumption-based national emission inventories. *Ecol. Econ.* **2008**, *65*, 13–23. [CrossRef]
3. Davis, S.J.; Caldeira, K. Consumption-based accounting of CO_2 emissions. *Proc. Natl. Acad. Sci. USA* **2010**, *107*, 5687–5692. [CrossRef] [PubMed]
4. Wiedmann, T. A review of recent multi-region input-output models used for consumption-based emission and resource accounting. *Ecol. Econ.* **2009**, *69*, 211–222. [CrossRef]
5. Hertwich, E.G. The Life Cycle Environmental Impacts Of Consumption. *Econ. Syst. Res.* **2011**, *23*, 27–47. [CrossRef]
6. Zhang, X.; Luo, L.; Skitmore, M. Household carbon emission research: An analytical review of measurement, influencing factors and mitigation prospects. *J. Clean. Prod.* **2015**, *103*, 873–883. [CrossRef]
7. Wiedenhofer, D.; Smetschka, B.; Akenji, L.; Jalas, M.; Haberl, H. Household time use, carbon footprints, and urban form: a review of the potential contributions of everyday living to the 1.5 °C climate target. *Curr. Opin. Environ. Sustain.* **2018**, *30*, 7–17. [CrossRef]
8. Ivanova, D.; Stadler, K.; Steen-Olsen, K.; Wood, R.; Vita, G.; Tukker, A.; Hertwich, E.G. Environmental Impact Assessment of Household Consumption. *J. Ind. Ecol.* **2016**, *20*, 526–536. [CrossRef]

9. IEA. *Key World Energy Statistics 2016*; IEA: Paris, France, 2016.
10. Shigetomi, Y.; Nansai, K.; Kagawa, S.; Tohno, S. Fertility-rate recovery and double-income policies require solving the carbon gap under the Paris Agreement. *Resour. Conserv. Recycl.* **2018**, *133*, 385–394. [CrossRef]
11. Shigetomi, Y.; Matsumoto, K.; Ogawa, Y.; Shiraki, H.; Yamamoto, Y.; Ochi, Y.; Ehara, T. Driving forces underlying sub-national carbon dioxide emissions within the household sector and implications for the Paris Agreement targets in Japan. *Appl. Energy* **2018**, *228*, 2321–2332. [CrossRef]
12. Nansai, K.; Kagawa, S.; Kondo, Y.; Suh, S.; Nakajima, K.; Inaba, R.; Oshita, Y.; Morimoto, T.; Kawashima, K.; Terakawa, T.; et al. Characterization of economic requirements for a "carbon-debt-free country". *Environ. Sci. Technol.* **2012**, *46*, 155–163. [CrossRef]
13. Shigetomi, Y.; Nansai, K.; Kagawa, S.; Tohno, S. Changes in the Carbon Footprint of Japanese Households in an Aging Society. *Environ. Sci. Technol.* **2014**, *48*, 6069–6080. [CrossRef] [PubMed]
14. Shigetomi, Y.; Nansai, K.; Kagawa, S.; Tohno, S. Influence of income difference on carbon and material footprints for critical metals: the case of Japanese households. *J. Econ. Struct.* **2016**, *5*, 1. [CrossRef]
15. Takase, K.; Kondo, Y.; Washizu, A. An analysis of sustainable consumption by the waste input-output model. *J. Ind. Ecol.* **2005**, *9*, 201–219. [CrossRef]
16. Kawajiri, K.; Tabata, T.; Ihara, T. Using a Rebound Matrix to Estimate Consumption Changes from Saving and its Environmental Impact in Japan. *J. Ind. Ecol.* **2015**, *19*, 564–574. [CrossRef]
17. Long, Y.; Dong, L.; Yoshida, Y.; Li, Z. Evaluation of energy-related household carbon footprints in metropolitan areas of Japan. *Ecol. Model.* **2018**, *377*, 16–25. [CrossRef]
18. Long, Y.; Yoshida, Y.; Meng, J.; Guan, D.; Yao, L.; Zhang, H. Unequal age-based household emission and its monthly variation embodied in energy consumption—A cases study of Tokyo, Japan. *Appl. Energy* **2019**, *247*, 350–362. [CrossRef]
19. Sovacool, B.K.; Kester, J.; Noel, L.; de Rubens, G.Z. The demographics of decarbonizing transport: The influence of gender, education, occupation, age, and household size on electric mobility preferences in the Nordic region. *Glob. Environ. Chang.* **2018**, *52*, 86–100. [CrossRef]
20. Ang, B.W.; Zhang, F.Q. A survey of index decomposition analysis in energy and environmental studies. *Energy* **2000**, *25*, 1149–1176. [CrossRef]
21. Rose, A.; Casler, S. Input-output structural decomposition analysis: a critical appraisal. *Econ. Syst. Res.* **1996**, *8*, 33–62. [CrossRef]
22. Ang, B.; Zhang, F.; Choi, K.-H. Factorizing changes in energy and environmental indicators through decomposition. *Energy* **1998**, *23*, 489–495. [CrossRef]
23. Xu, X.Y.; Ang, B.W. Index decomposition analysis applied to CO_2 emission studies. *Ecol. Econ.* **2013**, *93*, 313–329. [CrossRef]
24. Hoekstra, R.; van den Bergh, J.C.J.M. Comparing structural decomposition analysis and index. *Energy Econ.* **2003**, *25*, 39–64. [CrossRef]
25. Lenzen, M. Structural analyses of energy use and carbon emissions—An overview. *Econ. Syst. Res.* **2016**, *28*, 119–132. [CrossRef]
26. Donglan, Z.; Dequn, Z.; Peng, Z. Driving forces of residential CO_2 emissions in urban and rural China: An index decomposition analysis. *Energy Policy* **2010**, *38*, 3377–3383. [CrossRef]
27. Zang, X.; Zhao, T.; Wang, J.; Guo, F. The effects of urbanization and household-related factors on residential direct CO_2 emissions in Shanxi, China from 1995 to 2014: A decomposition analysis. *Atmos. Pollut. Res.* **2017**, *8*, 297–309. [CrossRef]
28. Xu, X.Y.; Ang, B.W. Analysing residential energy consumption using index decomposition analysis. *Appl. Energy* **2014**, *113*, 342–351. [CrossRef]
29. Guan, D.; Hubacek, K.; Weber, C.L.; Peters, G.P.; Reiner, D.M. The drivers of Chinese CO_2 emissions from 1980 to 2030. *Glob. Environ. Chang.* **2008**, *18*, 626–634. [CrossRef]
30. Feng, K.; Davis, S.J.; Sun, L.; Hubacek, K. Drivers of the US CO_2 emissions 1997–2013. *Nat. Commun.* **2015**, *6*, 7714. [CrossRef]
31. Yuan, B.; Ren, S.; Chen, X. The effects of urbanization, consumption ratio and consumption structure on residential indirect CO_2 emissions in China: A regional comparative analysis. *Appl. Energy* **2015**, *140*, 94–106. [CrossRef]

32. Chapman, A.; Fujii, H.; Managi, S. Key Drivers for Cooperation toward Sustainable Development and the Management of CO$_2$ Emissions: Comparative Analysis of Six Northeast Asian Countries. *Sustainability* **2018**, *10*, 244. [CrossRef]

33. Okamoto, S. Impacts of Growth of a Service Economy on CO$_2$ Emissions: Japan's Case. *J. Econ. Struct.* **2013**, 2. [CrossRef]

34. Timilsina, G.R.; Shrestha, A. Transport sector CO$_2$ emissions growth in Asia: Underlying factors and policy options. *Energy Policy* **2009**, *37*, 4523–4539. [CrossRef]

35. Matsumoto, K.; Shigetomi, Y.; Shiraki, H.; Ochi, Y.; Ogawa, Y.; Ehara, T. Addressing Key Drivers of Regional CO$_2$ Emissions of the Manufacturing Industry in Japan. *Energy J.* **2019**, *40*, 233–258. [CrossRef]

36. Miller, R.E.; Blair, P.D. *Input–Output Analysis*; Cambridge University Press: Cambridge, UK, 2009; ISBN 9780511626982.

37. Sun, J.W. Changes in energy consumption and energy intensity: A complete decomposition model. *Energy Econ.* **1998**, *20*, 85–100. [CrossRef]

38. Dietzenbacher, E.; Los, B. Structural Decomposition Techniques: Sense and Sensitivity. *Econ. Syst. Res.* **1998**, *10*, 307–324. [CrossRef]

39. Agency for Natural Resources and Energy of Japan Comprehensive Energy Statistics. Available online: https://www.enecho.meti.go.jp/statistics/total_energy/ (accessed on 20 July 2019).

40. MIC NSFIE; National Survey of Family Income and Expenditure. Available online: https://www.e-stat.go.jp/stat-search/database?page=1&toukei=00200564&survey (accessed on 1 May 2019).

41. Nansai, K.; Moriguchi, Y. Embodied Energy and Emission Intensity Data for Japan Using Input-Output Tables. Available online: http://www.cger.nies.go.jp/publications/report/d031/jpn/datafile/embodied/2005/403.htm (accessed on 20 July 2019).

42. Statistics Bureau of Japan Motor Vehicles Owned by Kind (F.Y.1936-2004). Available online: https://www.stat.go.jp/english/data/chouki/12.html?fbclid=IwAR1-62rFrm6hiZ-4nidMyIqEz1KX_W4YtvP-O4utpVG9mTEx3gURNhRkJvE (accessed on 20 July 2019).

43. Choi, Y.; Hirata, H.; Kim, S.H. Tax reform in Japan: Is it welfare-enhancing? *Jpn. World Econ.* **2017**, *42*, 12–22. [CrossRef]

44. MHLW; Ministry of Health Labour and Welfare Overview of Medical Service Regime in Japan. Available online: https://www.mhlw.go.jp/bunya/iryouhoken/iryouhoken01/dl/01_eng.pdf (accessed on 1 May 2019).

45. Chan, Y.C. How does retail sentiment affect IPO returns? Evidence from the Internet bubble period. *Int. Rev. Econ. Financ.* **2014**, *29*, 235–248. [CrossRef]

46. MHLW; Ministry of Health Labour and Welfare National Livelihood Survey. Available online: https://www.mhlw.go.jp/toukei/saikin/hw/k-tyosa/k-tyosa18/index.html (accessed on 20 July 2019).

Article

A Critical Review of CO_2 Capture Technologies and Prospects for Clean Power Generation

Najmus S. Sifat and Yousef Haseli *

Clean Energy and Fuel Lab, School of Engineering and Technology, Central Michigan University, Mount Pleasant, MI 48859, USA; sifat1n@cmich.edu
* Correspondence: hasel1y@cmich.edu

Received: 16 September 2019; Accepted: 28 October 2019; Published: 30 October 2019

Abstract: With rapid growth in global demand for energy, the emission of CO_2 is increasing due to the use of fossil fuels in power plants. Effective strategies are required to decrease the industrial emissions to meet the climate change target set at 21st Conference of the Parties (COP 21). Carbon capture and storage have been recognized as the most useful methods to reduce the CO_2 emissions while using fossil fuels in power generation. This work reviews different methods and updates of the current technologies to capture and separate CO_2 generated in a thermal power plant. Carbon capture is classified in two broad categories depending on the requirement of separation of CO_2 from the gases. The novel methods of oxy combustion and chemical looping combustion carbon capture have been compared with the traditional post combustion and precombustion carbon capture methods. The current state of technology and limitation of each of the processes including commonly used separation techniques for CO_2 from the gas mixture are discussed in this review. Further research and investigations are suggested based on the technological maturity, economic viability, and lack of proper knowledge of the combustion system for further improvement of the capture system.

Keywords: CO_2 capture; thermal power plants; oxyfuel combustion; allam cycle; post-combustion; pre-combustion

---

1. Introduction

Rapid industrialization over the past century has resulted in huge demand for power. The most common way to produce power is utilizing fossil fuels, but this causes emission of CO_2 which is the main component of greenhouse gas (GHG). The amount of CO_2 emitted by different power industries and the energy sector running on fossil fuels constitutes approximately 65% of the total emission of the GHG [1]. As the concern over climate change due to GHGs is increasing all over the world, the reduction in this emission has become an important area of research. Substantial reduction in the emission of CO_2 is necessary to follow the agreement of COP-21. The main outcome of the COP-21, held in Paris in 2015, was the agreement to maintain the global average temperature rise below 2 °C. Further efforts should be pursued to limit this temperature increase below 1.5 °C [2].

To reduce emissions, research is being conducted to use renewable resources instead of fossil fuels. One of the interesting methods is the conversion of CO_2 into organic compounds using photocatalytic reduction and producing fuel feedstock. Traditional fossil fuels can be used as renewable fuel following this process. Zhou et al. [3] illustrated the use of semiconductor ZnS during photoreduction of CO_2 to formate ($HCOO^-$). Sharma et al. [4] portrayed the use of sulfide-based photocatalysts for production of renewable fuel in this method. They used CU_3SnS_4 as a photocatalyst for reduction of CO_2 to CH_4. Excellent photocatalytic performance of the sulfide was observed along with good stability. However, no technology is advanced enough to compensate for the reduction in the use of conventional fuels. Though nuclear and renewable energy is predicted to play a significant role in low carbon power production, most of the future power demand is expected to be met by fossil fuels due to safety and other issues [5]. Therefore, it is necessary to think of a way to use conventional fuels for producing power while reducing the emission of CO_2. Carbon Capture and Storage (CCS) technology comes here as a rescue. It is the process of capturing the produced CO_2 and then storing it in a safe place so that it would not affect the environment. International Energy Agency (IEA) has a projection to abate the CO_2 emission; 17% of this abatement should be done by CCS by 2035 at the lowest cost [6].

To meet the expectation, research is ongoing all over the world to develop new technologies. Currently, the main obstacle in deploying CCS is the huge cost which in effect increases the price of electricity. The current estimated cost of capturing CO_2 with an established technology is at least \$60/t CO_2 [6]. This is too high to make CCS commercially attractive. This huge amount of compensation discourages investment in the energy market. To get rid of this barrier, research is ongoing mainly in developed countries to keep the cost of carbon capture around \$20/t CO_2 captured [6]. In addition to the increased cost associated with CCS, the environmental impact of the methods should also be considered. A reduction in CO_2 emission may lead to increasing other emissions affecting the environment [7].

Current technologies that are being developed for capturing CO_2 can be broadly categorized into the following divisions: (i) Carbon capture with separation, and (ii) Carbon capture without separation. This review focuses on the processes, the current state of the carbon capture technologies, and on identifying the area that demands further research.

2. Carbon Capture with Separation

This process requires technology to separate CO_2 from a mixture of different gases. The gas stream may form before or after the combustion. If the gas stream consisting of carbon dioxide is formed before combustion, then it is known as a precombustion carbon capture process. This gas stream consists of mainly CO_2 and H_2 in this case; otherwise, it is called post combustion carbon capture, where the main constituents of the gas stream are CO_2 and N_2. Several technologies are currently in use and under development for the separation of CO_2 from the gas mixture. Almost all the separation techniques can be applied for both processes.

2.1. Precombustion Carbon Capture

This method implies an alternative of combusting fuel directly in a combustor. At first, fuel is converted to a combustible gas. This gas is used for power generation [8]. CO_2 is separated and sequestered from this gas generated from fossil fuel before combustion [9]. A schematic diagram of the process is illustrated in Figure 1.

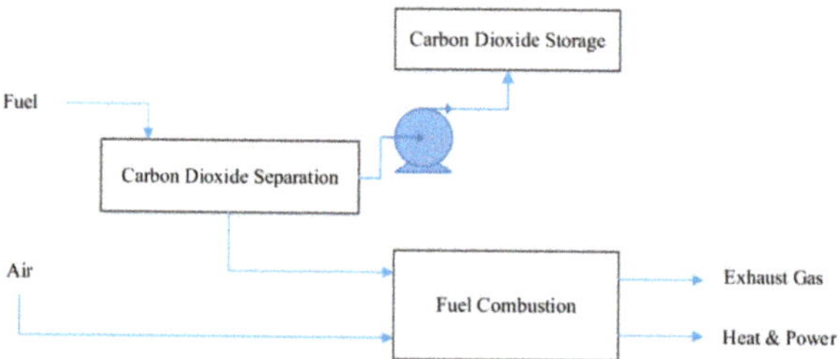

Figure 1. Schematic diagram of the pre combustion carbon capture process [10].

At first, synthesis gas (syngas), which is a mixture of mainly H_2 and CO with a trace of CO_2, is produced from a fossil fuel. It can be done by adding steam to the fossil fuel. This process is known as steam reforming [11]. Another way to produce syngas is by supplying pure oxygen after separating it from air to fossil fuel which is being used to produce power. This process is known as partial oxidation when it is applied to liquid or gaseous fuels. When it is applied for solid fuels, it is known as gasification. The reactions for this process are given below [11]:

Steam reforming: $C_xH_y + xH_2O \rightarrow xCO + \left(x + \frac{y}{2}\right)H_2$.

Partial oxidation: $C_xH_y + \frac{x}{2}O_2 \rightarrow xCO + \left(\frac{y}{2}\right)H_2$.

The syngas produced in this way is then converted to CO_2 from CO by water-gas shift reaction.

Water gas shift reaction: $CO + H_2O \rightleftharpoons CO_2 + H_2$.

The products of the water shift gas reaction remain under high pressure which facilitates the removal of CO_2. It is removed at ambient temperature. The remaining gas is mainly hydrogen with little impurities. This gas is used to generate power in a combined cycle power plant. High pressure (typically 2–7 MPa) and a high concentration of CO_2 (15–60% by volume) before the separation of CO_2/H_2 demand less energy for CO_2 separation and compression than post combustion carbon capture [6]. However, the energy requirement becomes high in this process due to the air separation and reforming or gasification processes. One way to reduce this energy penalty is to use Sorption Enhanced Water Gas Shift (SEWGS) technology. Water gas shift reaction and CO_2 separation can be integrated through this technology [12]. SEWGS increases the conversion rate of CO by removing CO_2 from the product of the water gas shift reaction. This results in an additional reduction of CO_2 emission [13]. The process is almost the same for any fossil fuel, but if any fuel other than the natural gas is used, then more refining stages should be included since more contaminants are produced [11].

Currently, the main research focus of precombustion carbon capture is to use this method in Integrated Gasification Combined Cycle (IGCC) power plants. A layout of the IGCC is shown in Figure 2.

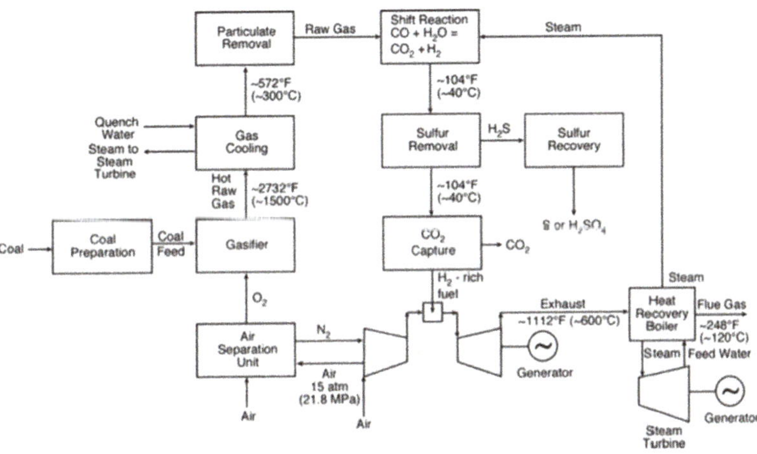

Figure 2. A schematic layout of an IGCC power plant using pre combustion carbon capture [14].

Here, oxygen is separated from air in a cryogenic air separation plant [14]. This oxygen is passed to a gasifier where coal is gasified at high pressure to produce syngas at high temperature. After cooling and preliminary cleaning, syngas is shifted through a water gas shift reaction in a water gas reactor and converted to H_2S, H_2, and CO_2. After several cleaning steps to remove sulphur, mercury, water, and other impurities, the syngas only consists of CO_2 and H_2. This gas mixture is passed through the CO_2 removal process where CO_2 is captured. Hydrogen is then used to produce power. Most commercially developed technologies employ physical solvents to separate CO_2 from syngas. A lot of work has been carried out for best performance of CO_2 separation from the syngas. Some of the important works using different separation technologies are summarized in Table 1.

Table 1. Summary of some important studies on precombustion carbon capture.

Author	Year	Method of Separation	Remarks
Romano et al. [15]	2010	Absorption	CO_2 capture using methyl diethanol amine solution was compared to other capture processes. They suggested avoiding more conservative assumption for greater efficiency.
Martin et al. [16]	2011	Adsorption	Hypercrosslinked polymers were evaluated for their adsorption capacity of CO_2. These polymers provided superior uptake of CO_2 than zeolite-based materials and commercial activated carbon. Also, they showed good selectivity towards CO_2 and low heat of adsorption.
Schell et al. [17]	2012	Adsorption	Three different materials of the Metal Organic Framework were tested as adsorption material. The USO-2-Ni Metal organic framework showed promising result when compared to the commercial activated carbon.
Garcia et al. [18]	2012	Adsorption	Partial pressure of CO_2 was found most influential when activated carbon was used for adsorption. Carbon capture capacity and breakthrough time were directly proportional to it and inversely proportional to temperature.
Casas et al. [19]	2013	Adsorption	Conducted a parametric analysis study on the PSA process for pre combustion carbon capture. They evaluated different process configurations and conditions for better separation performance. Separation improved with the decrease of operating temperature and desorption pressure. Adsorption pressure did not affect the separation.
Stefania et al. [20]	2014	Chemical absorption	They used air-blown gasification instead of oxygen-blown gasification. Efficiency penalty and carbon capture were competitive with air-blown gasification.
Jiang et al. [21]	2015	Adsorption	They used a mesoporous amine-TiO_2 as a sorbent. This inexpensive sorbent was stable and could be easily regenerated without loss of capacity and selectivity.
Park et al. [22]	2015	Absorption	Designed a two stage pre-combustion CO_2 capture process using three physical absorbents. Selexol was found as the most efficient pre combustion carbon capture process from an energy consumption point of view.
Dai et al. [23]	2016	Membrane absorption	Investigated separation performance of ionic-liquid based membrane contactor at high pressure and temperature. At high pressure, the membrane contactor became wetted which made it less efficient. The temperature did not have much effect on the process.
Ponnivalavan et al. [9]	2016	Hydrate based gas separation	Showed that tetrahydrofuran is better than other semiclathrate hydrate formers. Optimum concentration of THF was 5.56 mol% at 282.2 K and 6 MPa.
Mingjun et al. [24]	2016	Hydrate based gas separation	5% TBF + 10% TBAB was the most suitable choice for hydrate-based CO_2 capture.
Zheng et al. [25]	2017	Hydrate based gas separation	Proposed CO_2-H_2-TBAF semiclathrate hydrate process. Gas uptake was highest for CO_2 when it was used as promoter except for THF. Also, the process takes place at ambient temperature.
Muhammad et al. [26]	2018	Membrane contactor process	Cost of carbon capture using membrane contactor with PSA was too high to implement commercially.
Haibo et al. [27]	2018	Physical absorption using ionic liquid	They showed that using ionic liquid for absorption of carbon gives a similar result as the selexol process.

2.2. Post Combustion Carbon Capture

This technique is used in the existing power plants without a major modification of the plant. For this reason, it has the advantage of easier retrofitting compared to the other CCS processes [28–30]. It is the simplest technique to capture CO_2. In this method, CO_2 is removed from the exhaust flue gases of the power plants. Normally the flue gases exit at atmospheric pressure. The concentration of CO_2 in these gases is very low. A typical concentration of CO_2 in the flue gas is shown in Table 2.

Table 2. Amount of CO_2 in flue gases of power plants [31].

Method	Concentration of CO_2 (Vol. %)
Coal fired Boiler	14
Natural gas fired boiler	8
Natural gas combined cycle	4
Natural gas partial oxidation	40
Coal oxygen combustion	>80

Due to the low concentration of CO_2, the driving force is too low to capture it from the flue gas [31]. Large sized equipment and high capital cost are required to handle a huge volume of flue gases. Therefore, a cost-effective way to capture CO_2 from the flue gas needs to be identified. Also, the flue gas contains various types of contaminants such as SO_x, NO_x, fly ash, etc. They cause the separation process to become more costly with existing technologies [32].

Separation process for CO_2 from flue gas is challenging for some reasons. Equipment design is required to withstand the high temperature of the flue gas. The gas must be cleaned up before separating CO_2. Merkel et al. [32] has proposed a flow process to clean up the gas as shown in Figure 3. The hot exhaust gas leaving the boiler is passed through an electrostatic precipitator (ESP) that removes all the large particulates. After that, the sulphur products are removed through a flue gas desulphurization unit (FGD). Post combustion carbon capture technology is designed to treat the outlet gas of FGD. In this state, the gas mixture contains around 10–14% CO_2 mainly in a mixture of nitrogen. A schematic of a power plant using coal as fuel with solvent-based absorption post combustion carbon capture is shown in Figure 4.

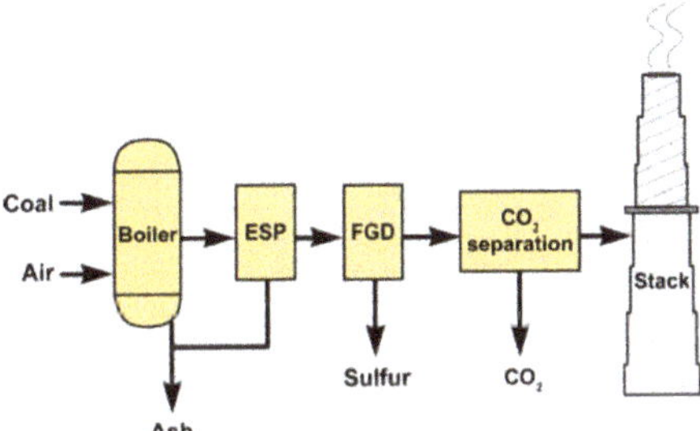

Figure 3. Schematic diagram of a simplified flue gas cleanup process for post combustion carbon capture [32].

Here, coal is pulverized and combusted with air to generate heat. This heat is used to produce steam which in turn produces power through three different steam turbines of various pressures. Low-quality exhaust steam is condensed in a condenser and sent back to the boiler. The exhaust flue gas from the boiler is passed through the cleaning process to remove sulphur, ash, NO_x and other impurities. After the final stage of cleaning, the gas is sent to the CO_2 capture process.

The complexity is much reduced when natural gas is used as fuel. A typical layout of a post combustion carbon capture combined cycle power plant using natural gas as fuel is shown in Figure 5. Natural gas is combusted with compressed air and the product is expanded through a gas turbine to produce power. The exhaust of the gas turbine remains at high temperature. This high-temperature flue gas is used to make steam. It produces additional power through a steam turbine. The cooled flue gas is then passed to the CO_2 capture process. Figure 5 shows a solvent-based CO_2 capture system using MEA. MEA scrubs CO_2 from the flue gas in the absorber column leaving clean gas to the exhaust. Later, the MEA is purified in the stripper column to use again in the absorber column. CO_2 is captured from the stripper column and compressed for storage. Using MEA is the most common method to separate CO_2 from flue gas. Other technologies are also used to separate CO_2 from the gas mixture.

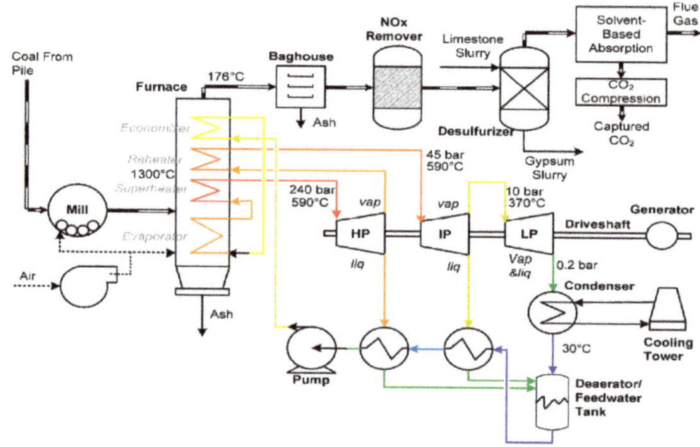

Figure 4. Layout of a post combustion carbon capture coal-fired power plant [33].

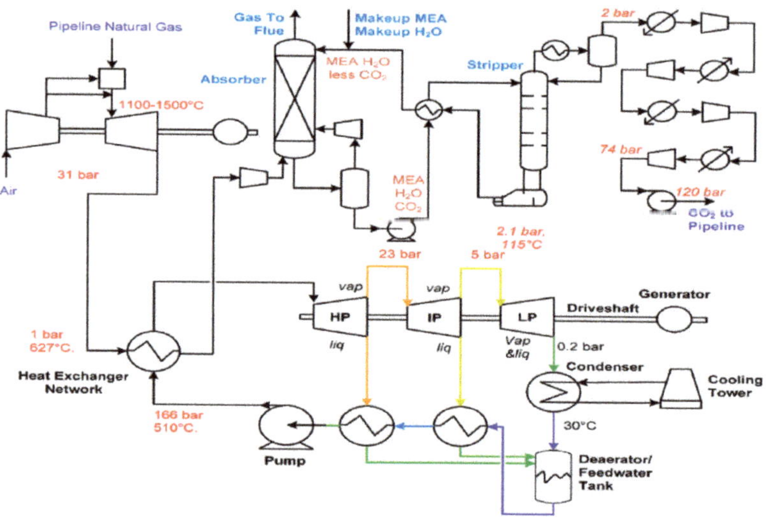

Figure 5. Layout of a post combustion carbon capture power plant operating with natural gas as the fuel [33].

Some of the important work on post combustion carbon capture using different separation technology is summarized in Table 3.

Table 3. Some of the important studies on post combustion carbon capture.

Author	Year	Method of Separation	Remarks
Tim et al. [32]	2010	Membrane separation	They developed and tested membranes with high permeance and applied a novel process by using incoming combustion air as sweep gas. More emphasis was given to high permeance of membrane rather than high selectivity.
Agarwal et al. [34]	2010	Adsorption	Presented a novel PSA superstructure to design and evaluate optimal Cycle configuration for CO_2 capture strategies. The superstructure could predict PSA cycles up to 98% purity.
Wappel et al. [35]	2010	Absorption	Different pure and diluted ionic liquids were tested as potential solvent and compared with a currently dominant solvent of MEA and water. One of the ionic liquids showed less energy demand than the reference solvent.
Jarad et al. [36]	2011	Adsorption	Two MOFs were evaluated in detail for their use as adsorbents in post combustion carbon capture via temperature swing adsorption. Mg_2 (dobdc) exhibited a better result in each case. They concluded that the presence of a strong CO_2 adsorption site is necessary for a MOF to be useful in post combustion carbon capture through temperature swing adsorption.
Savile et al. [37]	2011	Biotechnology	Showed the potential of carbonic anhydrase derived from thermophiles to accelerate the post combustion capture process.
Zhi et al. [38]	2012	Absorption	They proposed to use nanomaterials as absorbents due to their high surface area and adjustable properties and characteristics. The only problem was their high production cost and complicated synthesis process.
Scholes et al. [39]	2013	Membrane-cryogenic separation	Proposed modification of three membrane stages and cryogenic separation for increasing overall efficiency of post combustion carbon capture. The cost of capture remained comparable for the proposed process with current technologies.
Bae et al. [40]	2013	Adsorption	They evaluated a series of zeolite materials to use as adsorbents in post combustion carbon capture. Among the zeolites, Ca-A ($Na_{0.28}Ca_{0.36}AlSiO_4$) showed the highest uptake for CO_2.
Zhang et al. [41]	2014	Membrane absorption	Proposed a numerical model to investigate the effect of membrane properties on CO_2 absorption. Increased membrane length and number of fibers were found to have a positive effect on CO_2 capture. A decline in membrane thickness, inner fiber radius, and inner module radius showed improved performance in removal of CO_2 while it decreased the absorption performance.
Farid et al. [42]	2015	Absorption	A comparison was shown between an aqueous solution of amine and ammonia as a solvent. Aqueous ammonia was found to have higher absorption and loading capacity while requiring lower energy for regeneration but the absorption must be done at a lower temperature by cooling the flue gas.
Zhang et al. [43]	2016	Membrane absorption	Proposed a mathematical model to find optimum operating conditions for acid gas absorption in the hollow fiber membrane module. Chemical solvents were much better than physical solvents for CO_2 absorption. Piperazine showed the best performance among single solvents. Blended absorbent solution exhibited 20% higher efficiency for CO_2 removal.
Nabil et al. [44]	2017	Absorption	30 different aqueous amine solutions were characterized for better performance; 6 of them showed better performance than reference MEA. Among them, 2-ethylaminoethanol was considered best for its good CO_2 absorption, low heat of absorption and high kinetic reaction with CO_2.
Thompson et al. [45]	2017	Absorption	Two-stage stripping was evaluated for the amine absorber to lower the cost and increase the performance of the absorber. The secondary stripper used the heat rejected from primary stripper. Emissions of different substances were compared. Overall emission levels of amine, ammonia and aldehyde were comparable with other published results.
Thompson et al. [46]	2017	Absorption	Two-stage air stripping was evaluated for the amine absorber to lower the cost and increase the performance of the absorber. The secondary stripper used the heat rejected from the primary stripper. Degradation of amine due to the oxygen exposure in second stripper was tested. The second stripper showed negligible impact.
Zhang et al. [47]	2018	Membrane Absorption	Proposed a 2D model for a CO_2-piperazine membrane absorption system. They suggested optimum gas velocity, absorbent velocity, concentration of CO_2 and solvent for best performance.

A careful literature survey on post combustion carbon capture reveals the research is being directed lately to membrane absorption separation technology. This method combines the advantages of both absorption and membrane separation in a single technology [47].

2.3. CO$_2$ Separation Technology

A lot of separation technologies are available for separating CO_2 from gas mixtures. We can categorize the main technologies into five different sectors. They are absorption, adsorption, clathrate hydrate process, membrane technology, and calcium looping carbon capture. For the absorption process, research is mainly focused on the development and performance enhancement of different solvents. Adsorption technology emphasizes new and modified materials. The clathrate hydration separation process is being experimented with different thermodynamic promoters for better performance. Membrane-based research uses membrane of different materials including composite and hybrid membrane for enhanced performance [10].

2.3.1. Absorption

The method of absorbing CO_2 in a solvent to separate it from a gas stream has been in use on an industrial scale for more than 50 years [48], but the partial pressure of the gas streams is comparatively much higher in industrial applications. This process may, in general, be classified into physical absorption and chemical absorption. A detailed classification of the absorption-based CO_2 capture technique is shown in Figure 6.

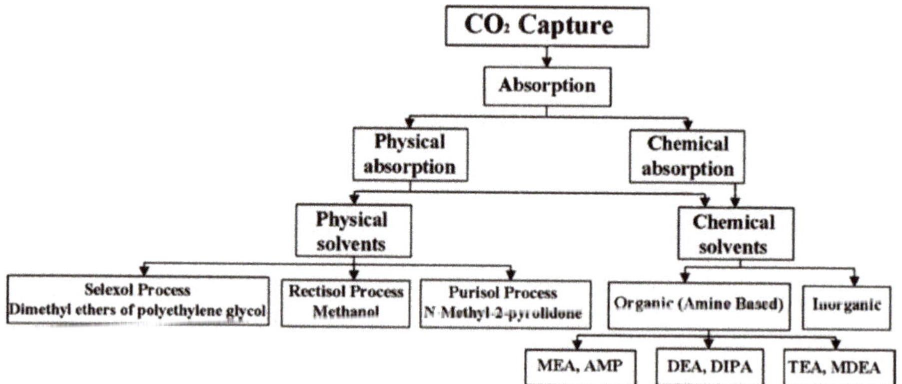

Figure 6. Classification of absorption processes for CO_2 capture [49].

If the solvent reacts with CO_2 and forms chemical compounds, then the process is known as chemical absorption. CO_2 is removed from the chemical compounds later. On the other hand, the solvent does not react with CO_2 if it is chemically inert. It soaks the CO_2 physically. This process is called physical absorption [48]. Chemical absorption of CO_2 is done in two stages. At first, the treated gas is brought into contact with the solvent stream in a counter flow. In this stage, the solvent absorbs CO_2 from the gas stream. This solvent is regenerated upon heating to desorb CO_2 in a stripping column. Pure CO_2 is collected from the top of the column [10]. It is then compressed and stored. The regenerated CO_2 lean solvent is sent back to the absorber [50]. The process is shown in Figure 7.

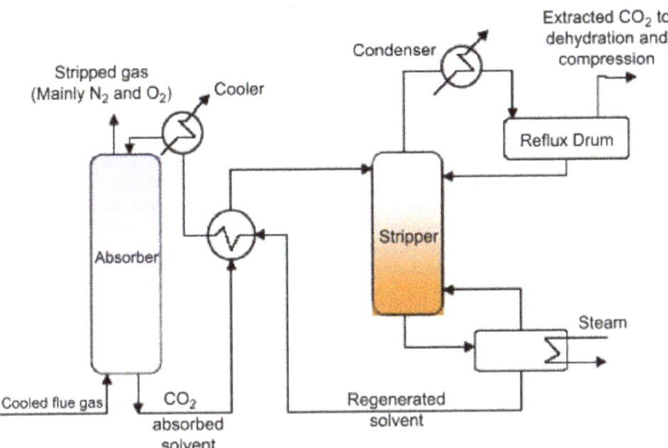

Figure 7. Schematic diagram of a CO_2 absorption plant [51].

The first stage of the process is optimal at high pressures and low temperatures whereas the second stage performs best at low pressures and high temperatures [10]. Chemical absorption is more favorable for capturing CO_2 at relatively low pressure. This is helpful for the post combustion process when amine or carbonate solutions are used as solvents [48].

In the case of physical absorption, organic or inorganic physical solvents are used. They do not react chemically with CO_2. This operation is based on Henry's law of vapor-liquid mixture equilibrium. According to this law, the amount of a gas dissolved in a unit volume of a solvent is proportional to the partial pressure of the gas in equilibrium with the solvent at any temperature [48]. Due to this pressure dependency of the physical absorption process; it shows better performance than chemical absorption at a higher partial pressure of CO_2 such as in an IGCC [11]. A physical absorption process is recommended to be used in IGCC due to the higher partial pressure of CO_2 in syngas which makes it more suitable for precombustion carbon capture. Physical solvents need lower energy for regeneration which is another advantage. [52].

The downside of this process is that the capacity of solvents is best at low temperatures. Therefore, the gas stream needs to be cooled before the absorption process. This causes a reduction in efficiency [52]. The processes that are being used commercially for physical absorption are known as Selexol, Rectisol, Purisol [11]. A comprehensive comparison using Aspen plus was done on these processes [49]. For capturing CO_2, Selexol was found more energy efficient than other investigated solvents. Lower consumption of energy to regenerate solvent and simple process configuration was the reason for this. David et al. [53] reported that the net efficiency would be greater than in the case with the selexol process if low-temperature CCS was applied in an IGCC.

The partial pressure of CO_2 in the flue gas stream is very low in the post combustion carbon capture process. For this reason, the focus of research on this process is to find a suitable solvent. A lot of research has been done on different processes and solvents to identify a cost-effective absorption method. A summary of the advantages and disadvantages of different processes is given Table 4.

Table 4. Advantages and disadvantages of different absorption technology [48].

Absorption Technology	Advantages	Disadvantages	Remarks
Fluor Process (Propylene Carbonate)	Well proven technology. Non-corrosive solvent. High selectivity for CO_2	Low H_2S tolerance in feed gas stream. Feed gas must be dehydrated due to high water solubility.	Physical absorption
Rectisol Process (Chilled Methanol)	Low solvent cost. Can remove a lot of contaminants in a single process. High selectivity for CO_2 and H_2S	High refrigeration energy cost. Higher selectivity for H_2S than CO_2.Feed gas must be dehydrated due to high water solubility	
Selexol Process (DMEPG)	Not subject to degradation. Very low vapor pressure. Dual-stage process can capture CO_2 and H_2S from syngas	Need high partial pressure of CO_2. High solvent viscosity. High solvent costs.	
Amine systems (MEA, DMEA etc.)	Well proven technology. Low capital cost. Benchmark for other solvent systems.	Limited loading capacity. High energy requirement for sorbent regeneration.Hazardous degradation of products. Solvent loss due to mist formation. Low tolerance to NO_x, SO_2 and O_2.	Chemical absorption
Bi-phasic liquid solvents	Reduced energy requirement. Lower corrosivity.	Higher pressure drops.	
Potassium carbonate system	Higher solvent loading capacity. Lower regeneration energy requirement. Low solvent cost. Lower corrosivity.	Slow reaction rate. High cost of additives. System fouling.	
Aqueous Ammonia	Lower regeneration energy requirement.Low solvent cost.	High solvent loss. Reduced operating temperature.	
Chilled Ammonia	Lower regeneration energy requirement. Lack of solvent degradation. Low solvent cost. Sellable by-product.	Near freezing operating conditions. Potential for fouling.	
Task specific and reversible Ionic Liquids	Very low volatility and solvent loss. High stability. Very low heat of absorption. Dual mode with high loading capacity.	High production cost. Reduced absorption efficiency with the presence of water. High viscosity reducing absorption rate.	
Sodium Hydroxide	Low cost and abundance of required chemicals. Proven technologies applied to other industries.	High energy requirement. High water and solvent loss.	

2.3.2. Adsorption

This is the process of removing a component from a mixture using a solid surface. Unlike absorption processes, the formation of physical or chemical bonds takes place between the solid phase adsorbent surface and CO_2. The intermolecular forces between the solid surfaces and gas are the driving force for adsorption [31]. Single or multiple layers of the gas can be absorbed based on adsorbent pore size, temperature, pressure and surface force [54].

At first, a column is filled up with the adsorbent. Then, the gas stream bearing CO_2 is passed through this column. The CO_2 adheres to the solid surface of the absorbent during the flow to the saturation of the adsorbent. When the surface becomes saturated with CO_2, it is removed and desorbed through different cycles for CO_2 adsorption [31].

Generally, four different regeneration cycles are used for single bed CO_2 adsorption. They are pressure swing adsorption (PSA), temperature swing adsorption (TSA), electrical swing adsorption (ESA), and vacuum swing adsorption (VSA). In temperature swinging adsorption, the temperature of the adsorbent is raised up to the point at which the chemical bonds are broken. CO_2 gets released at that point. An additional requirement of energy in this process makes this method costlier [54]. Also, this process is time consuming due to heating the adsorbent bed for desorption and cooling it again to make it ready for adsorption [48]. This can be done quickly using electrical swing adsorption. Here, low voltage electric current is passed through the adsorbent to heat up the adsorbent using joule effect. ESA makes it possible to regenerate the adsorbent fast, but it requires high-grade electrical energy instead of low-grade heat energy used in TSA [55].

In the case of pressure swinging adsorption, the pressure of the adsorbent is reduced to accomplish this consequence. Vacuum swing adsorption is a specialized PSA cycle which is used if the feed gas pressure is close to the ambient pressure. The extra energy required for achieving high pressure in PSA can be minimized using VSA [48]. Here, a partial vacuum is used at the downstream of the feed stage to draw the low-pressure feed gas. These cycles can be used in combination with one another. Plaza et al. [56] provided a model of VSA process using aspen plus for post combustion carbon capture. A schematic of the regeneration processes is shown in Figure 8.

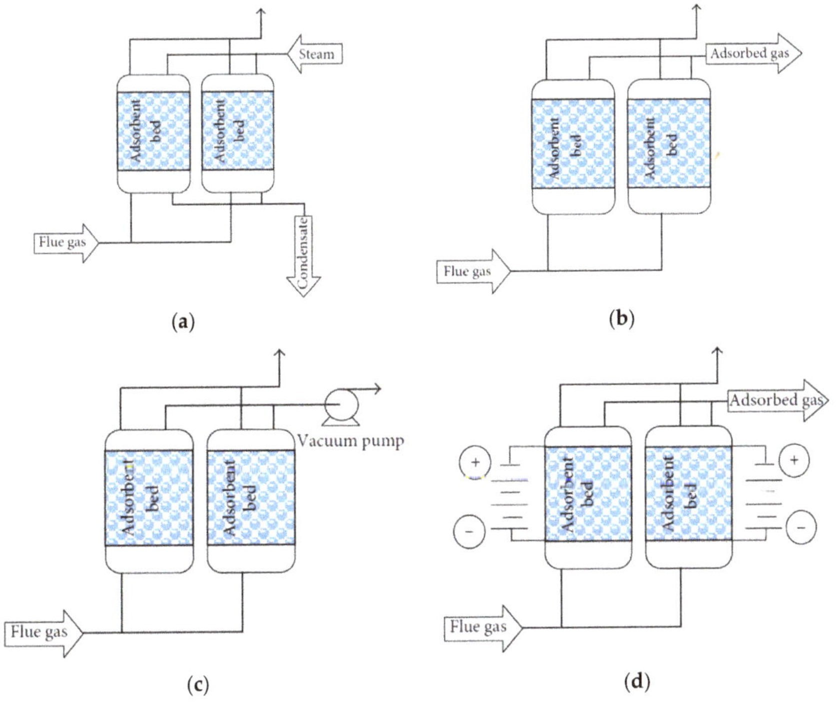

Figure 8. Schematic diagram of different adsorption regeneration cycles: (**a**) TSA (**b**) PSA (**c**) VSA (**d**) ESA [57].

Pressure swinging operation is favorable when the partial pressure of the CO_2 is high whereas temperature swinging adsorption is favorable if the concentration of CO_2 is low in the gas stream. PSA will take a much longer time if the concentration of CO_2 is low [58]. The adsorption process is more preferable because of its high adsorption capacity at normal pressure and temperature, long-term stability, low regeneration cost, high rate of adsorption, and lower energy requirement [59].

The focus of research on this process is to find a suitable sorbent to separate CO_2 from the gas stream. Various substances like zeolites, activated carbons, molecular sieves, hydrotalcites, and metal-organic framework materials have been investigated [60]. Garcia et al. showed [18] that the partial pressure of

CO_2 is the most influential variable when activated carbon is used as absorbent. Sorption-enhanced water gas shift combines adsorption of CO_2 with the water gas reaction. This method is more economical and more energy efficient than amine scrubbing in absorption [61]. Hydrotalcite-based materials are more suitable for adsorption at high temperatures. These materials exhibit improved result when used in a sorption-enhanced WGS reactor for better carbon capture [62].

2.3.3. Membrane Technology

Membranes are semi-permeable barriers of different materials which can separate different substances from a mixture by various mechanisms [31]. Membranes can be of organic or inorganic material. The solution-diffusion process takes place in non-facilitated membranes. The permeate diffuses through the membrane after being dissolved into it. The amount of CO_2 dissolved per unit volume is proportional to the partial pressure of CO_2 [63]. In the case of precombustion capture, the partial pressure of CO_2 remains comparatively high. Non-facilitated membrane separation technology has a greater use in this case.

Membranes can be used in a spiral wound, flat sheet, and hollow fiber modules. They can be selective or non-selective for a specific acidic gas [64]. Membrane technology can be classified into two categories for carbon capture: gas separation membrane and gas absorption membrane. In the gas separation membrane system, the CO_2 bearing gas is introduced at a high pressure into a membrane separator. The membrane separator typically consists of parallel cylindrical membranes. CO_2 passes through the membrane preferentially and it is recovered at a lower pressure at the other side of the membrane. A gas absorption system uses a microporous solid membrane to separate CO_2 from the gas stream. The removal rate of CO_2 is high for a gas absorption system due to minimization of flooding, foaming, channeling and entrainment. The equipment required is more compact than that for the membrane separator [53]. The two systems are shown in Figure 9.

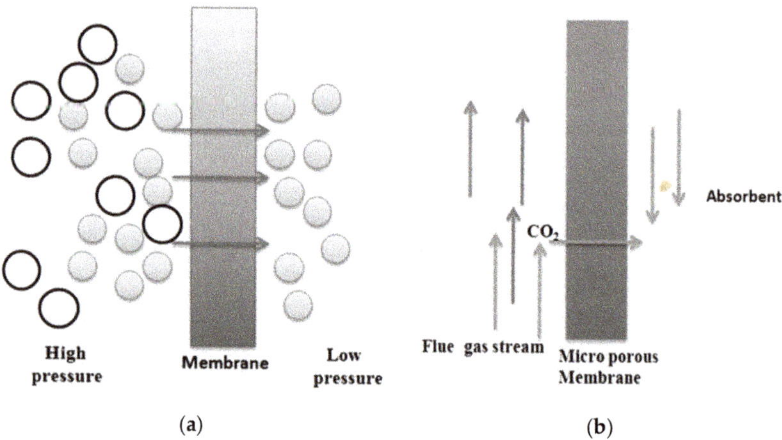

Figure 9. Principle of (a) gas separation membrane and (b) gas absorption membrane.

This technology has the advantages of operating without weeping, entrainment, foaming, and flooding which are common problems in operation with a packed column. They also have a higher surface area and better control of liquid and gas flow rates [65]. The main disadvantage of membranes is their reduced effectiveness at a lower concentration of CO_2. Membrane shows low flexibility and becomes unfeasible when the concentration of CO_2 in the gas stream is below 20% [66]. Therefore, this is not suitable for the post combustion capture process.

Membranes must be replaced periodically due to their limited lifetime. There is also a higher mass transfer resistance in the membrane fibers. Membrane pores should be filled up completely by the gas phase. When the liquid phase takes place in the membrane pores, resistance to mass transfer begins to build up through the membrane. Thus, the application of the membrane becomes economically unjustified. This phenomenon is known as wetting of a membrane. The desired condition is to fill the membrane pores completely with gas, but it is not always possible. Membrane pores become partially or fully wetted over long operational periods [64]. Several studies have been performed with different types of membranes and absorbents to investigate the wetting effect on mass transfer through the membrane [66–71]. The efficiency of absorption through membrane reduces significantly even for a low level of wetting. Using ionic liquid as absorbent can increase the efficiency by around 15% and 20% when compared with pure water in counter current and co current flows [66].

2.3.4. Clathrate Hydrate Process

Gas hydrate or clathrate hydrates are ice-like crystal compounds formed by water molecules and a number of other substances including CO_2, N_2, H_2, and O_2. These small gas molecules become trapped inside cavities of water molecules [72]. Concentrations of different gases in the crystals are different from their concentrations in the original gas mixture [73].

The main concept of separating CO_2 is the partition of the CO_2 selectively from a gas mixture between the solid hydrate crystal phase and the gaseous phase by forming a hydrate crystal. Thermodynamically, the minimum pressure to form hydrate at a temperature of 273.9 K is 5.56 MPa. The pressure of syngas after the water gas shift reaction is normally 2–7 MPa, whereas the flue gas in post combustion is almost at atmospheric pressure. Therefore, the gas stream requires compression to increase the rate of hydrate formation [74].

Different promoters have been tested to reduce the equilibrium condition to form hydrates. The most studied promoter is tetrahydrofuran (THF). Equilibrium of hydrate formation reduces with the addition of THF at any temperature. Increasing the concentration of THF causes a decrease in the hydrate formation pressure up to an optimum concentration ~1 mol% THF. It can be used in CO_2 separation industrially without compressing the flue gas significantly [75]. The equilibrium pressure to form hydrate may reduce by 50% if 3.2 mol% propane is added to a CO_2/O_2 mixture [76].

For the fuel gas mixture in the precombustion process, the hydrate phase equilibrium condition is reduced with the addition of tetra-n-butyl ammonium bromide (TBAB). Hydrate formation condition decreases with an increase in TBAB concentration up to the stoichiometric condition, beyond which the phase equilibrium increases with by increasing TBAB concentration [77]. Park et al. [78] investigated the effect of quaternary ammonium salts on hydrate formation. They showed that 95% of CO_2 can be captured from an IGCC using just one step of hydrate formation. TBAF showed a better result than TBAB but with a lower gas uptake.

Recent studies have also focused on the type of reactors. Zheng et al. [79] studied the impact of bed volume and bed reactor orientation for hydrate formation in precombustion carbon capture. Horizontal orientation performed better than vertical configuration. They also showed that low water saturation is preferable to form more hydrate.

2.3.5. Calcium Looping Technology

The calcium looping carbon capture system utilizes a different technique to capture CO_2 from a gas stream. In this method, a direct reaction takes place between CO_2 and CaO. This reaction produces solid calcium carbonate which is easily separable from the other gases. The main reversible reaction for this process is as follows [80].

$$CaO + CO_2 \leftrightharpoons CaCO_3 \tag{1}$$

The forward reaction, known as carbonation reaction, is exothermic. The reverse reaction is called calcination reaction, which is endothermic. The initial rate of a carbonation reaction is very fast but it comes to an abrupt slow rate after some time [81]. Due to the endothermic reaction at the calcination reactor, it needs a large amount of heat to be supplied at a high temperature. Often this heat is supplied by oxy combustion of coal or natural gas inside the calcination reactor [33]. After the recovery of CO_2 from the calcination reactor, it is compressed and stored. This process can be used for both precombustion and post combustion carbon capture. The following reaction is the desired key reaction in the gasifier of precombustion carbon capture [80]:

$$CO + H_2O + CaO \leftrightharpoons CaCO_3 + H_2 \tag{2}$$

Precombustion carbon capture has some advantages using calcium looping process. $CaCO_3$ and CaO increase the destruction rate of tar which is complex when hydrogen is used as fuel. The removal of CO_2 from the gas mixture also increases the rate of conversion from CH_4 and CO to H_2 [80].

The main use of this process is in post combustion carbon capture [82]. A schematic of the process is shown in Figure 10. Here, the limestone captures CO_2 from the exhaust flue gases of a power plant with the help of a circulating fluidized bed carbonator. The sorbent is then passed to a calciner which operates at a higher temperature. After regeneration, the sorbent is again passed to the carbonator. Coal or natural gas is burnt in an oxy fuel environment of the calciner to produce necessary heat.

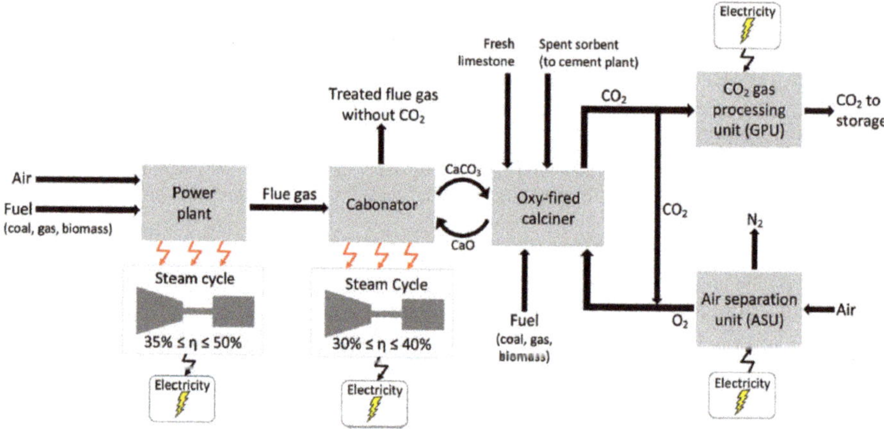

Figure 10. A schematic diagram of post combustion carbon capture using Calcium looping [82].

The overall reaction to form solid carbonate is exothermic. The high-grade heat produced at the carbonator can be supplied for a steam cycle in order to produce more power. This lowers the energy penalty from conventional post combustion capture [83]. The limestone is available in huge quantities and it is a non-hazardous substance. The price of the limestone is also much lower than the amines used for scrubbing in post combustion carbon capture. Used or spent sorbents can be further utilized for other purposes.

The sorbent is recycled and used repeatedly for CO_2 capture. The reversibility of the main reaction decreases with the increase of the number of cycles [84]. Therefore, the sorbent loses its carrying capacity with repeated use in the cycle. After the first cycle, the capacity of the sorbent is reduced by 1535% depending on favorable and unfavorable conditions. This loss of capacity decreases in each cycle [81]. A huge amount of makeup sorbent is required for this process.

3. Carbon Capture by Water Condensation

This method is comparatively novel in power generation. Here, instead of supplying air to the combustion chamber, pure oxygen is supplied for combustion. As a result, the combustion products consist of mainly CO_2 and steam. The CO_2 content of the mixture is captured by condensing steam. Thus, there is no need to apply any of the CO_2 separation technologies described in the previous section. Therefore, this method is economically more viable. If the oxygen is produced using an Air Separation Unit (ASU), the process is known as oxy-combustion carbon capture. Another way of supplying oxygen is to use a metal oxide with the help of a chemical looping known as chemical looping combustion [85,86].

3.1. Oxy-Combustion Carbon Capture

In this method, fuel is burnt in almost pure oxygen rather than air. Flue gas produced in this process is mainly a mixture of water and carbon dioxide. In a conventional power plant, fuel is combusted in air and the nitrogen of the air acts as a temperature moderator. As there is no N_2 present in the combustor of oxy fuel combustion, the flame temperature becomes too high. To keep the temperature within the limit, recycled CO_2 is passed to the combustor with pure oxygen. Another way to keep the flame temperature in the desired range is to inject steam in the combustion chamber [87].

After combustion, water is removed from the product by condensation [5]. The captured CO_2 is purified and compressed to supercritical condition for transporting or using again in the cycle. A flow sheet of the oxy fuel combustion concept is shown in Figure 11.

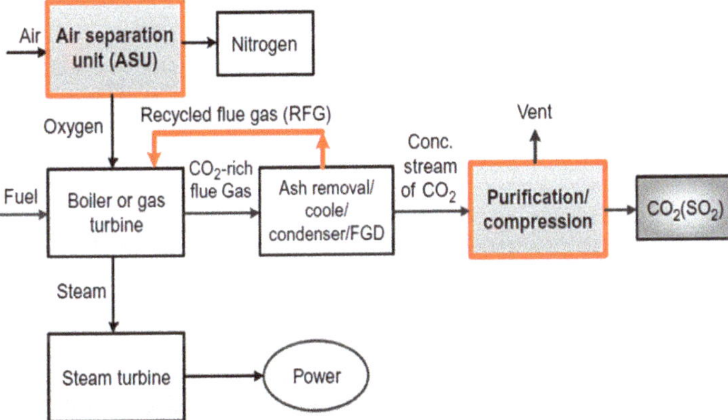

Figure 11. Flowsheet of oxy fuel combustion technology for power generation with CO_2 capture [5].

Since the properties of CO_2 and N_2 are different, the reaction pathway and combustion characteristics differ in oxy combustion from conventional air-fuel combustion [88]. These anomalies in the combustion characteristics demand in-depth research to understand and utilize this method.

Oxy fuel combustion has additional advantages compared to conventional combustion. There remains a large amount of N_2 in a conventional air firing system. Nitrogen consumes a lot of heat before being released to the environment, but in oxy combustion, this bulk N_2 is absent in the combustion environment. Due to the absence of nitrogen, there is no or much less production of NO_x in this process. There are no other significant pollutants in the products of combustion. The oxy fuel combustion technique is therefore a less expensive method compared to the previously discussed carbon capture technologies.

The main disadvantage of this method is the high operational cost for producing O_2 and pressurizing CO_2 after combustion [89]. One of the main challenges of this method is to produce oxygen with high purity at a reasonable expense. Wu et al. [90] summarized different work done on the separation process of oxygen from the air for use in oxy fuel combustion. They argue that membrane methods are more economical and simpler compared to the cryogenic method. The authors also suggested that adsorption technology is not yet updated to implement on large scales. Rather, a chemical looping air separation method is highly promising to become a more efficient and cost-effective technique for implementation in oxy fuel combustion.

A lot of research is ongoing to understand and improve the oxy-combustion method. Boiler design may be improved to enable more compact equipment with a deep understanding of oxy fuel combustion [91]. Knowledge of the combustion procedures will be necessary for this. With the use of a compact boiler, the cost of power generation can be reduced. This process helps reduce the amount of flue gas and thus the heat loss from the flue gas stream. It also reduces SO_x and NO_x emissions and improves combustibility. It has the potential to be more economical than other conventional processes.

The application of burning a fuel in pure oxygen was first done for different industrial processes. Later Abraham et al. [92] suggested this method as a solution to provide a large amount of CO_2 for oil recovery, but recycled flue gas was not used in that time. Later, the idea of using recycled flue gas was applied to produce a high purity CO_2 stream to use in oil recovery. It was also seen that this method could decrease the environmental impacts of fossil fuel power plants [93]. It can be applied with coal and natural gas. When this method is applied to coal, it can be classified as either an oxy-pulverized coal process or an oxy coal fired boiler process. In the case of natural gas, it can be classified as a CO_2-based cycle or water-based cycle [94].

For a conventional power plant using pulverized coal combustion, the concentration of CO_2 in the flue gas is relatively low (12–16%v dry basis). In the case of an oxy pulverized coal boiler, the air is replaced with pure oxygen. This results in a flue gas containing a high concentration of CO_2 (65–85%v dry basis) [94]. Rohan et al. [95] discussed the impact of Sulphur in oxy PC combustion. They reported that ash collection, furnace, CO_2 compression, transportation, and storage might be affected by Sulphur. If the recycle stream is taken before the flue gas clean up, it will increase the concentrations of the impurities, especially SO_x, in the furnace. The emission of SO_2 in oxy fuel combustion is lower than in air-combustion because of the retention of Sulphur in ash.

In comparison with other systems, Chen et al. [96] found that an oxy fuel system shows 1–5% less loss of efficiency than post combustion capture. The pressurized system gains around 3% more efficiency. Though the ASU requires more power in a pressurized system, it saves greater power during compression of CO_2. Chen et al. [96] also concluded that absorptivity and emissivity of the flue gases increase due to higher partial pressure in oxy fuel combustion. The optimized ratio of the recycle stream primarily depends on the type of fuel, the arrangement of the heat recuperator and the strategy of recycling. They found no influence of the oxy fuel environment on the devolatilization process of the solid fuel. Another finding is that ignition delay is longer in the combustion environment of an oxy fuel system than that of a conventional system.

Some researchers have investigated the effect of recycled CO_2 on the combustion environment. The burning velocity or speed of the propagation of a flame could decrease due to the incorporation of CO_2 instead of N_2 in the combustion environment [97]. Oh and Noh [98] examined the flame speed in an Oxy fuel environment in the atmospheric condition with a rich and lean fuel mixture. It was found that the speed of a methane flame in the oxy fuel environment is faster than that in an air-fuel environment which is contradictory to the results of Ref. [97]. Chen suggested that this discrepancy may result from the different prototypes used in these studies [97]. In previous research, the flame temperature was lower in the oxy fuel environment than the air-fuel environment, whereas in later research it was higher in the oxy fuel environment.

Mazas et al. [99] investigated the effect of water vapor on the speed of flame propagation [100]. They observed that with an increase in the molar fraction of steam, the flame velocity would decrease quasi-linearly, even at a high rate of dilution. The reduction in the burning velocity was larger for air combustion than oxy fuel combustion with the increase of the molar steam fraction. The effect of the equivalence ratio, CO_2 fraction, and pressure on the speed of oxy-methane flame was experimentally and numerically investigated by Xie et al. [100], who found that an increase in the CO_2 fraction would reduce the flame speed in $CH_4/O_2/CO_2$. Due to the presence of CO_2 in the oxy fuel environment, the radiation effect of CH_4 was much stronger.

Another important combustion parameter is the maximum flame temperature. Since the function of CO_2 in the combustor of oxy fuel environment is to control temperature, the temperature of the flame drops when CO_2 is added, but it has been shown that this decrease in temperature is not linear [97]. This is because the thermal effects are stronger at higher CO_2 concentrations. Wang et al. [88] examined the chemical and physical effects of CO_2 to find the dominant effect for determining the maximum flame temperature. They used different mole fractions of water and carbon dioxide in a counterflow oxy fuel combustion using methane as fuel. The maximum flame temperature was not affected because of the lower diffusion coefficient of CO_2, but a significant difference was observed in the temperature profile. They also noticed that the presence of H_2O did not have much influence on the maximum flame temperature. Pressure may also affect the maximum flame temperature. Seepana and Jayanti [101] numerically investigated the effect of pressure varying between 0.1 and 3 MPa in an oxy fuel environment. Their results indicated that the flame temperature would increase with an increase in the pressure, rapidly at first and then gradually. Low NO_x oxy fuel flame was found at high pressure with increased oxygen dilution.

Limited research has been conducted to improve understanding of the effect of ignition, flame stability, and flame extinction in an oxy combustion environment. Koroglu et al. [102] conducted a detailed experiment with methane to study ignition in an oxy fuel environment. They used a shock tube facility to do this experiment. The pressure was varied from 1 to 4 atm with temperature ranging from 1577 K to 2144 K. Their results revealed that the ignition delay would be longer when methane is burned in an O_2/CO_2 environment than in an O_2/N_2 environment due to the participation of CO_2 in chemical reactions, higher heat capacity, and different collision efficiency. A similar experiment was performed by Pryor et al. [103] at high pressure ranging from 6 to 31 atm and temperature varying between 1300 K and 2000 K. Using a sensitivity analysis, the authors showed that CO_2 could slow down the overall rate of reaction and increase the time of ignition delay.

Several research works have been done to determine the flammability of oxy methane flames. With the addition of CO_2, upper flammability decreases dramatically whereas it does not affect the lower flammability significantly [104]. An addition of steam also has an influence on the flammability limit of oxy combustion flame. The flammability limits become broader with the addition of CO_2 which is attributed to the lower heat capacity of steam compared to the supercritical CO_2 [97]. Some of the main studies on oxy fuel combustion are listed in Table 5.

Table 5. Some of the important studies on oxy combustion carbon capture.

Author	Year	Work
Said et al. [105]	2011	Integration of oxy fuel combustion process with carbonation process using $Mg(OH)_2$ to capture CO_2 was demonstrated.
Stanger et al. [95]	2011	Effect of Sulphur on oxy combustion CO_2 capture was shown
Riaza et al. [106]	2012	Investigated NO emission, ignition temperature and burnout with coal and biomass in an oxy fuel environment. Ignition temperature was increased when N_2 was replaced with CO_2. With the addition of biomass, this temperature subsequently decreased.
Allam et al. [107]	2013	Proposed a CO_2-based cycle using oxy combustion with near zero emission known as the Allam power cycle
Leckner et al. [108]	2014	Compared a ready to convert air fired CFB boiler with a newly designed oxy fuel CFB boiler for oxy combustion. The new designed one was more convenient.
Vellini et al. [109]	2015	An advanced supercritical steam cycle power plant with CO_2 capture based on oxy fuel combustion was studied. Oxygen transport membrane was used for a high oxygen production rate.
Scaccabarozzi et al. [110]	2016	Numerically investigated optimization of a NET power oxy combustion cycle
Falkenstein-Smith et al. [111]	2016	CO_2 selectivity and O_2 permeability were tested using a ceramic membrane catalytic reactor.
Climent et al. [112]	2016	Analyzed different oxy turbine power cycles with complete carbon capture for better performance.
Seon et al. [113]	2017	Studied the influence of recuperator performance on a semi-closed oxy combustion combined cycle.
Laumb et al. [114]	2017	Investigated different super critical CO_2 cycles for power production.

A novel approach for power generation using oxy fuel combustion was proposed by Allam et al. [107]. The proposed Allam cycle is basically a Brayton cycle operating pressurized supercritical CO_2 as the working fluid. The heat capacity of the high-pressure CO_2 is much higher than low-pressure CO_2. The Allam cycle uses this distinct thermodynamic property of CO_2. Since CO_2 is used as a working fluid in this cycle, there is no need to vaporize and condense water for the cycle. A high-pressure combustor burns the fuel in pure oxygen and produces a feed stream with pressure ranging from 200–400 bar. This stream is expanded in a single turbine with a pressure ratio ranging from 6 to 12. The heat of the high-temperature exhaust of the turbine is transferred to a high-pressure recycled CO_2 stream in a recuperator. The recycled stream is sent back to the combustor to control the turbine inlet temperature. With the help of theoretical analysis, the authors reported a (LHV-based) thermal efficiency of 59% for the cycle operating on natural gas and 52% when the cycle uses coal as the fuel while capturing CO_2 inherently. A schematic of the Allam cycle operating on natural gas and coal as fuel is shown Figures 12 and 13, respectively.

The construction of a 50 MWth demonstration plant operating on natural gas fired Allam cycle has just been completed in La Porte, Texas. A commercial 300 MW plant is in the planning stage to demonstrate the advantage of this cycle [114]. Since no extra measures are needed for capturing CO_2, this method is expected to produce electricity at a cost much lower than other conventional power plants using CCS.

As a new and potentially viable method to reduce carbon dioxide emissions from fossil fuel power plants, research is ongoing to optimize the parameters of the Allam cycle. The efficiency of this cycle is sensitive to the turbine inlet temperature and pressure, turbine outlet pressure, temperature difference on the hot side of the primary heat exchanger and the performance of the air separation unit of the cycle. Figure 14 depicts a distribution of the total power production of the turbine for the natural gas-fired Allam cycle. A similar graph is presented in Figure 15 which shows various losses as a percentage of the fuel thermal input. The figures are constructed using the simulation results of Mitchell et al. [116].

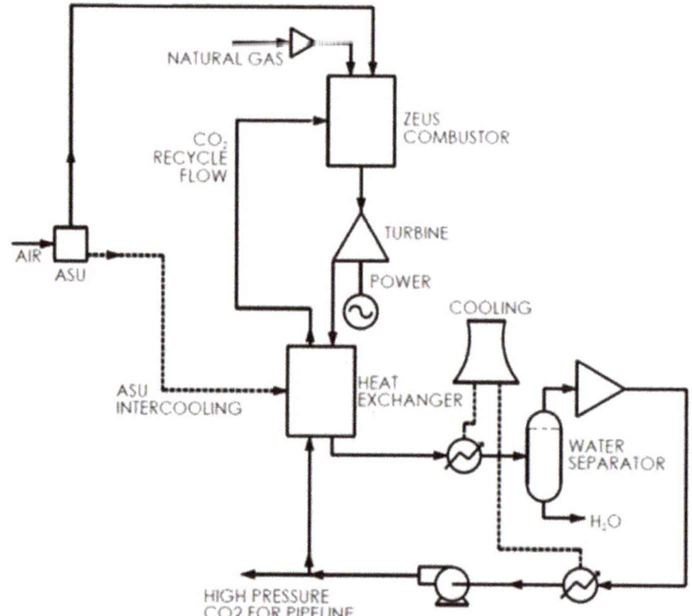

Figure 12. Schematic of the Allam power cycle operating on natural gas as fuel [107].

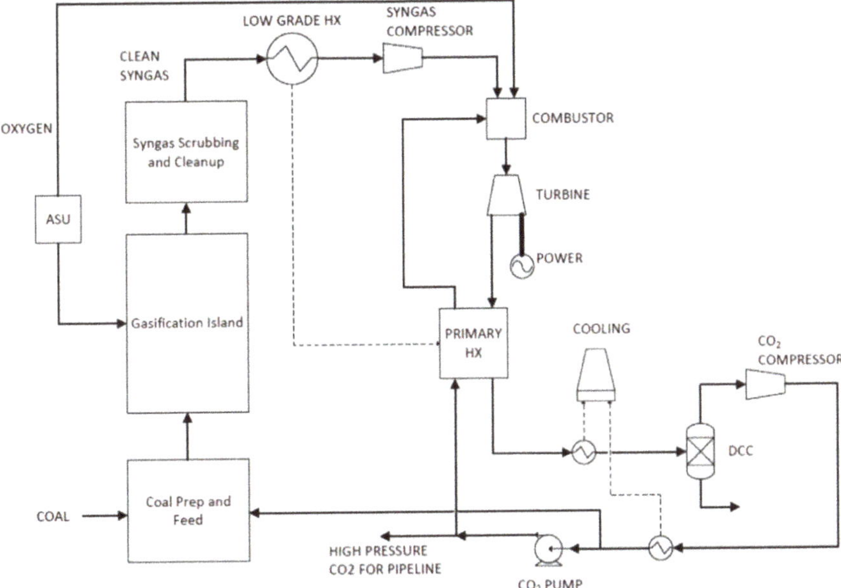

Figure 13. Schematic of the coal-fired Allam power cycle [115].

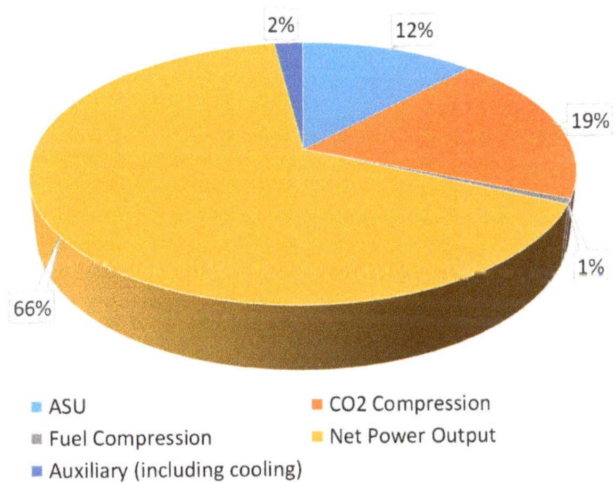

Figure 14. Distribution of the turbine power production in the natural gas-fired Allam cycle.

The power requirement of the CO_2 pressurization is the largest; it consumes about one-fifth of the turbine gross power. The second largest power-consuming component is the ASU, which requires 12% of the turbine power production. The fuel compressor requires ~1% of the turbine power. The total penalty is 34% so that 66% of the turbine gross power is transmitted to an electric generator as the net power output. From Figure 15, 12.4% of the fuel energy is discharged to the surroundings in the form of waste heat so the gross cycle efficiency is 87.6%. The total penalties account for 29.6% of the fuel thermal input which upon subtracting from the gross efficiency, the net cycle efficiency is found to be 58%. The cycle developers have also examined employing two turbines in the cycle. They have reported a noticeable increase in the power output while the capital cost increased a little.

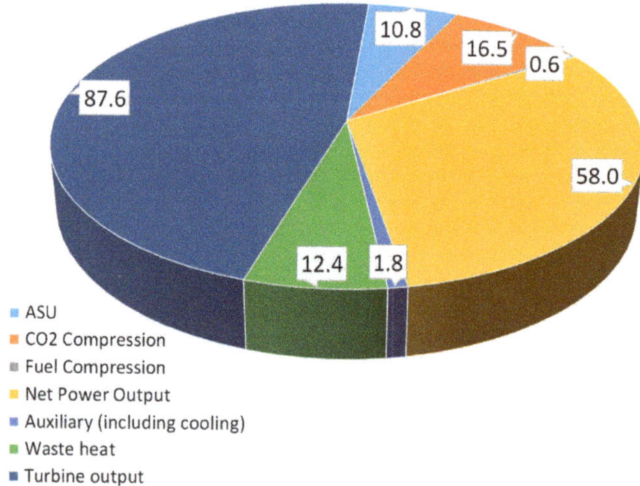

Figure 15. Distribution of the thermal energy input within the Allam cycle. The figures across the chart represent the percentage of the thermal input.

There are also some other power cycles that run on the oxy fuel combustion concept. Clean Energy Systems [117] proposed a cycle which uses water instead of recycled flue gas to control the turbine inlet temperature. Fuel is combusted with an oxidant in a high-pressure combustor in the presence of supercritical steam. The hot gas is then expanded in three different turbines with reheating between them. This cycle is considered by CES as a long-term solution to use in the oxy combustion process with natural gas. A schematic of the proposed cycle is shown in Figure 16.

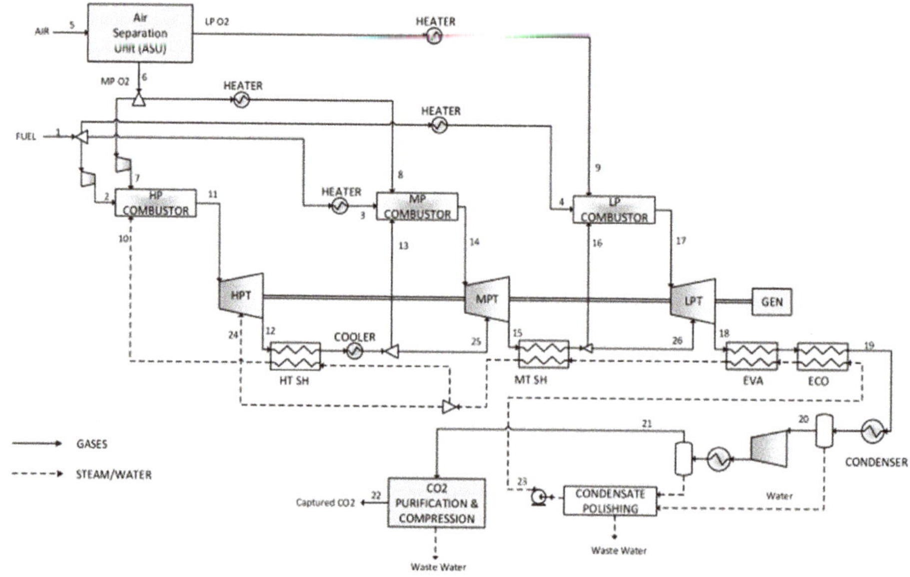

Figure 16. Schematic diagram of a supercritical CES cycle.

3.2. Chemical Looping Combustion (CLC)

Chemical looping combustion is a novel process in the field of carbon capture. This method has the potential to be the most efficient and a low-cost process for capturing carbon dioxide from fossil fuel power plants. IPCC identified this method as one of the cheapest technologies for carbon capture [93]. It has the inherent advantages of CO_2 separation with a minimum energy requirement.

There is no direct contact between the air and fuel which is why it is also known as unmixed combustion [118]. Instead of air, an appropriate oxygen carrier brings oxygen from the air to fuel [119]. Two fluidized bed reactors are used in this process. One is known as the air reactor and the other as the fuel reactor. A schematic of the process is shown in Figure 17. A solid oxygen carrier is circulated between these two reactors. The solid oxygen carrier is oxidized in the air reactor. After oxidation, the carrier goes to the fuel reactor. Fuel is oxidized in the fuel reactor while the oxygen carrier is being reduced. The corresponding chemical reaction can be written as follows [120]:

Fuel reactor: $(2n + m)M_yO_x + C_nH_{2m} \rightarrow (2n + 1)M_yO_{x-1} + mH_2O + nCO_2$

Air Reactor: $M_yO_{x-1} + \frac{1}{2}O_2 \rightarrow M_yO_x$

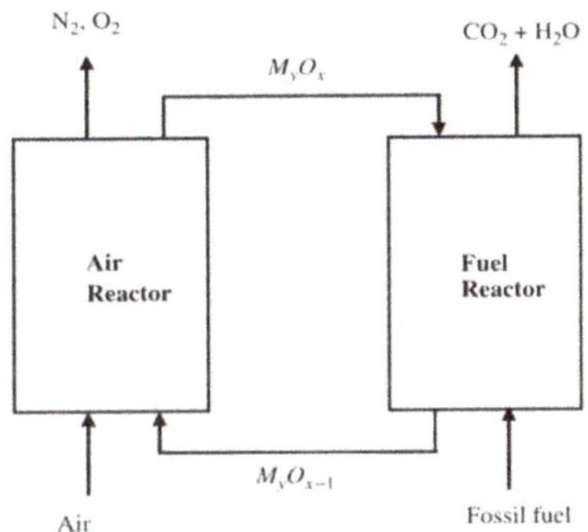

Figure 17. Schematic of the chemical looping combustion process [120].

After the completion of the fuel oxidation, the metal oxygen carrier is circulated back to the air reactor [120]. CO_2 and water are produced in the fuel reactor. CO_2 can be easily separated by condensing H_2O and then sequestered or used for other purposes. After oxidizing the oxygen carrier, the remaining air contains only nitrogen and unreacted oxygen. As they are not harmful to the environment, they can be released without further processing. Syngas produced from gasification of coal is used in the fuel reactor when coal is as the main fuel. A schematic layout of a power plant using chemical looping combustion with syngas from coal is shown in Figure 18.

Here, Ni reacts with high-pressure air in reactor 2 to form NiO and to remove oxygen from the air. Then, NiO is separated from the air. The hot and high-pressurized nitrogen-rich stream is passed through a turbine to generate power. NiO is passed to reactor 1 where it is reduced and the fuel is oxidized into CO_2 and H_2O. This high-pressure and high-temperature CO_2 and H_2O streams are used to produce steam for additional power generation. After that, CO_2 is captured by condensing H_2O from the stream. Reduced NiO forming Ni is returned to the air reactor to repeat the cycle.

Despite the novelty of the CLC process, this method has certain challenges. For example, the design of the reactor with two separated reaction zone is one of the main challenges. A CLC reactor system with two interconnected fluidized bed is shown in Figure 19. Here, the oxygen carrier should circulate between the zones, but the gas streams should not be leaked into one to another. The first continuous operation of this type of reactor was achieved with a 10 kW prototype at Chalmers University of Technology [122]. It showed a stable operation with 99.5% fuel conversion efficiency at ambient pressure. There was no significant gas leakage. There are also some other approaches in the design of the reactor. Shimomura [123,124] proposed a reactor with a rotating reactor wheel. Air and fuel were to be fed at separate compartments of the wheel. They considered an adsorbent wheel which would use Li_4SiO_4 based absorbent. Ivar et al. [125] worked on developing novel concepts of reactor. Their initial testing showed that the mixing of air and fuel could only be avoided partly. A steam stream was employed as a gas wall between the fuel and air. The flue gas in this experiment contained 85% CO_2 after steam removal.

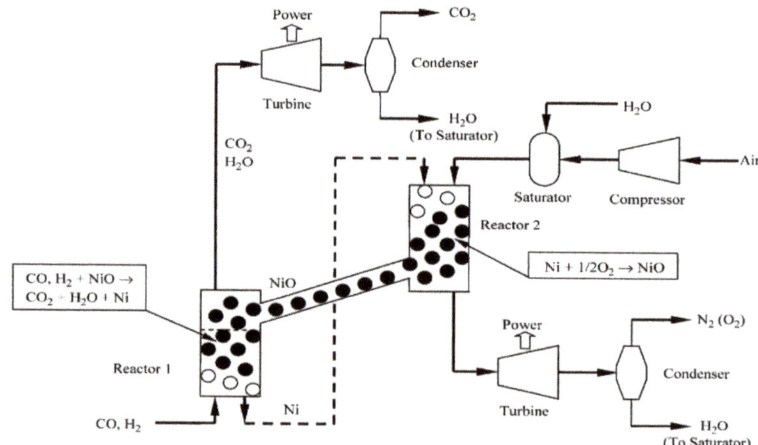

Figure 18. Layout of a power plant using chemical looping combustion with NiO and syngas [121].

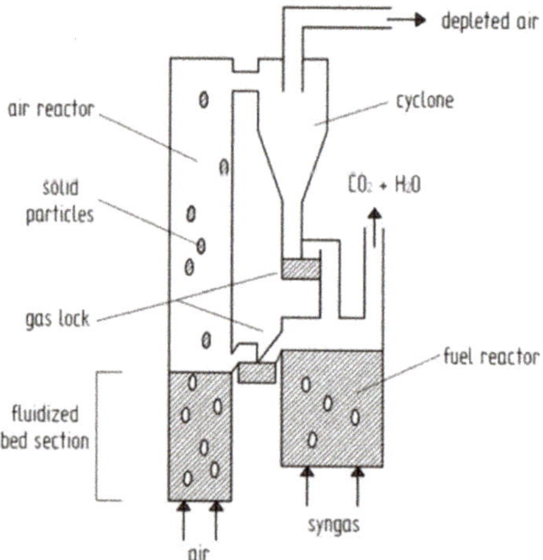

Figure 19. A CLC reactor with two interconnected fluidized beds [126].

The thermal efficiency of a combined cycle power plant equipped with CLC was found to be around 52–53% with respective operating temperature and pressure of 1200 °C and 13 atm in the air reactor. Implementation of CLC yields 3–5% higher efficiency than other carbon capture methods [127]. A 2.8% higher thermal efficiency was found for a CLC-IGCC power plant compared to an IGCC using physical absorption for carbon separation. Also, the CLC allowed for 100% capture of carbon dioxide while the physical absorption yielded 85% capture from the IGCC plant [122].

An important aspect of research in the CLC field is to find a suitable oxygen carrier that would have a high fuel conversion ratio, a good stability, and a high oxygen transport capacity [122]. Various materials are being tested for this purpose. Materials with reactivity above or near their melting point should not be used as the oxygen carrier because they would have to undergo a cyclic operation at a high temperature. Along with the reactivity, thermal stability, toxicity, and cost should be considered when choosing an oxygen carrier [128]. Some of the most likely elements to use as oxygen carrier are iron, copper, manganese, and nickel. They should be combined with inert materials like alumina, silica, titanium oxide etc. [122]. Lyngefelt et al. [129] tested more than 290 different particles as oxygen carrier including active oxides of copper, nickel, manganese, and iron. A conversion efficiency of 99.5% was attained in a 10 kW prototype reactor. In another experiment, 99% conversion efficiency was gained when $NiO/MgAl_2O_4$ was used as an oxygen carrier [130].

Another significant factor for better efficiency is the temperature of the air reactor. The temperature of the air reactor can be compared with the turbine inlet temperature of a conventional power plant [122]. The reaction that takes place in the air reactor is endothermic whereas the reaction at the fuel reactor can be endothermic or exothermic. Rehan et al. [131] examined a multi-stage chemical looping combustion for combined cycle. They found that at an oxidation temperature of 1200 °C, a CLC combined cycle would operate at achieved 52% efficiency without reheating. The same power plant exhibited 51% efficiency at 1000 °C and 53% efficiency at 1200 °C reactor temperature when a single stage reheat was employed. However, employing a double stage reheat did not improve the efficiency over the single reheat system. Zhu et al. [132] compared the performance of an IGCC with chemical looping and calcium looping processes. They concluded that the CLC based technology exhibits higher efficiency (39.78%) than the pre-combustion capture with physical absorption (36.21%) and calcium looping (37.72%). The payback period of the above three capture methods was estimated to be 13.45 years, 13.21 years and 17.25 years, respectively.

4. Comparison of the Methods

Carbon capture requiring separation of CO_2 is an age-old process and it has reached a certain maturity with various established full-scale application. A lot of experimental and numerical modeling studies have been done on these processes. The main advantage of post combustion capture is its easy integration capability with the existing power plants, but the partial pressure and concentration of CO_2 are very low in the flue gases. For transportation and storage of CO_2, a minimum concentration should be reached. The required extra energy and extra costs of carbon capture to attain a minimum required concentration are significantly high.

When using chemical absorption process for the separation, degradation of the solvent and severe corrosion of the used equipment take place. Therefore, a huge cost for solvents and other equipment becomes necessary for this process to make CO_2 ready for transportation and storage. These may increase the cost of producing electricity around 70% [118]. Research is ongoing for new solvents to reduce the cost of carbon capture. Large equipment size results in the high capital and operating cost in this method.

Pre-combustion carbon capture is mostly used in process industries. There are also full-scale CCS plants in some industries which use this method [11]. The amount of CO_2 is much higher in the gas mixture in this process than the conventional flue gas mixture. Due to the higher pressure and lower gas volume, less energy is required in this process compared to post combustion capture, but still, the energy penalty is high. Precombustion is mainly used in integrated gasification combined cycle technology. This technology demands a huge auxiliary system for smooth operation. Therefore, the capital cost of this system is too high compared to other systems.

On the other hand, carbon capture processes without requiring separation are comparatively novel in power generation. There is no full-scale operational plant based on these processes. There are some pilot scale operation and some subscale demonstration plants under development using oxy fuel combustion [10,30,94]. The most promising step regarding oxy fuel combustion is the 50 MWth demonstration power plant built in Texas by Net Power using the concept of Allam cycle. It will ensure near zero emission. This method has some other advantages like reduction in equipment size, compatibility with various kinds of coals, and no need for an onsite chemical plant [10].

The process, however, requires a large amount of high purity oxygen. Therefore, an energy-intensive ASU is needed for oxygen production. Membrane-based technology for air separation may compete with cryogenic ASU through a higher degree of integration into the power cycle [133]. Due to this ASU and CO_2 compression unit used in this process, net power output decreases significantly. Along with these, there are some technical uncertainties that demand more research to understand the full-scale operation. However, since no extra cost is required for CO_2 separation, this process remains a promising one to produce electricity at a lower cost while confirming near zero emission.

The CLC process for carbon capture is still in the preliminary stage. It has not been implemented on a commercial level yet. Further research is required to take advantage of this method. Due to the absence of flame, no NO_x is produced thermally and the outlet stream of air reactor is harmless for the environment [93]. Developing a proper oxygen carrier to use in CLC will make it more attractive than other processes.

A comparison of the thermal efficiency of power plants with different CO_2 capture processes is provided in Table 6. The efficiencies shown in the table are based on the lower heating value of the fuel. Bituminous coal is considered for coal-based power plants due to its extensive use in power production [30]. The Selexol process is taken into consideration for precombustion carbon capture in an IGCC GE type gasifier. In the case of chemical looping combustion, ilmenite was used as oxygen carrier for coal-based power production whereas Nickel Aluminum oxide was used for natural gas-based power production [134,135].

Table 6. Efficiency comparison of power generation with different carbon capture processes [30,107,134,135].

Fuel Type	Process	Net Efficiency	Net Power (MW)
Coal (Bituminous)	Without carbon capture	44	758
	Precombustion	31.5	676
	Post combustion	34.8	666
	Oxy Combustion	35.4	532
	Oxy Combustion (Allam Cycle)	51	226
	Chemical looping combustion	44	115.5
Natural Gas	Without carbon capture	55.6	776
	Pre combustion	41.5	690
	Post combustion	47.4	662
	Oxy Combustion	44.7	440
	Oxy Combustion (Allam Cycle)	59	303
	Chemical looping combustion	52.2	364

When coal is used as a fuel, the CLC exhibits the same efficiency as the base combustion technology using pulverized coal without any capture. Reduction in the efficiency is the highest in pre-combustion carbon capture. Post combustion and oxy combustion carbon capture show an almost similar drop in efficiency. An interesting observation in this comparison is the efficiency of the Allam cycle. The targeted efficiency of the Allam cycle is almost the same as the reference power plant efficiency without capture. If this cycle can be implemented commercially at a larger scale, the overall power generation efficiency will increase while ensuring total carbon capture.

When natural gas is used as fuel, the pre-combustion carbon capture shows a 14% drop in the efficiency from the reference powerplant whereas the post combustion carbon capture shows an 8% drop. The traditional oxy combustion process exhibits an efficiency of 44.7%. Chemical looping combustion indicates only a 4% drop in efficiency from the reference plant. The Allam cycle shows an extraordinary performance whose efficiency happens to be over 3 percentage points higher than that of the reference combined cycle without CO_2 capture.

From the efficiency comparison of Table 6, it may be concluded that the chemical looping combustion and the Allam cycle are expected to be the leading technologies in the near future for fossil fuel-based power generation. The 50 MWth Allam cycle provides the basis for deployment of large-scale facilities. Currently, 300 MW natural gas-fired plants are under development. The chemical looping method is not yet technologically ready to implement on an industrial basis. The method is still in the investigation stage. More experimental data are necessary before large-scale commercialization.

Conventional carbon capture process results in the reduction of efficiency. More fuel is burnt per unit of electricity production due to this inefficiency which leads to more production of CO_2. Also, the processes used for capturing carbon dioxide may affect the environment in different ways other than direct emission of CO_2. For example, different substances used for separating and capturing CO_2 may have undesired effect on the human body and environment. Using a solid sorbent covered with coating was experimented to reduce the formation of dust from the substance [136]. This could also reduce the capacity of the substance to capture carbon dioxide. Also, stripping of organic solvent from membranes and sorbents is suggested to prevent undesired odor. Before employing carbon capture, it should be ensured that reducing CO_2 is not being achieved at the cost of other environmental impacts.

Life cycle assessment of the plants is necessary to properly understand environmental impacts of the carbon capture methods. Schreibar et al. [7] used the life cycle assessment (LCA) methodology for post combustion carbon capture using MEA whose impact on the environment and human health was investigated for five power plants. The global warming potential (GWP), human toxicity potential (HTP), acidification potential (AP), photo oxidant formation potential, eutrophication potential (EP) were considered as impact categories. As expected, GWP was much lower with MEA compared to the power plants without capture whereas HTP was three times higher with MEA plants. Schreibar et al. [7] concluded that upstream and downstream processes such as emissions from fuel and material supply, waste disposal, and waste water treatment influence the environmental impact measures for power plants with carbon capture. Viebahn et al. [137] revealed about a 40% increase in AP, EP, HTP when post combustion carbon capture was implemented in a power plant.

A similar result was found by Veltman et al. [138]. They showed that a power plant with post combustion capture yields a 10 times increase in toxic impacts on freshwater compared to a plant without capture. Impacts on other categories were negligible. Degradation of MEA resulted in the emission of ammonia, acetaldehyde, and formaldehyde. Cuellar et al. [139] compared life cycle environmental impacts of carbon capture and storage with carbon capture and utilization. GWP with utilization was much greater than that with storage. The highest reduction of GWP was found for pulverized coal and IGCC plants employing the oxyfuel capture method as well as combined cycle gas turbine plants equipped with a post combustion capture technology.

Pehnt et al. [140] showed that a conventional power plant operating on coal with post combustion carbon capture would result in an increase in the environmental impact in almost all categories except GWP. Solvent degradation and energy penalty due to CO_2 capture process are the main reasons for this increase. Precombustion capture showed a decrease in all the environmental impact categories compared to a conventional power plant. They identified oxyfuel combustion as the most potential process to reduce all the environmental impact categories if co-capture of other pollutants can be achieved.

Nie et al. [141] investigated comparative environmental impacts of post combustion and oxy fuel combustion carbon capture. Their analysis showed that almost all environmental impact categories except GWP would increase with post combustion carbon capture. The same is true for oxyfuel combustion except GWP, AP and EP. However, the amount of increase of these impact categories was found to be less in oxyfuel combustion compared to the post combustion carbon capture. No LCA analysis was found for chemical looping combustion and the newly proposed Allam cycle based on oxyfuel technology.

5. Conclusions

Despite their harmful effects on the environment, fossil fuels will continue to be the primary source of power production. A comprehensive discussion of the technologies to reduce emissions from fossil fuels is presented in this review. Post-combustion carbon capture technology requiring separation is the best technology to retrofit the existing fossil fuel power plants. Pre-combustion carbon capture is more suitable in an integrated gasification combined cycle. A post combustion carbon capture process can be applied to the running power plants without much modification.

The separation of CO_2 from the gas stream in these processes requires additional energy which leads to an increase of the electricity price. Several technologies are available for separating CO_2 from the gas stream. Separation is less costly in precombustion processes because of the higher partial pressure of CO_2 in the gas stream. Among the separation technologies, the absorption process has become almost matured, but it requires attention to corrosion of equipment and high cost for regeneration of the solvent. An adsorption process for separation cannot be applied at large scales due to the low CO_2 adsorption and capacity and influence of gases on the adsorbents. New adsorbents should be developed to remove the barriers. Membrane technology is less energy intensive than other processes, though it is less effective at low concentration of CO_2. Also, more research is necessary to determine membrane behavior at high capacity. Formation of clathrate hydrate to separate CO_2 from gas mixture needs more attention. Proper additives or promoters should be developed to use in the process to make this process more competitive.

Carbon capture with water condensation is economically more viable to implement due to a simpler design and higher plant efficiency and better environmental aspects for the life cycle compared to the pre- and post-combustion capture methods. LCA shows that oxyfuel combustion has much less effect on the environment compared to other methods. The combustion characteristics are different in this case from traditional combustion. Due to the different characteristics, further research is necessary to make this technology competitive with others. There still is limited knowledge of many important aspects of oxy fuel combustion. Along with the theoretical achievement, it is necessary to find more experimental data in order to validate theoretical models.

Further research is necessary to properly understand the oxy fuel combustion-based near zero emission power cycles like the Allam cycle. Successful commercial implementation of these cycles will be a significant step in reducing emissions, thus meeting the challenge of global climate change. Chemical looping combustion has also the potential to become a cost-effective way to reduce emissions. It requires attention as a novel technology in this sector. The availability of suitable oxygen carriers and appropriate designing of the reactors can make this process comparable with others. LCA of these new methods is also necessary to properly understand their impact on the environment.

Author Contributions: Conceptualization, N.S.S. and Y.H.; methodology, Y.H.; investigation, N.S.S.; writing—original draft preparation, N.S.S.; writing—review and editing, Y.H.; supervision, Y.H.

Funding: This research received no external funding.

Acknowledgments: The research fund provided by Central Michigan University is gratefully acknowledged. The authors wish to express their gratitude to the anonymous reviewers for providing constructive feedback.

Conflicts of Interest: The authors declare no conflict of interest.

Abbreviations

CCS	Carbon Capture and Storage
IGCC	Integrated Gasification Combined Cycle
TSA	Temperature Swing Adsorption
PSA	Pressure Swing Adsorption
VSA	Vacuum Swing Adsorption
THF	Tetrahydrofuran
TBAB	Tetrabutylammonium Bromide

TBAF	Tetrabutylammonium Fluoride
ESP	Electrostatic Precipitator
FGD	Flue Gas Desulphurization
LCA	Life Cycle Assessment
MEA	Monoethanolamine
MOF	Metal Organic Framework
CLC	Chemical Looping Combustion
CFB	Circulating Fluidized Bed
CES	Clean Energy System

References

1. United States Environmental Protection Agency (EPA). *Global Greenhouse Gas Emissions Data.* Available online: https://www.epa.gov/ghgemissions/global-greenhouse-gas-emissions-data (accessed on 29 October 2019).
2. Rogelj, J.; Den Elzen, M.; Höhne, N.; Fransen, T.; Fekete, H.; Winkler, H.; Schaeffer, R.; Sha, F.; Riahi, K.; Meinshausen, M. Paris Agreement climate proposals need a boost to keep warming well below 2 °C. *Nature* **2016**, *534*, 631–639. [CrossRef]
3. Zhou, R.; Guzman, M.I. CO_2 reduction under periodic illumination of ZnS. *J. Phys. Chem. Chem. C* **2014**, *118*, 11649–11656. [CrossRef]
4. Sharma, N.; Das, T.; Kumar, S.; Bhosale, R.; Kabir, M.; Ogale, S. Photocatalytic activation and reduction of CO_2 to CH_4 over single phase nano Cu_3SnS_4: A combined experimental and theoretical study. *ACS Appl. Energy Mater.* **2019**, *2*, 5677–5685. [CrossRef]
5. Wall, T.; Liu, Y.; Spero, C.; Elliott, L.; Khare, S.; Rathnam, R.; Zeenathal, F.; Moghtaderi, B.; Buhre, B.; Changdong, S.; et al. An overview on oxyfuel coal combustion-state of the art research and technology development. *Chem. Eng. Res. Des.* **2009**, *87*, 1003–1016. [CrossRef]
6. Lockwood, T. A comparariitve review of next-generation carbon capture technologies for coal-fired power plant. *Energy Procedia* **2017**, *114*, 2658–2670. [CrossRef]
7. Schreiber, A.; Zapp, P.; Kuckshinrichs, W. Environmental assessment of German electricity generation from coal-fired power plants with amine-based carbon capture. *Int. J. Life Cycle Assess.* **2009**, *14*, 547–559. [CrossRef]
8. Pardemann, R.; Meyer, B. Pre-Combustion carbon capture. *Handb. Clean Energy Syst.* **2015**, 1–28. [CrossRef]
9. Babu, P.; Ong, H.W.N.; Linga, P. A systematic kinetic study to evaluate the effect of tetrahydrofuran on the clathrate process for pre-combustion capture of carbon dioxide. *Energy* **2016**, *94*, 431–442. [CrossRef]
10. Theo, W.L.; Lim, J.S.; Hashim, H.; Mustaffa, A.A.; Ho, W.S. Review of pre-combustion capture and ionic liquid in carbon capture and storage. *Appl. Energy* **2016**, *183*, 1633–1663. [CrossRef]
11. Jansen, D.; Gazzani, M.; Manzolini, G.; van Dijk, E.; Carbo, M. Pre-Combustion CO_2 capture. *Int. J. Greenh. Gas Control* **2015**, *40*, 167–187. [CrossRef]
12. Gazzani, M.; Macchi, E.; Manzolini, G. CO_2 capture in natural gas combined cycle with SEWGS. Part A: Thermodynamic performances. *Int. J. Greenh. Gas Control* **2013**, *12*, 493–501. [CrossRef]
13. Gazzani, M.; Macchi, E.; Manzolini, G. CO_2 capture in integrated gasification combined cycle with SEWGS—Part A: Thermodynamic performances. *Fuel* **2013**, *105*, 206–219. [CrossRef]
14. Davies, K.; Malik, A.; Li, J.; Aung, T.N. A meta-study on the feasibility of the implementation of new clean coal technologies to existing coal-fired power plants in an effort to decrease carbon emissions. *PAM Rev.* **2007**, 30–45. [CrossRef]
15. Romano, M.C.; Chiesa, P.; Lozza, G. Pre-Combustion CO_2 capture from natural gas power plants, with ATR and MDEA processes. *Int. J. Greenh. Gas Control* **2010**, *4*, 785–797. [CrossRef]
16. Martin, C.; Stockel, E.; Lowes, C.R.; Adams, D.; Cooper, A.; Pis, J.; Rubiera, F.; Piveda, C. Hypercrosslinked organic polymer networks as potential adsorbents for pre-combustion CO_2 capture. *J. Mater. Chem.* **2011**, *21*, 5475–5483. [CrossRef]
17. Schell, J.; Casas, N.; Blom, R.; Spjelkavik, A.I.; Andersen, A.; Cavka, J.H.; Mazzotti, M. MCM-41, MOF and UiO-67/MCM-41 adsorbents for pre-combustion CO_2 capture by PSA: Adsorption equilibria. *Adsorption* **2012**, *18*, 213–227. [CrossRef]

18. García, S.; Gil, M.V.; Martín, C.F.; Pis, J.J.; Rubiera, F.; Pevida, C. Breakthrough adsorption study of a commercial activated carbon for pre-combustion CO_2 capture. *Chem. Eng. J.* **2011**, *171*, 549–556.

19. Casas, N.; Schell, J.; Joss, L.; Mazzotti, M. A parametric study of a PSA process for pre-combustion CO_2 capture. *Sep. Purif. Technol.* **2013**, *104*, 183–192. [CrossRef]

20. Moioli, S.; Giuffrida, A.; Gamba, S.; Romano, M.C.; Pellegrini, L.; Lozza, G. Pre-Combustion CO_2 capture by MDEA process in IGCC based on air-blown gasification. *Energy Procedia* **2014**, *63*, 2045–2053. [CrossRef]

21. Jiang, G.; Huang, Q.; Kenarsari, S.D.; Hu, X.; Russell, A.G.; Fan, M.; Shen, X. A new mesoporous amine-TiO_2 based pre-combustion CO_2 capture technology. *Appl. Energy* **2015**, *147*, 214–223. [CrossRef]

22. Park, S.H.; Lee, S.J.; Lee, J.W.; Chun, S.N.; Lee, J.B. The quantitative evaluation of two-stage pre-combustion CO_2 capture processes using the physical solvents with various design parameters. *Energy* **2015**, *81*, 47–55. [CrossRef]

23. Dai, Z.; Deng, L. Membrane absorption using ionic liquid for pre-combustion CO_2 capture at elevated pressure and temperature. *Int. J. Greenh. Gas Control* **2016**, *54*, 59–69. [CrossRef]

24. Yang, M.; Jing, W.; Zhao, J.; Ling, Z.; Song, Y. Promotion of hydrate-based CO_2 capture from flue gas by additive mixtures (THF (tetrahydrofuran) + TBAB (tetra-n-Butyl ammonium bromide)). *Energy* **2016**, *106*, 546–553. [CrossRef]

25. Zheng, J.; Zhang, P.; Linga, P. Semiclathrate hydrate process for pre-combustion capture of CO_2 at near ambient temperatures. *Appl. Energy* **2017**, *194*, 267–278. [CrossRef]

26. Usman, M.; Hillestad, M.; Deng, L. Assessment of a membrane contactor process for pre-combustion CO_2 capture by modelling and integrated process simulation. *Int. J. Greenh. Gas Control* **2018**, *71*, 95–103. [CrossRef]

27. Zhai, H.; Rubin, E.S. Systems analysis of physical absorption of CO_2 in ionic liquids for pre-combustion carbon capture. *Environ. Sci. Technol.* **2018**, *52*, 4996–5004. [CrossRef]

28. Hammond, G.P.; Spargo, J. The prospects for coal-fired power plants with carbon capture and storage: A UK perspective. *Energy Convers. Manag.* **2014**, *86*, 476–489. [CrossRef]

29. Yoro, K.; Sekoai, P. The potential of CO_2 capture and storage technology in South Africa's coal-fired thermal power plants. *Environments* **2016**, *3*, 24. [CrossRef]

30. Leung, D.Y.C.; Caramanna, G.; Maroto-Valer, M.M. An overview of current status of carbon dioxide capture and storage technologies. *Renew. Sustain. Energy Rev.* **2014**, *39*, 426–443. [CrossRef]

31. Mondal, M.K.; Balsora, H.K.; Varshney, P. Progress and trends in CO_2 capture/separation technologies: A review. *Energy* **2012**, *46*, 431–441. [CrossRef]

32. Merkel, T.C.; Lin, H.; Wei, X.; Baker, R. Power plant post-combustion carbon dioxide capture: An opportunity for membranes. *J. Membr. Sci.* **2010**, *359*, 126–139. [CrossRef]

33. Adams, T.A., II; Hoseinzade, L.; Madabhushi, P.B.; Okeke, I.J. Comparison of CO_2 capture approaches for fossil-based power generation: Review and meta-study. *Processes* **2017**, *5*, 1–37.

34. Agarwal, A.; Biegler, L.T.; Zitney, S.E. A superstructure-based optimal synthesis of PSA cycles for post-combustion CO_2 capture. *AIChE J.* **2010**, *56*, 1813–1828. [CrossRef]

35. Wappel, D.; Gronald, G.; Kalb, R.; Draxler, J. Ionic liquids for post-combustion CO_2 absorption. *Int. J. Greenh. Gas Control* **2010**, *4*, 486–494. [CrossRef]

36. Mason, J.A.; Sumida, K.; Herm, Z.R.; Krishna, R.; Long, J.R. Evaluating metal-organic frameworks for post-combustion carbon dioxide capture via temperature swing adsorption. *Energy Environ. Sci.* **2011**, *4*, 3030–3040. [CrossRef]

37. Savile, C.K.; Lalonde, J.J. Biotechnology for the acceleration of carbon dioxide capture and sequestration. *Curr. Opin. Biotechnol.* **2011**, *22*, 818–823. [CrossRef]

38. Lee, Z.H.; Lee, K.T.; Bhatia, S.; Mohamed, A.R. Post-Combustion carbon dioxide capture: Evolution towards utilization of nanomaterials. *Renew. Sustain. Energy Rev.* **2012**, *16*, 2599–2609. [CrossRef]

39. Scholes, C.A.; Ho, M.T.; Wiley, D.E.; Stevens, G.W.; Kentish, S.E. Cost competitive membrane-cryogenic post-combustion carbon capture. *Int. J. Greenh. Gas Control* **2013**, *17*, 341–348. [CrossRef]

40. Bae, T.H.; Hudson, M.R.; Mason, J.A.; Queen, W.L.; Dutton, J.J.; Sumida, K.; Micklash, K.J.; Kaye, S.S.; Brown, C.M.; Long, J.R. Evaluation of cation-exchanged zeolite adsorbents for post-combustion carbon dioxide capture. *Energy Environ. Sci.* **2013**, *6*, 128–138. [CrossRef]

41. Zhang, Z.; Yan, Y.; Zhang, L.; Chen, Y.; Ju, S. CFD investigation of CO_2 capture by methyldiethanolamine and 2-(1-piperazinyl)-ethylamine in membranes: Part B. Effect of membrane properties. *J. Nat. Gas Sci. Eng.* **2014**, *19*, 311–316. [CrossRef]

42. Shakerian, F.; Kim, K.H.; Szulejko, J.E.; Park, J.W. A comparative review between amines and ammonia as sorptive media for post-combustion CO_2 capture. *Appl. Energy* **2015**, *148*, 10–22. [CrossRef]

43. Zhang, Z. Comparisons of various absorbent effects on carbon dioxide capture in membrane gas absorption (MGA) process. *J. Nat. Gas Sci. Eng.* **2016**, *31*, 589–595. [CrossRef]

44. El Hadri, N.; Quang, D.V.; Goetheer, E.L.V.; Abu Zahra, M.R.M. Aqueous amine solution characterization for post-combustion CO_2 capture process. *Appl. Energy* **2017**, *185*, 1433–1449. [CrossRef]

45. Thompson, J.G.; Combs, M.; Abad, K.; Bhatnagar, S.; Pelgen, J.; Beaudry, M.; Rochelle, G.; Hume, S.; Link, D.; Figueroa, J.; et al. Pilot testing of a heat integrated 0.7 MWe CO_2 capture system with two-stage air-stripping: Emission. *Int. J. Greenh. Gas Control* **2017**, *64*, 267–275. [CrossRef]

46. Thompson, J.G.; Combs, M.; Abad, K.; Bhatnagar, S.; Pelgen, J.; Beaudry, M.; Rochelle, G.; Hume, S.; Link, D.; Figueroa, J.; et al. Pilot testing of a heat integrated 0.7 MWe CO_2 capture system with two-stage air-stripping: Amine degradation and metal accumulation. *Int. J. Greenh. Gas Control* **2017**, *64*, 23–33. [CrossRef]

47. Zhang, Z.; Chen, F.; Rezakazemi, M.; Zhang, W.; Lu, C.; Chang, H.; Quan, X. Modeling of a CO_2-piperazine-membrane absorption system. *Chem. Eng. Res. Des.* **2017**, *131*, 375–384. [CrossRef]

48. Rackley, S.A. Absorption capture systems. In *Carbon Capture and Storage*, 2nd ed.; Elsevier: Amsterdam, The Netherlands, 2017.

49. Padurean, A.; Cormos, C.-C.; Agachi, P.-S. Pre-Combustion carbon dioxide capture by gas–liquid absorption for integrated gasification combined cycle power plants. *Int. J. Greenh. Gas Control* **2012**, *7*, 1–11. [CrossRef]

50. Yu, C.H.; Huang, C.H.; Tan, C.S. A review of CO_2 capture by absorption and adsorption. *Aerosol Air Qual. Res.* **2012**, *12*, 745–769. [CrossRef]

51. Mofarahi, M.; Khojasteh, Y.; Khaledi, H.; Farahnak, A. Design of CO_2 absorption plant for recovery of CO_2 from flue gases of gas turbine. *Energy* **2008**, *33*, 1311–1319. [CrossRef]

52. Figueroa, J.D.; Fout, T.; Plasynski, S.; McIlvried, H.; Srivastava, R.D. Advances in CO_2 capture technology—The U.S. Department of Energy's Carbon Sequestration Program. *Int. J. Greenh. Gas Control* **2008**, *2*, 9–20. [CrossRef]

53. Berstad, D.; Anantharaman, R.; Nekså, P. Low-Temperature CCS from an IGCC power plant and comparison with physical solvents. *Energy Procedia* **2013**, *37*, 2204–2211. [CrossRef]

54. Meisen, A.; Shuai, X. Research and development issues in CO_2 capture. *Energy Convers. Manag.* **1997**, *38*, S37–S42. [CrossRef]

55. Lillia, S.; Bonalumi, D.; Grande, C.; Manzolini, G. A comprehensive modeling of the hybrid temperature electric swing adsorption process for CO_2 capture. *Int. J. Greenh. Gas Control* **2018**, *74*, 155–173. [CrossRef]

56. Plaza, M.G.; Durán, I.; Rubiera, F.; Pevida, C. Adsorption-Based process modelling for post-combustion CO_2 capture. *Energy Procedia* **2017**, *114*, 2353–2361. [CrossRef]

57. Songolzadeh, M.; Soleimani, M.; Takht Ravanchi, M.; Songolzadeh, R. Carbon dioxide separation from flue gases: A technological review emphasizing reduction in greenhouse gas emissions. *Sci. World J.* **2014**, *2014*, 828131. [CrossRef] [PubMed]

58. Grande, C.A. Advances in pressure swing adsorption for gas separation. *ISRN Chem. Eng.* **2012**. [CrossRef]

59. Rashidi, N.A.; Yusup, S. An overview of activated carbons utilization for the post-combustion carbon dioxide capture. *J. CO2 Util.* **2016**, *13*, 1–16. [CrossRef]

60. Langlois, P.; Pentchev, I.; Hinkov, I.; Lamari, F.D.; Langlois, P.; Dicko, M.; Chilev, C.; Pentchev, I. Carbon dioxide capture by adsorption. *J. Chem. Tech. Matall.* **2016**, *51*, 609–627.

61. Wright, A.; White, V.; Hufton, J.; Selow, E.; van Hinderink, P. Reduction in the cost of pre-combustion CO_2 capture through advancements in sorption-enhanced water-gas-shift. *Energy Procedia* **2009**, *1*, 707–714. [CrossRef]

62. Cobden, P.D.; van Beurden, P.; Reijers, H.T.J.; Elzinga, G.D.; Kluiters, S.C.A.; Dijkstra, J.W.; Jansen, D.; van den Brink, R.W. Sorption-Enhanced hydrogen production for pre-combustion CO_2 capture: Thermodynamic analysis and experimental results. *Int. J. Greenh. Gas Control* **2007**, *1*, 170–179. [CrossRef]

63. Khalilpour, R.; Mumford, K.; Zhai, H.; Abbas, A.; Stevens, G.; Rubin, E.S. Membrane-Based carbon capture from flue gas: A review. *J. Clean. Prod.* **2015**, *103*, 286–300. [CrossRef]

64. Ibrahim, M.H.; El-Naas, M.H.; Zhang, Z.; Van der Bruggen, B. CO_2 capture using hollow fiber membranes: A review of membrane wetting. *Energy Fuels* **2018**, *32*, 963–978. [CrossRef]

65. El-Naas, M.H.; Al-Marzouqi, M.; Marzouk, S.A.; Abdullatif, N. Evaluation of the removal of CO_2 using membrane contactors: Membrane wettability. *J. Membr. Sci.* **2010**, *350*, 410–416. [CrossRef]

66. Rostami, S.; Keshavarz, P.; Raeissi, S. Experimental study on the effects of an ionic liquid for CO_2 capture using hollow fiber membrane contactors. *Int. J. Greenh. Gas Control* **2018**, *69*, 1–7. [CrossRef]

67. Kreulen, H.; Smolders, C.A.; Versteeg, G.F.; Van Swaaij, W.P.M. Determination of mass transfer rates in wetted and non-wetted microporous membranes. *Chem. Eng. Sci.* **1993**, *48*, 2093–2102. [CrossRef]

68. Mavroudi, M.; Kaldis, S.P.; Sakellaropoulos, G.P. Reduction of CO_2 emissions by a membrane contacting process. *Fuel* **2003**, *82*, 2153–2159. [CrossRef]

69. Kreulen, H.; Smolders, C.A.; Versteeg, G.F.; van Swaaij, W.P.M. Microporous hollow fibre membrane modules as gas-liquid contactors part 2. Mass transfer with chemical reaction. *J. Membr. Sci.* **1993**, *78*, 217–238. [CrossRef]

70. Dindore, V.Y.; Brilman, D.W.F.; Geuzebroek, R.H.; Versteeg, G.F. Membrane-Solvent selection for CO_2 removal using membrane gas-liquid contactors. *Sep. Purif. Technol.* **2004**, *40*, 133–145. [CrossRef]

71. Malek, A.; Li, K.; Teo, W.K. Modeling of microporous hollow fiber membrane modules operated under partially wetted conditions. *Ind. Eng. Chem. Res.* **1997**, *36*, 784–793. [CrossRef]

72. Kumar, R.; Linga, P.; Englezos, P. Pre and Post Combustion Capture of Carbon Dioxide via Hydrate Formation. In Proceedings of the 2006 IEEE EIC Climate Change Technology Conference, EICCCC 2006, Ottawa, ON, Canada, 10–12 May 2006; pp. 1–7.

73. Zhong, D.; Sun, D.; Lu, Y.; Yan, J.; Wang, J. Adsorption—Hydrate hybrid process for methane separation from a $CH_4/N_2/O_2$ gas mixture using pulverized coal particles. *Ind. Eng. Chem. Res.* **2014**, *53*, 15738–15746. [CrossRef]

74. Babu, P.; Linga, P.; Kumar, R.; Englezos, P. A review of the hydrate based gas separation (HBGS) process forcarbon dioxide pre-combustion capture. *Energy* **2015**, *85*, 261–279. [CrossRef]

75. Lee, H.J.; Lee, J.D.; Linga, P.; Englezos, P.; Kim, Y.S.; Lee, M.S.; Kim, Y.D. Gas hydrate formation process for pre-combustion capture of carbon dioxide. *Energy* **2010**, *36*, 2729–2733. [CrossRef]

76. Kumar, R.; Wu, H.J.; Englezos, P. Incipient hydrate phase equilibrium for gas mixtures containing hydrogen, carbon dioxide and propane. *Fluid Ph. Equilib.* **2006**, *244*, 167–171. [CrossRef]

77. Kim, S.M.; Lee, J.D.; Lee, H.J.; Lee, E.K.; Kim, Y. Gas hydrate formation method to capture the carbon dioxide for pre-combustion process in IGCC plant. *Int. J. Hydrog. Energy* **2011**, *36*, 1115–1121. [CrossRef]

78. Park, S.; Lee, S.; Lee, Y.; Seo, Y. CO_2 capture from simulated fuel gas mixtures using semiclathrate hydrates formed by quaternary ammonium salts. *Environ. Sci. Technol.* **2013**, *47*, 7571–7577. [CrossRef]

79. Zheng, J.; Lee, Y.K.; Babu, P.; Zhang, P.; Linga, P. Impact of fixed bed reactor orientation, liquid saturation, bed volume and temperature on the clathrate hydrate process for pre-combustion carbon capture. *J. Nat. Gas Sci. Eng.* **2016**, *35*, 1499–1510. [CrossRef]

80. Blamey, J.; Anthony, E.J.; Wang, J.; Fennell, P.S. The calcium looping cycle for large-scale CO_2 capture. *Prog. Energy Combust. Sci.* **2010**, *36*, 260–279. [CrossRef]

81. Silaban, A.; Harrison, D.P. High temperature capture of carbon dioxide: Characteristics of the reversible reaction between CaO(s) and CO_2 (G). *Chem. Eng. Commun.* **1995**, *137*, 177–190. [CrossRef]

82. Bui, M.; Adjiman, C.S.; Bardow, A.; Anthony, E.J.; Boston, A.; Brown, S.; Fennell, P.S.; Fuss, S.; Galindo, A.; Hackett, L.A.; et al. Carbon capture and storage (CCS): The way forward. *Energy Environ. Sci.* **2018**, *11*, 1062–1176. [CrossRef]

83. Hanak, D.P.; Manovic, V. Calcium looping with supercritical CO_2 cycle for decarbonisation of coal-fired power plant. *Energy* **2016**, *102*, 343–353. [CrossRef]

84. Barker, R. The reversibility of the reaction $CaCO_3 \leftrightarrows CaO + CO_2$. *J. Chem. Technol. Biotechnol.* **1973**, *23*, 733–742. [CrossRef]

85. Buhre, B.J.P.; Elliott, L.K.; Sheng, C.D.; Gupta, R.P.; Wall, T.F. Oxy-Fuel combustion technology for coal-fired power generation. *Prog. Energy Combust. Sci.* **2005**, *31*, 283–307. [CrossRef]

86. Yin, C.; Yan, J. Oxy-Fuel combustion of pulverized fuels: Combustion fundamentals and modeling. *Appl. Energy* **2016**, *162*, 742–762. [CrossRef]

87. Seepana, S.; Jayanti, S. Steam-Moderated oxy-fuel combustion. *Energy Convers. Manag.* **2010**, *51*, 1981–1988. [CrossRef]

88. Wang, L.; Liu, Z.; Chen, S.; Zheng, C.; Li, J. Physical and chemical effects of CO_2 and H_2O additives on counterflow diffusion flame burning methane. *Energy Fuels* **2013**, *27*, 7602–7611. [CrossRef]

89. Edge, P.; Gharebaghi, M.; Irons, R.; Porter, R.; Porter, R.T.J.; Pourkashanian, M.; Smith, D.; Stephenson, P.; Williams, A. Combustion modelling opportunities and challenges for oxy-coal carbon capture technology. *Chem. Eng. Res. Des.* **2011**, *89*, 1470–1493. [CrossRef]

90. Wu, F.; Argyle, M.D.; Dellenback, P.A.; Fan, M. Progress in O_2 separation for oxy-fuel combustion—A promising way for cost-effective CO_2 capture: A review. *Prog. Energy Combust. Sci.* **2018**, *67*, 188–205. [CrossRef]

91. Huang, X.; Guo, J. Opportunities and challenges of oxy-fuel combustion. In *Oxy-Fuel Combustion*; Elsevier: Amsterdam, The Netherlands, 2018; pp. 1–12.

92. Abraham, B.M.; Asbury, J.G.; Lynch, E.P.; Teotia, A.P.S. Coal-Oxygen process provides CO_2 for enhanced recovery. *Oil Gas J.* **1982**, *80*(11), 68–75.

93. Adanez, J.; Abad, A.; Garcia-Labiano, F.; Gayan, P.; De Diego, L.F. Progress in chemical-looping combustion and reforming technologies. *Prog. Energy Combust. Sci.* **2012**, *38*, 215–282. [CrossRef]

94. Stanger, R.; Wall, T.; Spörl, R.; Paneru, M.; Grathwohl, S.; Weidmann, M.; Scheffknecht, G.; McDonald, D.; Myöhänen, K.; Ritvanen, J.; et al. Oxyfuel combustion for CO_2 capture in power plants. *Int. J. Greenh. Gas Control* **2015**, *40*, 55–125. [CrossRef]

95. Stanger, R.; Wall, T. Sulphur impacts during pulverised coal combustion in oxy-fuel technology for carbon capture and storage. *Prog. Energy Combust. Sci.* **2011**, *37*, 69–88. [CrossRef]

96. Chen, L.; Yong, S.Z.; Ghoniem, A.F. Oxy-Fuel combustion of pulverized coal: Characterization, fundamentals, stabilization and CFD modeling. *Prog. Energy Combust. Sci.* **2012**, *38*, 156–214. [CrossRef]

97. Chen, S. Fundamentals of oxy-fuel combustion. In *Oxy-Fuel Combustion*; Elsevier: Amsterdam, The Netherlands, 2018; pp. 13–30.

98. Oh, J.; Noh, D. Laminar burning velocity of oxy-methane flames in atmospheric condition. *Energy* **2012**, *45*, 669–675. [CrossRef]

99. Mazas, A.N.; Lacoste, D.A.; Schuller, T. Experimental and Numerical Investigation on the Laminar Flame Speed of CH_4/O_2 Mixtures Diluted with CO_2 and H_2O. In Proceedings of the ASME Turbo Expo 2010: Power for Land, Sea, Air, Glasgow, UK, 14–18 June 2010; pp. 411–421.

100. Xie, Y.; Wang, J.; Zhang, M.; Gong, J.; Jin, W.; Huang, Z. Experimental and numerical study on laminar flame characteristics of methane oxy-fuel mixtures highly diluted with CO_2. *Energy Fuels* **2013**, *27*, 6231–6237. [CrossRef]

101. Seepana, S.; Jayanti, S. Flame structure and NO generation in oxy-fuel combustion at high pressures. *Energy Convers. Manag.* **2009**, *50*, 1116–1123. [CrossRef]

102. Koroglu, B.; Pryor, O.M.; Lopez, J.; Nash, L.; Vasu, S.S. Shock tube ignition delay times and methane time-histories measurements during excess CO_2 diluted Oxy-Methane Combustion. *Combust. Flame* **2016**, *164*, 152–163. [CrossRef]

103. Pryor, O.; Barak, S.; Lopez, J.; Ninnemann, E.; Koroglu, B.; Nash, L.; Vasu, S.; Ninnemann, E.; Koroglu, B.; Nash, L.; et al. High pressure shock tube ignition delay time measurements during oxy-methane combustion with high levels of CO_2 dilution. *J. Energy Resour. Technol.* **2017**, *139*. [CrossRef]

104. Hu, X.; Yu, Q.; Sun, N.; Qin, Q. Experimental study of flammability limits of oxy-methane mixture and calculation based on thermal theory. *Int. J. Hydrog. Energy* **2014**, *39*, 9527–9533. [CrossRef]

105. Said, A.; Eloneva, S.; Fogelholm, C.J.; Fagerlund, J.; Nduagu, E.; Zevenhoven, R. Integrated carbon capture and storage for an oxyfuel combustion process by using carbonation of $Mg(OH)_2$ produced from serpentinite rock. *Energy Procedia* **2011**, *4*, 2839–2846. [CrossRef]

106. Riaza, J.; Gil, M.V.; Álvarez, L.; Pevida, C.; Pis, J.J.; Rubiera, F. Oxy-Fuel combustion of coal and biomass blends. *Energy* **2012**, *41*, 429–435. [CrossRef]

107. Allam, R.J.; Palmer, M.R.; Brown, G.W.; Fetvedt, J.; Freed, D.; Nomoto, H.; Itoh, M.; Okita, N.; Jones, C. High efficiency and low cost of electricity generation from fossil fuels while eliminating atmospheric emissions, including carbon dioxide. *Energy Procedia* **2013**, *37*, 1135–1149. [CrossRef]

108. Leckner, B.; Gómez-Barea, A. Oxy-Fuel combustion in circulating fluidized bed boilers. *Appl. Energy* **2014**, *125*, 308–318. [CrossRef]

109. Vellini, M.; Gambini, M. CO_2 capture in advanced power plants fed by coal and equipped with OTM. *Int. J. Greenh. Gas Control* **2015**, *36*, 144–152. [CrossRef]

110. Scaccabarozzi, R.; Gatti, M.; Martelli, E. Thermodynamic optimization and part-load analysis of the NET power cycle. *Energy Procedia* **2017**, *114*, 551–560. [CrossRef]

111. Falkenstein-Smith, R.; Zeng, P.; Ahn, J. Investigation of oxygen transport membrane reactors for oxy-fuel combustion and carbon capture purposes. *Proc. Combust. Inst.* **2017**, *36*, 3969–3976. [CrossRef]

112. Climent Barba, F.; Martínez-Denegri Sánchez, G.; Soler, S.B.; Gohari Darabkhani, H.; Anthony, E.J. A technical evaluation, performance analysis and risk assessment of multiple novel oxy-turbine power cycles with complete CO_2 capture. *J. Clean. Prod.* **2016**, *133*, 971–985. [CrossRef]

113. Seon, B.; Jae, M.; Ho, J.; Seop, T. Influence of a recuperator on the performance of the semi-closed oxy-fuel combustion combined cycle. *Appl. Therm. Eng.* **2017**, *124*, 1301–1311.

114. Laumb, J.D.; Holmes, M.J.; Stanislowski, J.J.; Lu, X.; Forrest, B.; McGroddy, M. Supercritical CO_2 cycles for power production. *Energy Procedia* **2017**, *114*, 573–580. [CrossRef]

115. Allam, R.J.; Martin, S.; Forrest, B.; Fetvedt, J.; Lu, X.; Freed, D.; Brown, G.W.; Sasaki, T.; Itoh, M.; Manning, J. Demonstration of the allam cycle: An update on the development status of a high efficiency supercritical carbon dioxide power process employing full carbon capture. *Energy Procedia* **2017**, *114*, 5948–5966. [CrossRef]

116. Mitchell, C.; Avagyan, V.; Chalmers, H.; Lucquiaud, M. An initial assessment of the value of Allam Cycle power plants with liquid oxygen storage in future GB electricity system. *Int. J. Greenh. Gas Control* **2019**, *87*, 1–18. [CrossRef]

117. Ferrari, N.; Mancuso, L.; Davison, J.; Chiesa, P.; Martelli, E.; Romano, M.C. Oxy-Turbine for Power Plant with CO_2 Capture. *Energy Procedia* **2017**, *114*, 471–480. [CrossRef]

118. Zhang, Z.; Borhani, T.N.G.; El-Naas, M.H. Carbon capture. In *Exergetic, Energetic and Environmental Dimensions*; Elsevier: Amsterdam, The Netherlands, 2017; pp. 997–1016.

119. Yang, H.; Xu, Z.; Fan, M.; Gupta, R.; Slimane, R.B.; Bland, A.E.; Wright, I. Progress in carbon dioxide separation and capture: A review. *J. Environ. Sci.* **2008**, *20*, 14–27. [CrossRef]

120. Hossain, M.M.; De Lasa, H.I. Chemical-Looping combustion (CLC) for inherent CO_2 separations—A review. *Chem. Eng. Sci.* **2008**, *63*, 4433–4451. [CrossRef]

121. Jin, H.; Ishida, M. A new type of coal gas fueled chemical-looping combustion. *Fuel* **2004**, *83*, 2411–2417. [CrossRef]

122. Erlach, B.; Schmidt, M.; Tsatsaronis, G. Comparison of carbon capture IGCC with pre-combustion decarbonisation and with chemical-looping combustion. *Energy* **2011**, *36*, 3804–3815. [CrossRef]

123. Shimomura, Y. The CO_2 wheel: A revolutionary approach to carbon dioxide capture. *Modern Power Syst.* **2003**, *23*, 15–17.

124. Quinn, R.; Kitzhoffer, R.J.; Hufton, J.R.; Golden, T.C. A high temperature lithium orthosilicate-based solid absorbent for post combustion CO_2 capture. *Ind. Eng. Chem. Res.* **2012**, *51*, 9320–9327. [CrossRef]

125. Dahl, I.M.; Bakken, E.; Larring, Y.; Spjelkavik, A.I.; Håkonsen, S.F.; Blom, R. On the development of novel reactor concepts for chemical looping combustion. *Energy Procedia* **2009**, *1*, 1513–1519. [CrossRef]

126. Johansson, M.; Mattisson, T.; Lyngfelt, A. Investigation of Mn_3O_4 with stabilized ZrO_2 for chemical-looping combustion. *Chem. Eng. Res. Des.* **2006**. [CrossRef]

127. Li, J.; Zhang, H.; Gao, Z.; Fu, J.; Ao, W.; Dai, J. CO_2 capture with chemical looping combustion of gaseous fuels: An overview. *Energy Fuels* **2017**, *31*, 3475–3524. [CrossRef]

128. Tang, M.; Xu, L.; Fan, M. Progress in oxygen carrier development of methane-based chemical-looping reforming: A review. *Appl. Energy* **2015**, *151*, 143–156. [CrossRef]

129. Lyngfelt, A.; Kronberger, B.; Adanez, J.; Morin, J.X.; Hurst, P. 2005, The Grace Project: Development of Oxygen Carrier Particles for Chemical-Looping Combustion. Design and Operation of a 10 KW Chemical-Looping Combustor. In Proceedings of the 7th International Conference on Greenhouse Gas Control Technologies, Vancouver, BC, Canada, 5 September 2004; pp. 115–123.

130. Johansson, E.; Mattisson, T.; Lyngfelt, A.; Thunman, H. Combustion of syngas and natural gas in a 300 W chemical-looping combustor. *Chem. Eng. Res. Des.* **2006**, *36*, 3804–3815. [CrossRef]

131. Naqvi, R.; Bolland, O. Multi-Stage chemical looping combustion (CLC) for combined cycles with CO_2 Capture. *Int. J. Greenh. Gas Control* **2007**, *1*, 19–30. [CrossRef]

132. Zhu, L.; Jiang, P.; Fan, J. Comparison of carbon capture IGCC with chemical-looping combustion and with calcium-looping process driven by coal for power generation. *Chem. Eng. Res. Des.* **2015**, *104*, 110–124. [CrossRef]

133. Pfaff, I.; Kather, A. Comparative thermodynamic analysis and integration issues of CCS steam power plants based on oxy-combustion with cryogenic or membrane based air separation. *Energy Procedia* **2009**, *1*, 495–502. [CrossRef]

134. Naqvi, R.; Wolf, J.; Bolland, O. Part-Load analysis of a chemical looping combustion (CLC) combined cycle with CO_2 capture. *Energy* **2007**, *32*, 360–370. [CrossRef]

135. Fan, J.; Zhu, L.; Hong, H.; Jiang, Q.; Jin, H. A thermodynamic and environmental performance of in situ gasification of chemical looping combustion for power generation using ilmenite with different coals and comparison with other coal-driven power technologies for CO_2 capture. *Energy* **2017**, *119*, 1171–1180. [CrossRef]

136. Wilberforce, T.; Baroutaji, A.; Soudan, B.; Al-Alami, A.H.; Olabi, A.G. Outlook of carbon capture technology and challenges. *Sci. Total Environ.* **2019**, *657*, 56–72. [CrossRef]

137. Viebahn, P.; Nitsch, J.; Fischedick, M.; Esken, A.; Schüwer, D.; Supersberger, N.; Zuberbühler, U.; Edenhofer, O. Comparison of carbon capture and storage with renewable energy technologies regarding structural, economic, ecological aspects in Germany. *Int. J. Greenh. Gas Control* **2007**, *1*, 121–133. [CrossRef]

138. Veltman, K.; Singh, B.; Hertwich, E.G. Human and environmental impact assessment of postcombustion CO_2 capture focusing on emissions from amine-based scrubbing solvents to air. *Environ. Sci. Technol.* **2010**, *44*, 1496–1502. [CrossRef]

139. Cuéllar-Franca, R.M.; Azapagic, A. Carbon capture, storage and utilisation technologies: A critical analysis and comparison of their life cycle environmental impacts. *J. CO2 Util.* **2015**, *9*, 82–102. [CrossRef]

140. Pehnt, M.; Henkel, J. Life cycle assessment of carbon dioxide capture and storage from lignite power plants. *Int. J. Greenh. Gas Control* **2009**, *3*, 49–66. [CrossRef]

141. Nie, Z.; Korre, A.; Durucan, S. life cycle modelling and comparative assessment of the environmental impacts of oxy-fuel and post-combustion CO_2 capture, transport and injection processes. *Energy Procedia* **2011**, *4*, 2510–2517. [CrossRef]

Article

Macroeconomic Effects of EU Energy Efficiency Regulations on Household Dishwashers, Washing Machines and Washer Dryers

Paola Rocchi *, José Manuel Rueda-Cantuche, Alicia Boyano and Alejandro Villanueva

European Commission, Joint Research Centre (JRC), Spain. C/Inca Garcilaso, 3, 41092 Sevilla, Spain;
Josem.RCANTUCHE@ec.europa.eu (J.M.R.-C.); Alicia.BOYANO-LARRIBA@ec.europa.eu (A.B.);
Alejandro.VILLANUEVA@ec.europa.eu (A.V.)
* Correspondence: Paola.ROCCHI@ec.europa.eu

Received: 5 August 2019; Accepted: 8 November 2019; Published: 12 November 2019

Abstract: Testing the relationship between economic performance and energy consumption is of utmost importance in nearly all countries. Taking the European Union as scope, this paper analyses the impacts of energy efficiency legislation on a selection of household appliances. In particular, it analyses the employment and value added impacts of the stricter energy efficiency requirements for dishwashers, washing machines, and washer dryers. To do so, this paper combines a bottom-up stock model with a macro-econometric dynamic general equilibrium model (FIDELIO) to quantify the direct and indirect value added and employment impacts in the European Union. The analysis shows that stricter energy efficiency requirements on household dishwashers, washing machines, and washer dryers have a net negative macroeconomic impact on value added (roughly 0.01 % of the total European Union value added) and a slightly net positive impact on employment. In fact, the regulations cause a shift in the composition of the household consumption basket that seems to favor labor-intensive industries.

Keywords: energy efficiency policy; household appliances; eco-design; energy labelling; indirect impacts; general equilibrium model; FIDELIO model

---

1. Introduction

This paper (The views expressed are purely those of the authors and may not in any circumstances be regarded as stating an official position of the European Commission) focuses on the relationship between energy efficiency policies and macroeconomic performance. A better understanding of this relationship is of key importance due to the increasing volume of regulated energy related products, as well as the number of countries implementing energy efficiency standards. In particular, the aim of the analysis is to add new empirical evidence regarding the macroeconomic impact of specific energy efficiency regulations on three household appliances: dishwashers, washing machines, and washer dryers in the European Union (EU). These appliances contribute approximately 2.8% to the residential energy consumption of the EU [1].

Nowadays, energy efficiency policies are one of the key measures to reduce the impact of human activities on the environment and the climate. Looking at the EU current growth strategy, climate change and energy sustainability priorities establish, for instance, the need of a 32.5% energy efficiency improvement for 2030. The improvement of energy efficiency has key impacts on the efforts that the EU undertakes to reduce energy consumption and greenhouse gas (GHG) emissions [2].

Theoretically speaking, energy efficiency policies are expected to have impacts not only on the environmental footprint of human activities, but also on the economic system through different channels [3]. In a first instance, these policies create incentives for certain economic sectors to introduce new production technologies, with possible consequent changes in the costs and product prices. On the other hand, energy efficiency technologies imply a reduction in energy consumption and, therefore, a consequent reduction in energy expenditure. These two direct effects will then have repercussions in the entire economy, through changes in the distribution of productive resources or changes in the composition of end-user spending.

The growing attention and use of energy efficiency policies has given rise to a growing body of literature that seeks to provide scientific evidence on the economic impact of these policies. These studies propose different methodologies, and can be classified into two main approaches: analyses that use partial models, and analyses that instead use general equilibrium models.

When considering the first approach, in recent years a growing number of analyses [4–7] focus on the effects of the implementation of energy efficiency measures on household appliances by using engineering approaches and bottom-up stock models. For example, there are some scientific studies analyzing and projecting the electricity use, water use, pollutant emissions, and consumer welfare implications for household appliances in Europe [8,9]. McNeil et al. (2013) [10] use a bottom-up stock accounting model that predicts the energy consumption of different equipment in 11 EU countries according to engineering-based estimates of the annual unit energy consumption. The authors model a high-efficiency policy scenario and conclude that significant energy savings can be achieved by adopting the current best practices of appliance energy efficiency policies. Braungardt et al. (2016) [11] investigate the impact of eco-innovations on the EU residential electricity demand (excluding heating) while using a detailed bottom-up modelling approach and find out that energy savings are achieved through the development of technologies with efficiencies beyond the status quo. Studies on the market penetration of high energy efficiency appliances in the residential sector are carried out by Radpout et al. (2017) [12]. These authors develop a model based on econometrics and time series analysis combined with the cost models. They find that government incentives to encourage people to buy higher energy efficient appliances are more effective than electricity price policies. Yilmaz et al. (2019) [6] publish a study that describes the development and application of a stock model that allows for quantifying the changes in the number of household appliances in stock, the related evolution of energy efficiency, as well as the changes/projections of electricity consumption between 2000 and 2035 in Switzerland. From the methodological point of view, this bottom-up stock model addresses the limitations of other previous models [13,14] that offer insight into demand dynamics but provide scarce information regarding the evolution of key parameters, such as new technologies, systems, or practices or the forecasting of the sales and stock. Specifically, for washing machines, washer-dryers, and dishwashers, Boyano et al. (2017) [15,16] analyze the environmental, employment, and economic impacts in the EU of stricter energy efficiency requirements for these appliances. The authors built-up a bottom-up stock model that projects stock sales, lifespans, and related evolution of energy efficiency classes. They estimate the electricity consumption, water consumption, and consumer welfare between 2015 and 2030. The authors find a decrease in the employment, but they were unable to quantify indirect, supply-chain, or cross-sectorial impacts.

While using a general equilibrium framework, other studies focus instead on the economic and environmental impact of energy efficiency policy while considering all interlinkages and dependences of the economic system. Hanson and Laitner (2004) [17] use the All Modular Industry Growth Assessment (AMIGA) system to analyse the policies that increase investment in energy-efficient technologies in the United States (U.S.) economy. They find that the policies would lead to substantial domestic reduction of carbon emissions and a net positive impact on the economy. Rose and Wei (2012) [18] analyse the impacts of the Florida Energy and Climate Change Action Plan that was introduced in 2001. Using the econometric general equilibrium REMI model, they find that most of the recommended options individually—as well as the combined recommendations—have positive impacts on the state's economy. Barker et al. (2016) [19] analyse four different policies that the International Energy Agency suggests in order to close the 2020 emissions gap. They estimate the GDP and employment effect while using the econometric general equilibrium model Global Energy-Environment-Economy Model

(E3MG), finding that the policies are not enough to reach the required emissions reduction, although presenting positive impacts on GDP and employment.

There is finally a third group of studies that combine the two approaches using hybrid models. These analyses integrate detailed bottom-up technical descriptions of specific industries affected by the policies with a broader economic perspective provided by the macroeconomic framework. Barker et al. (2007) [20] analyze the UK Climate Change Agreements and related energy efficiency policies for energy-intensive industrial sectors. They combine bottom-up estimates of the effects of these policies and the dynamic econometric model of the UK economy Multisectoral Dynamic Model-Energy-Environment-Economy (MDM-E3). They find final energy reduction and a slight increase in economic growth through improved international competitiveness. Ringel et al. (2016) [21] use a hybrid approach—bottom-up model together with the Assessment of Transport Strategies (ASTRA) model—to analyze the environmental and socio-economic impacts of Germany's latest energy efficiency and climate strategies for the year 2020. They find that enhanced green energy policies bring about economic benefits in terms of GDP and employment, even in the short term. Additionally, the European Commission uses a hybrid approach in the impact assessment of the proposal for the revision of the energy efficiency directive [22]. The analysis uses bottom-up models— Price-Induced Market Equilibrium System (PRIMES), Greenhouse gas - Air pollution Interactions and Synergies (GAINS), Global Biosphere Management Model - Global Forest Model (GLOBIOM-G4M), Prometheus, Common Agricultural Policy Regional Impact (CAPRI)—together with two different general equilibrium macroeconomic models: the computational general equilibrium model GEM-E3 and the dynamic econometric global model E3ME. While the E3ME model presents positive impacts GDP and employment in all analyzed scenarios, the results for the GEM-E3 models are mixed. Finally, Hartwig et al. (2017) [3] present a case study for Germany, where a scenario including ambitious energy efficiency measures for building, household appliances, industry, and the service sector is compared to a reference scenario with respect to the macroeconomic impacts. Connecting the energy demand models Forecast and invert/EE-lab with the macroeconomic model ASTRA-D is undertaken to analyze the effects of the policy scenarios. The authors conclude that the macroeconomic effects of ambitious energy efficiency policies in Germany have considerable positive impacts on employment (particularly in those that produce energy efficiency technologies and construction and manufacturing sector, as well as in real-state and consulting) and GDP.

In this paper, we follow the third approach to use the technical information that is offered by bottom-up models, while going beyond the direct impacts on employment and value added of the energy efficiency regulations to analyze their impacts across industries and countries. In particular, we use a hybrid framework combining the detailed bottom-up energy demand stock based model that Boyano et al. developed [15,16], together with the macro-econometric dynamic general equilibrium model FIDELIO [23]. The resulting hybrid model is used to provide new empirical evidence on a specific energy efficiency policy, which is the revision of the energy efficiency regulatory framework for dishwashers, washing machines, and washer dryers. In particular, the overarching regulatory framework for energy efficiency products is the combination of two policies: The Energy Labelling Regulation (EU) 2017/1369 [24], which defines the process of determination of energy label to be displayed in new appliances, and the Eco-design Directive 2009/125/EC [25], which specifies the process of defining minimum energy performance levels. Acting in combination as a pull-push effect, these regulatory measures have improved the average energy efficiency of the household appliance stock over the years in the EU [26]. Regarding the analyzed appliances, their specific regulations are: (1) for washing machines, Regulation EU No 1061/2010 on energy label and Regulation EU No 1015/2010 on Eco-design requirements, (2) for washer dryers Regulation EU No 96/60/EC on energy label, and (3) for dishwashers, Regulation EU No 1060/2010 on energy label and Regulation EU No 1016/2010 on Eco-design requirements. These regulations have been revised in 2014–2018, and adopted in 2019. The revision introduces stricter energy efficiency requirements of the products that enter the EU market from March 2021, and a rescaling of the energy efficiency classes. Additionally, some changes in the testing programs are introduced to bring the energy efficiency developments of the products closer to the end-user behavior. This paper quantifies the EU-wide economic (in terms of value added) and employment impacts between 2020 and 2030 of such proposed changes to the EU energy efficiency regulations on the aforementioned household's appliances. The analysis provides results by countries and for most of the economic sectors in the EU.

The paper is organized as follows. Section 2 provides an overview of the methods and materials that were used in the study. Section 3 presents the impacts from changes in the EU regulations estimated using FIDELIO. Section 4 discusses a range of implications of the empirical results, while Section 5 draws some concluding remarks.

2. Methods and Materials

The modelling tool that is proposed in this analysis has two main components: a bottom-up approach, used together with the top-down macro-economic Fully Interregional Dynamic Econometric Long-term Input-Output model (FIDELIO). In this section, we first describe the two models (in Sections 2.1 and 2.2, respectively). Next, Section 2.3 describes the necessary further steps to link the two approaches, and it offers a graphic representation of the methodological proposal.

2.1. Bottom-Up Approach

We use a bottom-up energy demand model covering all EU countries. In this model, the EU electricity consumption and sale prices of dishwashers, washing machines, and washer dryers are based on the related energy efficiency technological improvements, which are estimated while using an engineering approach.

The technological improvements that were triggered by the implementation of the new regulations have an effect on the sales price of the appliances as well as in the electricity consumption of the overall stock at EU level. The sales price of each machine is estimated based on the manufacturing costs, manufacturers and retailers' mark-ups, the value added tax, and, wherever appropriate, the additional costs of the improvement options that are added to the basic models to achieve a better energy efficiency and, therefore, a better energy efficiency label classification. The manufacturing costs are provided by the manufacturers and assumed to decrease over time according to the experience curve [27], experience gained by the manufacturer in producing the machines. This correction is applied to the sale prices beyond 2015.

The annual electricity consumption of the overall stock at the EU level is estimated based on the average unitary electricity consumption of one appliance. Data regarding the energy consumption of the appliances are based on the performance data provided by the manufacturers, data from consumer surveys on how the appliances are used, and the evolution of technology. This unitary electricity consumption also depends on the sales distribution over the energy efficiency classes of the year when the appliance is purchased. The annual market share of each energy efficiency class is based on the historical data series and the influence that labelling has on the investment decisions of consumers, directing preference towards more energy efficiency appliances [28].

The way that this bottom-up model calculates the energy consumption is similar to that used by Yilmaz et al. [6], as both models estimate the number of replaced machines throughout a Weibull distribution. However, Yilmaz et al. [6] disregard the effect of the users when using the appliances and assumed the testing program energy consumption as the average value. For a complete description of the calculations and the assumptions that were considered in this bottom-up stock model, see [15,16].

A variety of studies have addressed the rebound effects for appliances, including an increase in operation hours, appliance size, or ownership rate [11,29,30]. The direct rebound effect related to dishwashers, washing machines, and washer dryers has been estimated to be negligible. The most recent user surveys [10,13] report that consumers remain using these appliances in the same way, regardless of their energy efficiency class, and this behavior has been stable in the last years. For example, the annual number of cycles of use of the appliances is stable and it mainly depends on the household size. The laundry load has remained constant through the last years at approx. 3.4 kg laundry/cycle, and the number of washing machines per household has remained constant, which indicates a saturated market.

2.2. Top-Down Model: FIDELIO

FIDELIO is based on a neo-Keynesian demand-driven non-optimization macroeconomic framework in the line of the E3ME (Cambridge Econometrics) model. This family of models is frequently compared to another set of macroeconomic models that are often used for policy and environmental analysis that is computational general equilibrium (CGE) models. One of the main broad differences between the two types of models is that CGE models are based on neoclassical assumptions in line with economic theory of optimization. Prices adjust to market clearing, aggregate demand adjusts to meet potential supply, and output is determined by available capacity. Instead, macro-econometric models assume that agents lack perfect knowledge and do not optimize their decisions. They provide a more empirically grounded approach and the alternative assumption ruling agents' choices is represented by econometric estimations. The parameters are estimated from time-series databases; therefore, they are validated against historical relationship: agents behave as they did in the past. Differently from CGE models, market imperfections exist and the economy is not assumed to be in equilibrium. There is no guarantee that all available resources are used. The level of output is a function of the level of demand and it might be less than potential supply.

Besides offering a relatively strong empirical grounding, the use of FIDELIO offers two additional advantages for the analysis carried out. First, the model offers a fairly high level of geographical and sectorial disaggregation. FIDELIO covers 35 regions (the 28 EU Member States plus Brazil, China, India, Japan, Russia, Turkey, and the United States), with each of them being disaggregated in 56 industries and products (see in Appendix A, Table A1, the list of industries available in FIDELIO).

Second, the model offers a useful instrument to analyze policies that influence household consumption. In fact, while the supply side is described in a relatively simple way—it is characterized by an input-output core enlarged with nested constant-elasticity-of-substitution production function—the household block is modelled with relatively high detail. In FIDELIO, households receive three sources of income: wages, a share of the firms' gross operating surplus, and some government transfers. This income, after taxes, is either used for consumption or saved. In particular, households consume different categories of products: durable products (housing rents and vehicles) and non-durable products, such as appliances, electricity, heating, fuel for private transport, public transport, food, clothing, furniture and equipment, health, communication, recreation and accommodation, financial services, and other products. For almost all consumption categories—including also appliances and electricity consumption—the demand is characterized through econometric estimations with different consumption categories modelled with different functional forms. For a complete description of the characteristics, the assumptions and equations of the FIDELIO model, see [23]. Appendix B offers a short description of the data sources that are needed to build the FIDELIO database.

2.3. Bridging Bottom-Up and Top-Down Approaches

Introducing the shock values into the FIDELIO model requires additional information. In fact, both shocks—to the sale prices of appliances and to the electricity requirements—are separately estimated for each specific appliance: dishwashers, washing machines, and washer dryers. However, the FIDELIO model only operates with one single household consumption category, which includes these and all other appliances together. Therefore, additional information is required to weigh the estimated shocks and calculate the corresponding equivalent shocks that are to be introduced in the FIDELIO model.

As regards the shock of the sale prices of appliances, we use the penetration rates that were estimated in [31] for dishwashers, washing machines, and washer dryers to weigh each (exogenously estimated) sales price shock and compute the single weighted equivalent sales price shock that includes all three household appliances. Next, we use information from the 2010 Household Budget Survey

(HBS) micro-data produced by Eurostat in the COICOP (Classification of Individual Consumption According to Purpose) classification at the five-digit level. In particular, the survey provides information on the household total consumption of appliances and the household consumption of "clothes washing machines, clothes drying machines, and dish washing machines". Using this information, for each EU country, we compute the share of "clothes washing machines, clothes drying machines, and dish washing machines" over the broader category "household appliances". Eventually, we compute the sales price variations that are to be used in the FIDELIO model by using the weighted equivalent price shock for washing machines, washer dryers, and dishwashers, and the weights based on the HBS data. These price variations are introduced as shock parameters into the endogenously determined prices of appliances of FIDELIO. We implicitly assume that the shock in the sales price affects both domestic and imported products since the price of appliances in FIDELIO is computed as an average of the price of the domestic products and the price of the imported products. This is how the policy is actually expected to operate.

We use a similar approach to combine the electricity consumption shock related to each appliance into an aggregate shock in the value of the household total electricity consumption. First, we use the weights based on the penetration rates previously described to compute a weighted variation in the value of electricity consumed for dishwasher, washing machine and washer-dryer appliances. Next, to know the share of electricity consumption that households use for dishwashing, washing machine and washer-dryer appliances over the household total consumption of electricity we use data from the European Environment Agency [32] and from the ODYSSEE database [33]. These databases distinguish among different uses of household electricity (for electrical large appliances, other appliances, lighting, space heating, water heating, cooking, and air cooling). These shares are used to compute the final variations in household total electricity consumption used as a shock in FIDELIO. In FIDELIO, household electricity consumption depends on the stock of appliances, the electricity price, an exogenous index capturing the efficiency of appliances, the previous year's electricity consumption and stock of appliances, and the demand for energy that is needed for heating. We impose a shock in the efficiency parameter that would be equivalent, *ceteris paribus*, to the exogenously computed shock in the electricity consumed to simulate the electricity consumption variation computed through the bottom-up model. Appendix C—Tables A2–A6—presents the cost variations and the household total electricity consumption variations that were introduced in FIDELIO, and the weights used to compute them.

Given how consumers' choices are described in FIDELIO, the model takes indirect rebound effects into account. By reducing their energy consumption, households might use their additional savings to buy other goods and services that require additional use of electricity, partially offsetting the initial electricity reduction.

Figure 1 provides a graphical description of the two models used for the analysis and of the input (in yellow) and outputs (in green) flows between the two models. As the figure shows, the revised ecodesign requirements and the new energy efficiency classes are inputs for the bottom-up model that computes the shock in the appliances sale prices and household electricity consumption. These outputs of the bottom-up model are inputs for FIDELIO that simulates the policy revision impacts on employment and value added.

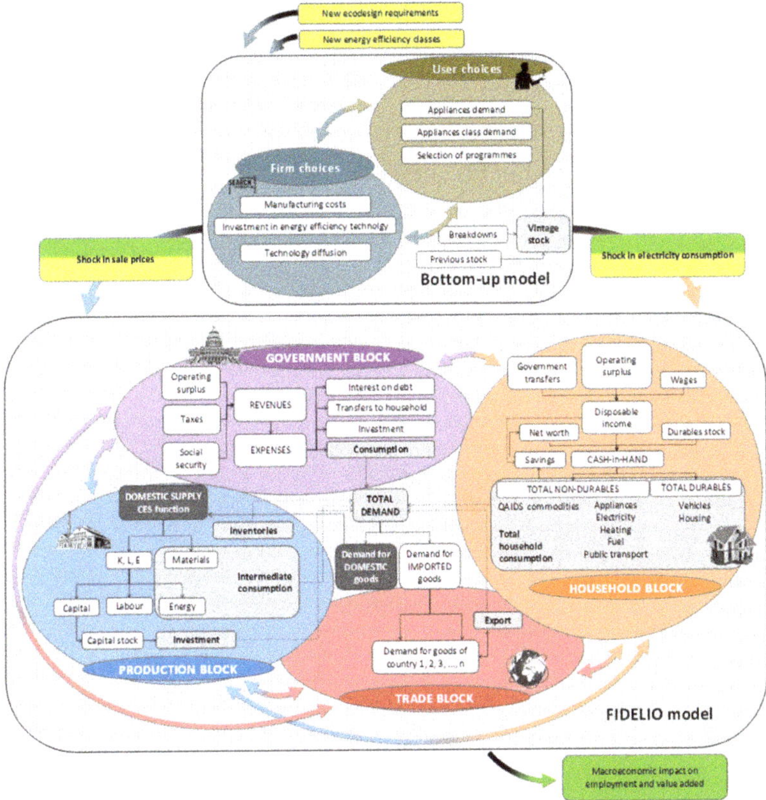

Figure 1. Graphical representation of the hybrid model used in the analysis. For more details, see [15,16,23].

3. Results

We shock the model in 2020 and run it up to 2030 for a baseline scenario (no new regulatory measures) and for a scenario with the proposed stricter EU energy efficiency regulations. The results that are presented hereafter correspond to variations of value added and employment with respect to the baseline scenario, in 2030. The results are very similar for the other years. While Section 3.1 focuses on the macroeconomic impact of the policies, Section 3.2 provides an overview of the environmental impact in terms of CO_2-equivalent emissions.

3.1. Economic Impact

Table 1 shows the variations in value added and in employment in the EU economy as a whole.

Table 1. Absolute and relative variation of value added and employment.

Value Added		Employment	
Absolute Variation (Million Euros)	**Relative Variation (%)**	**Absolute Variation (Thousand Jobs)**	**Relative Variation (%)**
−1901.5	−0.01	23.9	0.01

While the value added decreases by 1.9 billion euros, employment increases in around 24,000 jobs. In relative terms, with 0.01%, none of these two results represents a significant share of the total value added and the total employment in the EU.

Sections 3.1.1 and 3.1.2 provide more detailed results at the industry and country level, respectively, in order to give more insight into the EU (negative) value added effects and the EU (positive) employment effects.

3.1.1. Industry Level Analysis

Input-output based models typically provide results that are broken down by economic activities or industries (also sometimes denoted as sectors). Even if the analyzed policies aim at influencing the energy efficiency of some specific household appliances, produced by some specific industries, FIDELIO simulates the indirect impacts that these policies would also have on other industries. These impacts can be caused, for example, by an indirect effect of regulations—such as the reduction of electricity consumption, or by changes in the quantity of intermediate inputs necessary to produce the appliances, or by changes in the households' bundle of goods and services consumed.

Table 2 shows the absolute and relative variations in the EU value added, broken down by industry. The left hand side of Table 2 shows industries with value added increases, while the right hand side shows industries that are worse off. In both sides, the industries are ranked based on their share over the total variation in value added, in decreasing order.

Table 2. Value added variation in the European Union (EU) by industry.

Increase in Value Added				Decrease in Value Added			
Industry	Variation (Million Euros)	Relative Variation (%)	Share over VA Increase (%)	Industry	Variation (Million Euros)	Relative Variation (%)	Share over VA Decrease (%)
Accommodation	360.3	0.09	17	Electricity, gas	−1945.6	−0.69	49
Retail trade	342.6	0.06	16	Mining and quarrying	−484.5	−0.34	12
Other services	290.7	0.07	14	Construction	−271.7	−0.04	7
Food, beverages	253.2	0.09	12	Public administration	−146.3	−0.01	4
Crop and animal	175.1	0.09	8	Sewerage; waste collection	−103.7	−0.1	3
Telecommunications	97.8	0.04	5	Forestry and logging	−96.7	−0.42	2
Electrical equipment	89.7	0.08	4	Machinery and equipment	−79.2	−0.04	2
Financial service activities	87.8	0.02	4	Chemicals	−72	−0.05	2
Human health activities	78.4	0.01	4	Coke and refined petroleum	−71.4	−0.24	2
Activities of households	77	0.12	4	Architectural and engineering	−66.4	−0.03	2
Insurance and pension	69.1	0.06	3	Land transport	−64.4	−0.02	2
Auxiliary to financial	33.7	0.03	2	Fabricated metal products	−62.1	−0.03	2
Motion picture production	30.6	0.04	1	Rental and leasing	−59.7	−0.01	1
Textiles	21.2	0.03	1	Legal and accounting activities	−53.9	−0.01	1
Publishing activities	18.2	0.02	1	Repair	−53.8	−0.06	1
Furniture	13.6	0.01	1	Real estate activities	−53.6	0	1
Wholesale trade	11.9	0	1	Warehousing	−40.8	−0.02	1
Wholesale and retail (vehicles)	9	0	0	Basic metals	−31.8	−0.04	1
Fishing	7.1	0.09	0	Other non-metallic products	−31.7	−0.04	1
Printing and reproduction	5.1	0.01	0	Computer, electronic products	−30.8	−0.03	1
Other professional	4.9	0.01	0	Computer programming	−25.2	−0.01	1
Paper and paper products	3.1	0.01	0	Other transport equipment	−22.7	−0.03	1
Advertising	2.7	0	0	Wood and of products of wood	−20.6	−0.05	1
Postal activities	2.4	0	0	Motor vehicles	−18.9	−0.01	0
Pharmaceutical products	0	0	0	Scientific research	−16.5	−0.02	0
				Education	−16	0	0
				Water collection and supply	−15.1	−0.05	0
				Rubber and plastic products	−13.5	−0.01	0
				Air transport	−6.3	−0.02	0
				Water transport	−6.2	−0.01	0
				Pharmaceutical products	−5.7	0	0
TOTAL	2085.2	0.02	100	TOTAL	−3986.7	−0.03	100

Even when the total value added decreases, it does not shrink in all industries. The value added of the household appliances producer industry (electrical equipment) increases by 90 million euros.

However, the positive impact in other industries is even greater. In fact, 60% of the value added growth takes place in the accommodation and food services (360 million euros), retail trade (343 million euros), food and beverages (253 million euros), and other services (290 million euros), including activities, such as repairing services, art, entertainment, and recreation services, among others.

A possible reason why these sectors are increasing their production and, consequently, their value added, can be found in the way consumers' choices are modelled in FIDELIO. Households increase their demand of these products mainly due to savings made in electricity consumption. In fact, 50% of the value added reduction takes place in the electricity production industry (and corresponds to around two-billion euros). However, this decrease only represents 0.69% of the value added of the electricity industry.

Regarding employment, the results present a positive effect in the EU economy, in contrast to the value added decrease. The reason lies in the fact that the industries that show an increase in production, value added, and employment are more labor intensive than the industries that are worse off. Table 3 shows the absolute and relative variations in employment at the EU level, broken down by industry (with the same structure as Table 2).

Table 3. Employment variation in the EU by industry.

Increase in Employment				Decrease in Employment			
Industry	Variation (Thousand Jobs)	Relative Variation (%)	Share Over Total Increase (%)	Industry	Variation (Thousand Jobs)	Relative Variation (%)	Share Over Total Decrease (%)
Crop and animal	19	0.14	26	Electricity, gas,	−12.2	−0.75	25
Accommodation	12.1	0.09	17	Construction	−7.4	−0.04	15
Retail trade	10	0.04	14	Mining and quarrying	−5	−0.41	10
Other services	8.1	0.07	11	Forestry and logging	−3.1	−0.52	6
Food, beverages	6	0.1	8	Public administration	−2.9	−0.02	6
Activities of households	5.8	0.13	8	Land transport	−1.9	−0.03	4
Human health activities	2.3	0.01	3	Fabricated metal products	−1.6	−0.04	3
Textiles	1.7	0.05	2	Rental and leasing	−1.6	−0.01	3
Electrical equipment	1.6	0.09	2	Sewerage; waste collection	−1.5	−0.1	3
Financial service activities	0.9	0.02	1	Architectural and engineering	−1.4	−0.04	3
Insurance and pension	0.8	0.07	1	Machinery and equipment	−1.3	−0.04	3
Telecommunications	0.7	0.05	1	Repair	−1	−0.06	2
Furniture	0.6	0.02	1	Legal and accounting activities	−1	−0.01	2
Auxiliary to financial	0.6	0.03	1	Wood and of products of wood	−0.7	−0.05	1
Motion picture production	0.4	0.04	0	Other non-metallic products	−0.7	−0.04	1
Publishing activities	0.3	0.03	0	Chemicals	0.6	−0.04	1
Fishing	0.2	0.09	0	Basic metals	−0.6	−0.05	1
Printing and reproduction	0.2	0.02	0	Warehousing	−0.6	−0.02	1
Postal activities	0.2	0.01	0	Coke and refined petroleum	−0.4	−0.24	1
Paper and paper products	0.1	0.01	0	Computer programming	−0.4	−0.01	1
Wholesale and retail (vehicles)	0.1	0	0	Rubber and plastic products	−0.3	−0.01	1
Pharmaceutical products	0	0	0	Computer, electronic products	−0.3	−0.02	1
Wholesale trade	0	0	0	Motor vehicles	−0.3	−0.01	1
Advertising	0	0	0	Other transport equipment	−0.3	−0.03	1
Other professional	0	0	0	Scientific research	−0.2	−0.02	1
				Water collection and supply	−0.2	−0.05	0
				Real estate activities	−0.2	−0.01	0
				Air transport	−0.1	−0.02	0
				Pharmaceutical products	0	0	0
				Water transport	0	−0.01	0
				Education	0	0	0
TOTAL	71.7	0.03	100	TOTAL	−47.8	0.02	100

The total positive impact on employment (ca. 72,000 jobs) in some sectors more than compensates the negative impact (ca. 48,000 jobs) in others. The positive impacts mainly come from the agricultural sector, accommodation and food services, retail trade, and other services. On the negative side, the electricity industry is the one that suffers most in terms of employment (25% of the total impact), although much less than in terms of value added (50% of the total impact). Other industries that show a decrease in employment are, for example, construction, mining and quarrying, and forestry and logging.

3.1.2. Country-Level Analysis

In addition to the analysis at the industry level, in this section we analyze how the impact is distributed among the different EU countries. Figure 2 shows the absolute variation in value added and employment by country.

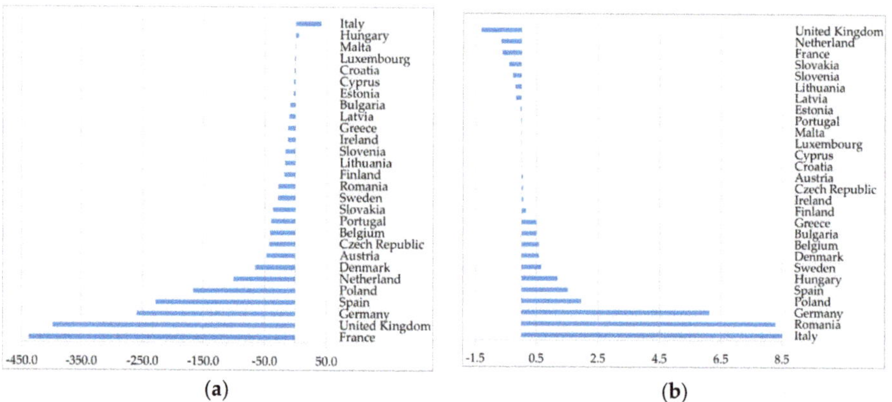

Figure 2. Impact at a country level: (**a**) Variation in value added (million euros); and, (**b**) Variation in employment (thousand jobs).

All EU countries would see their total value added reduced, except Hungary and Italy. However, these reductions would represent very small values in relative terms with respect to their total national levels, with most of them being close to zero. The maximum value is −0.05% for Lithuania, Latvia and Slovakia

In absolute terms, the countries absorbing most of the value added decrease are Germany, Spain, France, United Kingdom, and Poland. For these countries, between 31% and 54% of the value added decrease occurs in the electricity industry. However, this just corresponds to a 1% value added of the electricity industry in France, Poland, Spain, and United Kingdom. Other industries contributing to the overall value added decrease are the mining sector and the construction sector.

Although not representing relevant variations in absolute terms, there are other industries that show higher value added decreases, between 1% and 3%—the relatively most affected sectors, these are the manufacture of transport equipment in Cyprus (1%); the electricity industry in Denmark (1.1%), Lithuania (1.3%), Slovakia (1.3%), and Greece (1.6%); the forestry and logging industry in Greece (3%), Romania (1.9%), Lithuania (1.1%), Slovenia (1.1%), and Italy (1%); and, the refined oil products industry in Slovenia (1.3%).

The positive impact in Italy and Hungary is driven by industries, such as accommodation, repairing, retail trade, agriculture, and manufacture of food products.

The employment effects are positive in most of the countries. The three countries showing the biggest employment increase are Germany (ca. 6100 jobs), Italy (ca. 8600 jobs), and Romania (ca. 8300 jobs). In Germany and Italy, the industries that mostly drive the employment increase are accommodation, retail trade, agriculture, and manufacture of food products. For Romania, 70% of the employment increase is in the agricultural sector, followed by the manufacture of food products and the manufacture of textiles.

For some countries, such Estonia, France, Lithuania, Latvia, the Netherlands, Slovenia, and Slovakia, employment (slightly) decreases around less than 1000 jobs in all cases. The exception is United Kingdom, where the employment decrease was around 1300 jobs.

3.2. Environmental Impact

In FIDELIO, the monetary value of energy that is consumed by firms and households is linked to energy consumption and is then used to compute emissions of carbon dioxide (CO_2), methane (CH_4), and nitrous oxide (N_2O) related to energy production and consumption (see [23] for a description of the conversion factors from primary energy consumption to emissions). To describe the environmental impact of the proposed stricter energy efficiency requirements, we look at the variation of GHG emissions. In particular, for a synthetic measure of the GHG effect, the emissions are converted into CO_2-equivalent units while using the Global Warming Potential (GWP), as in [34]. Conversion factors are 1 for CO_2, 265 for N_2O and 28 for CH_4.

The emissions increase is, in absolute terms, smaller than the emission reduction that is driven by stricter energy efficiency requirements (as expected), and the net effect is a decrease in GHG emissions equal to 1.5 million tonnes. This reduction is driven by the industry producing electricity that is responsible for the 70% of the total reduction, followed by mining and quarrying (15%), sewerage and waste collection and treatment (10%), and the manufacture of coke and refined petroleum products (3%).

Figure 3 shows the industries that are responsible for the resulting increase in emissions and those industries that are responsible for emission reductions.

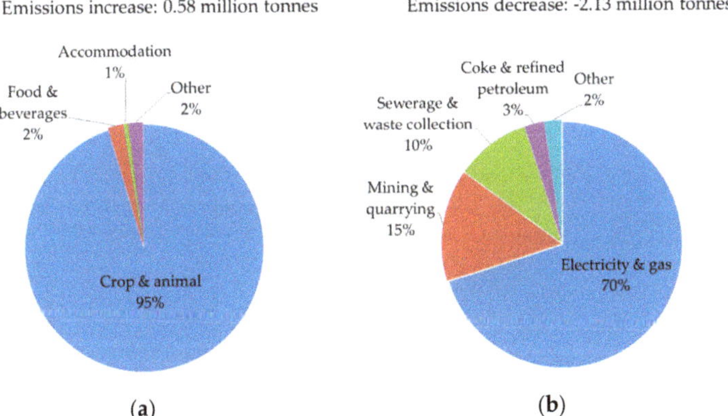

Figure 3. CO_2-equivalent emissions variation: (**a**) Distribution of emissions increase by industry; and, (**b**) Distribution of emissions decrease by industry (unit: percentage).

The three main industries showing the biggest shares of emission increases are the agriculture industry, the manufacture of food and beverages, and accommodation services. Together, they would cause GHG emissions to increase in around half a million tonnes, out of which 95% would correspond to the agriculture industry. This effect might be considered as rebound effects, a reduction in the expected environmental gains of the regulation caused by behavioral responses.

Figure 4 shows the distribution of emissions decrease among EU countries. Germany, Poland, and the United Kingdom are responsible of half of the total reduction of GHG emissions that are caused by the proposed stricter energy efficiency requirements on dishwashers, washing machines, and washer dryers.

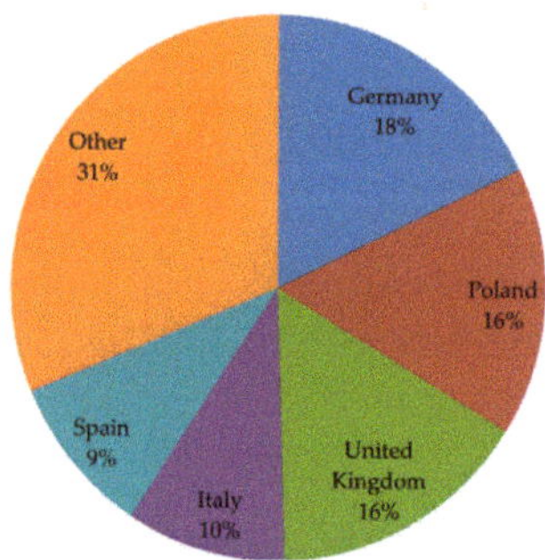

Figure 4. Distribution of CO_2-equivalent emissions decrease among EU countries (unit: percentage).

4. Discussion

One of the main findings of our analysis is that energy efficiency regulations on household dishwashers, washing machines, and washer dryers have a net negative macroeconomic impact on value added (roughly 0.01 % of the total EU value added) and a slightly net positive impact on employment (ca. 24,000 jobs). In both cases, energy efficiency regulations have positive and negative economic effects, depending on the industry sector analyzed. Most of the negative impact comes from the reduction in the consumption of energy due to the implementation of more energy-efficient technologies in household appliances. Positive impacts are derived from the new investments on more efficient technologies, and the shift in the composition of the household consumption basket, while using less energy in favor of goods and services that are produced by more labor-intensive industries.

While the most part of studies analyzed in the literature review find positive impacts of energy efficiency policies on both GDP and employment, these analyses usually consider the impact of a bundle of policies together. On the contrary, our results are consistent with the findings of Barker et al. (2016) [19]. The authors test whether energy efficiency measures contribute to closing the 2020 emissions gap without a loss in GDP and employment, and they offer the results disaggregated by different policy measures. One of the policy measures that they analyze is the use of more efficient appliances and lighting in residential and commercial buildings. For this specific measure, in line with our results, they find negative economic impacts for the EU, particularly in Germany and positive economic impacts in Italy. These authors also reported net positive employment effects. Therefore, it seems useful to specifically analyze individual energy efficiency policies, given that different measures may have different impacts on the economy.

In any case, it would be important to validate the results that were obtained using alternative theoretical frameworks to check the sensitivity of the results that were obtained to the assumptions underlying the used approach. For instance, following the findings of [11,29,30], in the bottom-up approach, we assume no relevant direct rebound effects induced by the revised regulations. A further extension of the analysis might corroborate how the results change relaxing this assumption. It is also important to note that one of the strengths of the FIDELIO model is its capacity to re-allocate household consumption across different goods and services in reaction to price changes, based on a relatively simple description of the electricity market and the electricity production function. Thus, the model is neither able to consider strategic choices that the electricity industry can make to accommodate the new energy efficiency policies nor the possible incentives towards innovative business models. Therefore, it would be interesting to deepen the analysis that was carried out with complementary approaches, for instance through microeconomic analyses of the electricity sector, or through energy models.

The revised energy efficiency regulations on household appliances that are studied in this paper are part of the EU initiatives to reach the EU targets on energy efficiency and GHG emissions reduction.

Even if the main aim of the regulation is environmental, in the political discussion it is necessary to add information regarding economic and social impacts.

In particular, the energy efficiency policies force producers' and consumers' intertemporal choices. Whenever technological improvements are already available, but not used, stricter regulations create incentives for firms to anticipate investments that bring future revenues. For consumers, the revision of the regulations implies an increase in the purchase cost of the appliances, but a decrease in the future spending in electricity. Further research could dwell on the alternative investments not realized by producers and consumers, as a consequence of forced investments in energy efficiency.

The approach that is used to address the quantification of indirect impacts of product energy efficiency policies has the potential to expand and nuance the policy-making discussion. Assuming that the proposed changes are required to reach the EU environmental targets therefore they are appropriate per se, the analysis shows that they are expected to have negative but small economic impact (with a small 0.01% decrease in the EU GDP) and a small positive social impact on employment. Indeed, the economic brake that a reduction in the use of energy could cause is compensated by other trade-off in the economic system, such as a change in the household consumption basket or the economic growth that is driven by the new investments induced by the revised regulations. The presented hybrid approach provides quantitative information that complements the policy discussion of energy saving policy, enriching this discussion with knowledge of the indirect impacts and tradeoffs between the economy sectors and countries/regions in the EU. Being able to highlight which sectors and countries are most benefited and which bear the weight of the policy the most adds relevant information. This information can indeed be used during the policy process to adjust the reform, for example by introducing some compensation mechanism for countries or sectors that are more disadvantaged.

Further research might look more in depth to distributional issues across the EU countries and regions, as well as non-EU trade issues. Additionally, another important aspect to be investigated could be the impacts of these regulations not only on the energy efficiency requirements, but also on the lifetime of the appliances. In fact, some papers demonstrate that an extension of the lifetime of durable goods has a positive effect on the overall lifecycle energy consumption and GHG emissions, as long as the production and end of life stages require less energy than the use phase [35–38]. Moreover, in a background of progressive greening and de-carbonization of the economy, the incentives behind energy efficiency policy are transforming. In a new metric of social welfare, carbon footprint and intensity might deserve further analysis to articulate the narrative for and/or against stricter product efficiency policies.

5. Conclusions

Many studies support a positive correlation between income growth and energy consumption [39]. This implies that policies aimed at improving energy efficiency and decreasing energy consumption can have a negative impact on economic growth. With respect to this topic, the study that is reported in this article provides additional insights and granularity on the energy-growth relationship, at EU level, and it suggests two main outcomes.

First, the analysis shows that stricter energy efficiency requirements of three household appliances—washing machines, dishwashers, and washer-dryers—have a negative macroeconomic impact on value added, a decrease of around two-billion euros. The order of magnitude of the changes that the regulations introduce is small as compared to the whole EU economy since these appliances constitute around 20% of total household appliance energy use. In terms of value added, with 0.01% of the total EU value added, the impact is very small. The reasons behind this result are manifold. Firstly, the stricter requirements are expected to cause an increase in the cost of manufacturing and, consequently, the sales price of appliances, which increases by ca. 10%. Secondly, the result is also due to a change in the composition of household spending: households' savings resulting from a reduction in energy demand can be used to consume other goods and services. Although the energy market is a key industry for economic development, the negative impact that the proposed changes in the EU regulations are expected to have on this industry is partially compensated by the increase in other industries' production.

The second main outcome is that while the impact on value added is negative, the impact on employment is small but positive. Again, the shift in the composition of the household consumption basket seems to mainly favor industries that use relatively more labor than the industries that are negatively affected by the analyzed policies.

In terms of industry distribution, the sector that has the greatest negative impact on value added is the electricity industry, and this effect is quite homogenous among European countries. This result is straightforward: policies that aim at controlling emissions by improving energy efficiency have a negative impact on the electricity production, which is one of the main sectors responsible for GHG emissions. According to the European Environment Agency, in 2017 electricity production generated the largest share of GHG emissions (23.3% of total GHG emissions) [40].

Finally, our results suggest that the proposed changes in the EU regulations would cause a reduction in emissions of around 1.5 million tonnes of CO_2-equivalent emissions. As expected, this reduction is driven by the industry producing electricity, but the 30 % of total GHG emissions reduction comes from other industries, such as mining and quarrying, sewerage and waste collection and treatment, or manufacture of coke and refined petroleum products.

Author Contributions: Conceptualization, P.R., J.M.R.-C., A.B. and A.V.; methodology, P.R., J.M.R.-C. and A.B.; software, P.R.; formal analysis, P.R. and A.B.; data curation, P.R. and A.B.; writing—original draft preparation, P.R., J.M.R.-C., and A.B.; writing—review and editing, P.R., J.M.R.-C., A.B. and A.V.

Funding: This research received no external funding.

Acknowledgments: We would like to thank Frédéric Reynes and Jinxue Hu for their fruitful comments and support.

Conflicts of Interest: The authors declare no conflict of interest.

Appendix A

Table A1. List of NACE Rev. 2 industries in FIDELIO.

Sector	Description
A01	Crop and animal production, hunting and related service activities
A02	Forestry and logging
A03	Fishing and aquaculture
B	Mining and quarrying
C10T12	Manufacture of food products, beverages and tobacco products
C13T15	Manufacture of textiles, wearing apparel and leather products
C16	Manufacture of wood and of products of wood and cork, except furniture; manufacture of articles of straw and plaiting materials
C17	Manufacture of paper and paper products
C18	Printing and reproduction of recorded media
C19	Manufacture of coke and refined petroleum products
C20	Manufacture of chemicals and chemical products
C21	Manufacture of basic pharmaceutical products and pharmaceutical preparations
C22	Manufacture of rubber and plastic products
C23	Manufacture of other non-metallic mineral products
C24	Manufacture of basic metals
C25	Manufacture of fabricated metal products, except machinery and equipment
C26	Manufacture of computer, electronic and optical products
C27	Manufacture of electrical equipment
C28	Manufacture of machinery and equipment n.e.c.
C29	Manufacture of motor vehicles, trailers and semi-trailers
C30	Manufacture of other transport equipment
C31_32	Manufacture of furniture; other manufacturing
C33	Repair and installation of machinery and equipment
D35	Electricity, gas, steam and air conditioning supply
E36	Water collection, treatment and supply
E37T39	Sewerage; waste collection, treatment and disposal activities; materials recovery; remediation activities and other waste management services
F	Construction

Table A1. *Cont.*

Sector	Description
G45	Wholesale and retail trade and repair of motor vehicles and motorcycles
G46	Wholesale trade, except of motor vehicles and motorcycles
G47	Retail trade, except of motor vehicles and motorcycles
H49	Land transport and transport via pipelines
H50	Water transport
H51	Air transport
H52	Warehousing and support activities for transportation
H53	Postal and courier activities
I	Accommodation; food and beverage service activities
J58	Publishing activities
J59_60	Motion picture, video and television program production, sound recording and music publishing activities; programming and broadcasting activities
J61	Telecommunications
J62_63	Computer programming, consultancy and related activities; information service activities
K64	Financial service activities, except insurance and pension funding
K65	Insurance, reinsurance and pension funding, except compulsory social security
K66	Activities auxiliary to financial services and insurance activities
L68	Real estate activities
M69_70	Legal and accounting activities; activities of head offices; management consultancy activities
M71	Architectural and engineering activities; technical testing and analysis
M72	Scientific research and development
M73	Advertising and market research
M74_75	Other professional, scientific and technical activities; veterinary activities
N	Administrative and support service activities
O84	Public administration and defense; compulsory social security
P85	Education
Q	Human health and social work activities
R-S	Arts, entertainment and recreation. Other service activities
T	Activities of households as employers; undifferentiated goods and services producing activities of households for own use
U	Activities of extraterritorial organizations and bodies

Appendix B

To build the FIDELIO database, many different data sources are used. The core of the database is the IO data, which is the main source of information that feed the production block and the trade block. The IO core is built mainly using the World Input Output Database [41] (WIOD, 2016 release). In particular, the model uses the WIOD international and national supply and use tables. Whenever the WIOD database does not provide all the necessary information, other databases, such as Eurostat supply and use tables or OECD data are used.

Besides the IO core, other databases are necessary in order to compile data for the other blocks of the model. For the household block, the main data sources come from Eurostat. The main datasets used are "non-financial transactions of households and non-profit institutions serving households" (nasa_10_nf_tr), "final consumption expenditure of households by consumption purpose—COICOP 3 digit" (nama_10_co3_p3), "heating degree-days by NUTS 2 regions—annual data under Energy statistics" (nrg_esdgr_a) and "financial balance sheets of households and non-profit institutions serving households" (nasa_10_f_bs). Other datasets come from the OECD—"simplified non-financial accounts" (no. 13), "final consumption expenditure of households" (no. 5) and "financial balance sheets—consolidated" (no. 710)—and from the National Statistical Institutes of Belgium, China, Czech Republic, Hungary, India, Slovakia, Turkey and the United Kingdom. Some data on household energy consumption are taken from the EU Reference Scenario 2016 on energy, transport and GHG emissions containing trends to 2050 [42]; population demographics from United Nations projections [43]. As regards the government block, the main sources come from Eurostat datasets—the main datasets used are: "non-financial transactions" (nasa_10_nf_tr), "government revenue, expenditure and main aggregates" (gov_10a_main) and "government deficit/surplus, debt and associated data" (gov_10dd_edpt1)—, WIOD data, the OECD dataset on general government debt—"general government debt – Maastricht" (no. 750)—and data from the World Bank. The labor market is described using the WIOD social accounts, and other data from Eurostat, World Bank, and CEDEFOP. Finally, for the energy block data comes from the WIOD (2019) energy accounts (https://europa.eu/!Un47Cp), the POLES model, the Eurostat table on "air emissions accounts by NACE Rev. 2 activity" (env_ac_ainah_r2), the ODYSSEE database [33], the EXIOBASE database [44] and the KLEMS database [45]. A detailed description of all the data sources and methods used to build the FIDELIO database can be found in [23].

Appendix C

This appendix show the information used to weight the exogenous shocks in the price of each appliance and the exogenous shocks in the total electricity consumption as described in Section 2.2.

Table A2. Penetration rates (Unit: percentage).

	Dishwashers	**Washing Machines**	**Washer Dryers**
2020	52.5	92.0	6.9
2021	53.8	92.0	6.8
2022	55.2	92.0	6.8
2023	56.5	92.0	6.8
2024	57.8	92.0	6.8
2025	59.0	92.0	6.8
2026	60.1	92.0	6.7
2027	61.2	92.0	6.7
2028	62.2	92.0	6.7
2029	63.2	92.0	6.7
2030	64.1	92.0	6.7

Table A3. Share of washing and drying machines over total appliances from Eurostat HBS (Unit: percentage).

Country	**Share**
Austria	20
Belgium	21
Bulgaria	2
Croatia	16
Cyprus	20
Czech republic	9
Denmark	35
Estonia	20
Finland	18
France	28
Germany	20
Greece	19
Hungary	19
Ireland	39
Italy	12
Latvia	22
Lithuania	4
Luxembourg	15
Malta	3
Netherland	20

Table A3. *Cont.*

Country	Share
Poland	19
Portugal	19
Romania	3
Slovakia	5
Slovenia	19
Spain	19
Sweden	20
United Kingdom	12

Table A4. Share of electricity used by large appliances over total electricity use from EEA and ODYSSEE databases (Unit: percentage).

Country	Share
Austria	28
Belgium	26
Bulgaria	19
Croatia	22
Cyprus	22
Czech republic	15
Denmark	29
Estonia	14
Finland	17
France	21
Germany	23
Greece	24
Hungary	22
Ireland	20
Italy	32
Latvia	23
Lithuania	32
Luxembourg	27
Malta	19
Netherland	34
Poland	28
Portugal	15
Romania	38
Slovakia	36
Slovenia	14
Spain	29
Sweden	23
United Kingdom	20

Table A5. Price variation for the aggregate household consumption category "appliances" (unit: percentage).

	2020	2021	2022	2023	2024	2025	2026	2027	2028	2029	2030
Austria	1.3	1.4	1.5	1.6	1.7	1.9	1.8	1.9	1.9	2.0	2.0
Belgium	1.3	1.5	1.6	1.7	1.8	1.9	1.9	2.0	2.0	2.1	2.1
Bulgaria	0.1	0.1	0.1	0.1	0.1	0.2	0.1	0.2	0.2	0.2	0.2
Croatia	1.0	1.1	1.2	1.3	1.3	1.5	1.4	1.5	1.5	1.6	1.6
Cyprus	1.3	1.4	1.5	1.6	1.7	1.8	1.8	1.9	1.9	2.0	2.0
Czech Republic	0.6	0.6	0.7	0.7	0.8	0.8	0.8	0.9	0.8	0.9	0.9
Denmark	2.2	2.4	2.5	2.8	2.9	3.1	3.1	3.3	3.2	3.4	3.4
Estonia	1.2	1.3	1.4	1.6	1.6	1.8	1.8	1.8	1.8	1.9	1.9
Finland	1.1	1.2	1.3	1.4	1.5	1.6	1.6	1.7	1.7	1.7	1.8
France	1.8	1.9	2.0	2.3	2.3	2.5	2.5	2.6	2.6	2.7	2.8
Germany	1.3	1.4	1.5	1.6	1.7	1.9	1.8	1.9	1.9	2.0	2.0
Greece	1.2	1.3	1.4	1.5	1.6	1.7	1.7	1.8	1.7	1.8	1.9
Hungary	1.2	1.3	1.4	1.6	1.6	1.8	1.7	1.8	1.8	1.9	1.9
Ireland	2.5	2.7	2.9	3.2	3.3	3.6	3.5	3.7	3.6	3.8	3.9
Italy	0.8	0.8	0.9	1.0	1.0	1.1	1.1	1.2	1.1	1.2	1.2
Latvia	1.4	1.5	1.6	1.8	1.9	2.1	2.0	2.1	2.1	2.2	2.2
Lithuania	0.2	0.3	0.3	0.3	0.3	0.3	0.3	0.4	0.3	0.4	0.4
Luxembourg	0.9	1.0	1.1	1.2	1.2	1.4	1.3	1.4	1.4	1.4	1.5
Malta	0.2	0.2	0.2	0.2	0.2	0.3	0.3	0.3	0.3	0.3	0.3
Netherland	1.3	1.4	1.5	1.6	1.7	1.9	1.8	1.9	1.9	2.0	2.0
Poland	1.2	1.3	1.4	1.5	1.5	1.7	1.7	1.7	1.7	1.8	1.8
Portugal	1.2	1.3	1.4	1.5	1.6	1.7	1.7	1.8	1.8	1.8	1.9
Romania	0.2	0.2	0.2	0.2	0.2	0.2	0.2	0.3	0.3	0.3	0.3
Slovakia	0.3	0.4	0.4	0.4	0.5	0.5	0.5	0.5	0.5	0.5	0.5
Slovenia	1.3	1.4	1.5	1.6	1.7	1.9	1.8	1.9	1.9	2.0	2.0
Spain	1.2	1.3	1.4	1.6	1.6	1.8	1.7	1.8	1.8	1.9	1.9
Sweden	1.2	1.3	1.4	1.5	1.6	1.7	1.7	1.8	1.8	1.8	1.9
United Kingdom	0.8	0.8	0.9	1.0	1.0	1.1	1.1	1.2	1.1	1.2	1.2

Table A6. Shock in household total electricity consumption (unit: percentage).

	2020	2021	2022	2023	2024	2025	2026	2027	2028	2029	2030
Austria	3.1	−0.4	−3.6	−3.9	−4.4	−4.6	−4.7	−4.9	−5.0	−5.2	−5.5
Belgium	−3.0	−3.2	−3.4	−3.8	−4.2	−4.4	−4.5	−4.7	−4.8	−5.0	−5.2
Bulgaria	−2.2	−2.3	−2.5	−2.7	−3.0	−3.2	−3.3	−3.4	−3.5	−3.6	−3.8
Croatia	−2.5	−2.7	−2.9	−3.2	−3.5	−3.7	−3.8	−4.0	−4.1	−4.2	−4.4
Cyprus	−2.4	−2.7	−2.8	−3.1	−3.4	−3.6	−3.7	−3.8	−3.9	−4.1	−4.3
Czech Republic	−1.7	−1.9	−2.0	−2.2	−2.4	−2.6	−2.6	−2.7	−2.8	−2.9	−3.1
Denmark	−3.3	−3.6	−3.8	−4.2	−4.6	−4.9	−5.0	−5.2	−5.3	−5.5	−5.8
Estonia	−1.6	−1.7	−1.8	−2.0	−2.2	−2.3	−2.4	−2.5	−2.6	−2.6	−2.8
Finland	−2.0	−2.1	−2.3	−2.5	−2.7	−2.9	−3.0	−3.1	−3.2	−3.3	−3.5
France	−2.3	−2.6	−2.7	−3.0	−3.3	−3.4	−3.5	−3.7	−3.8	−3.9	−4.1
Germany	−2.6	−2.8	−3.0	−3.2	−3.6	−3.8	−3.9	−4.0	−4.1	−4.3	−4.5
Greece	−2.7	−3.0	−3.1	−3.4	−3.8	−4.0	−4.1	−4.2	−4.4	−4.5	−4.8
Hungary	−2.5	−2.7	−2.9	−3.2	−3.5	−3.7	−3.8	−3.9	−4.0	−4.2	−4.4
Ireland	−2.3	−2.5	−2.7	−2.9	−3.2	−3.4	−3.5	−3.6	−3.7	−3.8	−4.0
Italy	−3.6	−3.9	−4.2	−4.6	−5.1	−5.3	−5.5	−5.7	−5.8	−6.1	−6.4
Latvia	−2.6	−2.9	−3.1	−3.3	−3.7	−3.9	−4.0	−4.1	−4.3	−4.4	−4.6
Lithuania	−3.6	−4.0	−4.2	−4.6	−5.1	−5.3	−5.5	−5.7	−5.9	−6.1	−6.4
Luxembourg	−3.1	−3.4	−3.6	−3.9	−4.4	−4.6	−4.7	−4.9	−5.0	−5.2	−5.5
Malta	−2.2	−2.4	−2.5	−2.8	−3.1	−3.2	−3.3	−3.4	−3.5	−3.7	−3.8
Netherland	−3.8	−4.1	−4.4	−4.8	−5.3	−5.6	−5.8	−6.0	−6.1	−6.4	−6.7
Poland	−3.2	−3.5	−3.7	−4.0	−4.4	−4.7	−4.8	−5.0	−5.1	−5.3	−5.6
Portugal	−1.7	−1.9	−2.0	−2.2	−2.4	−2.5	−2.6	−2.7	−2.8	−2.9	−3.0
Romania	−4.3	−4.6	−4.9	−5.4	−5.9	−6.2	−6.4	−6.7	−6.9	−7.1	−7.5
Slovakia	−4.1	−4.5	−4.8	−5.2	−5.8	−6.1	−6.2	−6.5	−6.7	−6.9	−7.2
Slovenia	−2.6	−2.9	−3.0	−3.3	−3.7	−3.9	−4.0	−4.1	−4.2	−4.4	−4.6
Spain	−3.3	−3.6	−3.8	−4.1	−4.6	−4.8	−4.9	−5.1	−5.3	−5.4	−5.7
Sweden	−1.5	−1.7	−1.8	−2.0	−2.2	−2.3	−2.3	−2.4	−2.5	−2.6	−2.7
United Kingdom	−2.3	−2.5	−2.6	−2.9	−3.2	−3.3	−3.4	−3.6	−3.7	−3.8	−4.0

References

1. Eurostat. Energy Consumption in Households. Available online: https://ec.europa.eu/eurostat/statistics-explained/index.php/Energy_consumption_in_households (accessed on 1 August 2019).
2. European Commission. Communication from the Commission, Ecodesign Working Plan 2016–2019. (COM(2016) 773 Final). 2016. Available online: https://ec.europa.eu/energy/sites/ener/files/documents/com_2016_773.en_.pdf (accessed on 1 August 2019).
3. Hartwig, J.; Kockat, J.; Schade, W.; Braundardt, S. The macroeconomic effect of ambitious energy efficiency policy in Germany—Combining bottom-up energy modelling with a non-equilibrium macroeconomic model. *Energy* **2017**, *124*, 510–520. [CrossRef]
4. Jollands, N.; Waide, P.; Ellis, M.; Onoda, T.; Lautsen, J.; Tanaka, K.; de T'Serclaes, P.; Barnsley, I.; Bradley, R.; Meier, A. The 25 IEA energy efficiency policy recommendations to plan of action. *Energy Policy* **2010**, *38*, 6409–6418. [CrossRef]
5. Lund, P.D. Exploring past energy changes and their implications for the pace of penetration of new energy technologies. *Energy* **2010**, *35*, 647–656. [CrossRef]
6. Yilmaz, S.; Majcen, D.; Heidari, M.; Mahmoodi, J.; Brosch, T.; Patel, M.K. Analysis of the impact of energy efficiency labelling and potential changes on electricity demand reduction of white goods using a stock model: The case of Switzerland. *Appl. Energy* **2019**, *239*, 117–132. [CrossRef]
7. Bertoldi, P.; Atanasiu, B. *Electricity Consumption and Efficiency Trends in the Enlarged European Union- Status Report 2006*; Publications Office of the European Union: Luxembourg, 2007; ISBN 978-92-79-05558-4.
8. Bucher, M.; Koch, S.; Andersson, G. A dynamic household appliance stock model for load management introduction strategies. In Proceedings of the 8th International Conference on the EU Energy Market (EEM), Zagreb, Croatia, 25–27 May 2011; pp. 717–722.
9. Cabeza, L.F.; Urge-Vorsatz, D.; Urge, D.; Palacios, A.; Barreneche, C. Household appliances penetration and ownership trends in residential buildings. *Renew. Sustain. Energy Rev.* **2018**, *98*, 1–8. [CrossRef]
10. McNeil, M.A.; Letschert, V.E.; De la Rue du Can, S.; Ke, J. Bottom-up energy analysis (BUENAS)—An international appliance efficiency policy tool. *Energy Effic.* **2013**, *6*, 191–217. [CrossRef]
11. Braungardt, S.; Elsland, R.; Eichhammer, W. The environmental impact of eco-innovations: The case of the EU residential electricity use. *Environ. Eco Policy Stud.* **2016**, *18*, 213–228. [CrossRef]
12. Radpour, S.; Mondal, M.A.H.; Kumar, A. Market penetration modeling of high energy efficiency appliances in the residential sector. *Energy* **2017**, *134*, 951–961. [CrossRef]
13. O'Doherty, J.; Lyons, S.; Tol, R. Energy-using appliances and energy-saving features: Determinants of ownership in Ireland. *Appl. Energy* **2008**, *85*, 650–662. [CrossRef]
14. Zhan, T.; Yaun, Z.; Bi, J.; Huang, L. Estimating future generation of obsolete household appliances in China. *Waste Manag. Resour.* **2012**, *30*, 11.
15. Boyano Larriba, A.; Cordella, M.; Espinosa Martinez, M.; Villanueva Krzyzaniak, A.; Graulich, K.; Rüdinauer, I.; Alborzi, F.; Hook, I.; Stamminger, R. *Ecodesign and Energy Label for Household Washing Machines and Household Washer-Dryers*; EUR 28809 EN, JRC108604; Publications Office of the European Union: Luxembourg, 2017; ISBN 978-92-79-74183-8.
16. Boyano, A.; Moons, H.; Villanueva, A.; Graulich, K.; Rüdenauer, I.; Alborzi, F.; Hook, I.; Stamminger, R. *Follow-Up for the Preparatory Study for Ecodesign and Energy Label for Household Dishwashers*; Publications Office of the European Union: Luxembourg, 2017; ISBN 978-92-79-73895-1.
17. Hanson, D.; Laitner, J.A.S. An integrated analysis of policies that increase investments in advanced energy-efficient/low-carbon technologies. *Energy Econ.* **2004**, *26*, 739–755. [CrossRef]
18. Rose, A.; Wei, D. Macroeconomic impact of the Florida Energy and Climate Change Action Plan. *Clim. Policy* **2012**, *12*, 50–69. [CrossRef]
19. Barker, T.; Alexandri, E.; Mercure, J.F.; Ogawa, Y.; Pollitt, H. GDP and employment effects of policies to close the 2020 emissions gap. *Clim. Policy* **2016**, *16*, 393–414. [CrossRef]
20. Barker, T.; Ekins, P.; Foxon, T. Macroeconomic effects of efficiency policies for energy-intensive industries: The case of the UK Climate Change Agreement, 2000–2010. *Energy Econ.* **2007**, *29*, 760–778. [CrossRef]
21. Ringel, M.; Schlomann, B.; Krail, M.; Rohde, C. Towards a green economy in Germany? The role of energy efficiency. *Appl. Energy* **2016**, *179*, 1293–1303. [CrossRef]

22. European Commission. Impact Assessment Accompanying the Document Proposal for a Directive of the European Parliament and of the Council Amending Directive 2012/27/EU on Energy Efficiency (COM (2016) 761 Final). 2016. Available online: https://eur-lex.europa.eu/legal-content/EN/TXT/?uri=CELEX%3A52016SC0405 (accessed on 22 October 2019).

23. Rocchi, P.; Salotti, S.; Reynès, F.; Hu, J.; Bulavskaya, T.; Rueda Cantuche, J.M.; Valderas Jaramillo, J.M.; Velázquez Afonso, A.; Amores, A.F.; Corsatea, T. *FIDELIO 3 Manual: Equations and Data Sources*; Publications Office of the European Union: Luxembourg, 2019; ISBN 978-92-79-98872-1.

24. European Parliament and Council. Regulation 2017/1369 of the EU Parliament and of the Council of 4 July 2017 Setting a Framework for Energy Labelling and Repealing Directive 2010/30/EU: 2017. Available online: https://eur-lex.europa.eu/legal-content/EN/TXT/PDF/?uri=CELEX:32017R1369 (accessed on 1 August 2019).

25. European Parliament and Council. Directive 2009/125/EC of the EU Parliament and of the Council of 21 October 2009 Establishing a Framework for the Setting of Ecodesing Requirements for Energy-Related Products (Recast) 2009. Available online: https://eur-lex.europa.eu/LexUriServ/LexUriServ.do?uri=OJ:L:2009:285:0010:0035:en:PDF (accessed on 1 August 2019).

26. VHK. Ecodesign Impact Accounting, Overview Report. 2016. Available online: https://ec.europa.eu/energy/sites/ener/files/documents/eia_ii_-_overview_report_2016_rev20170314.pdf (accessed on 1 August 2019).

27. McNeil, M.; Bojda, N. Cost effectiveness of high efficiency appliances in the US residential sector: A case study. *Energy Policy* **2012**, *45*, 33–42. [CrossRef]

28. Bull, J. Loads of green washing—Can behavioral economics increase willingness-to-pay for efficient washing machines in UK? *Energy Policy* **2012**, *50*, 242–252. [CrossRef]

29. Antal, M.; van den Bergh, J.C.J.M. Is there an energy efficiency gap? *J. Econ. Perspect.* **2012**, *26*, 3–28.

30. Barker, T.; Dagoumas, A.; Rubin, J. The macroeconomic rebound effect and the world economy. *Energy Effic.* **2009**, *2*, 411–427. [CrossRef]

31. CLASP. Estimating Potential Additional Energy Savings from Upcoming Revisions to Existing Regulations Under the Ecodesign and Energy Labelling Directives: A Contribution to the Evidence Base. 2013. Available online: http://www.clasponline.org/en/Resources/Resources/PublicationLibrary/2013/CLASP-and-eceee-Point-To-Additional-Savings-from-Ecodesign-and-Energy-Labelling.aspx (accessed on 8 April 2019).

32. European Environmental Energy Agency. Final Household Electricity Consumption by Use. Available online: https://www.eea.europa.eu/data-and-maps/daviz/final-household-electricity-consumption-by-use#tab-chart (accessed on 1 August 2019).

33. ODYSSEE Database. Available online: https://www.odyssee-mure.eu/project.html (accessed on 1 August 2019).

34. IPCC. *Climate Change 2014: Synthesis Report. Contribution of Working Groups I, II and III to the Fifth Assessment Report of the Intergovernmental Panel on Climate Change*; Pachauri, R.K., Meyer, L.A., Eds.; IPCC: Geneva, Switzerland, 2014; p. 151.

35. Kagawa, S.; Kudoh, Y.; Nansai, K.; Tasaki, T. The Economic and Environmental Consequences of Automobile Lifetime Extension and Fuel Economy Improvement: Japan's Case. *Econ. Syst. Res.* **2008**, *20*, 3–28. [CrossRef]

36. Kagawa, S.; Nansai, K.; Kondo, Y.; Hubacek, K.; Suh, S.; Minx, J.; Kudoh, Y.; Tasaki, T.; Nakamura, S. Role of Motor Vehicle Lifetime Extension in Climate Change Policy. *Environ. Sci. Technol.* **2011**, *45*, 1184–1191. [CrossRef] [PubMed]

37. Nishijima, D. Product Lifetime, Energy Efficiency and Climate Change: A Case Study of Air Conditioner in Japan. *J. Environ. Manag.* **2016**, *181*, 582–589. [CrossRef] [PubMed]

38. Nakamoto, Y.; Nishijima, D.; Kagawa, S. The Role of Vehicle Lifetime Extensions of Countries on Global CO_2 Emissions. *J. Clean. Prod.* **2019**, *207*, 1040–1046. [CrossRef]

39. Belke, A.; Dobnik, F.; Dreger, C. Energy consumption and economic growth: New insights into the cointegration relationship. *Energy Econ.* **2011**, *33*, 782–789. [CrossRef]

40. European Environment Agency. Greenhouse Gas Emissions. Available online: https://www.eea.europa.eu/data-and-maps/data/data-viewers/greenhouse-gases-viewer#tab-based-on-data (accessed on 1 August 2019).

41. WIOD. World Input Output Database Project (2016 Release). Available online: http://www.wiod.org/home (accessed on 1 August 2019).

42. Capros, P.; De Vita, A.; Tasios, N.; Siskos, P.; Kannavou, M.; Petropoulos, A.; Evangelopoulou, S.; Zampara, M.; Papadopoulos, D.; Paroussos, L.; et al. *EU Reference Scenario 2016—Energy, Transport and GHG Emissions Trends to 2050*; Publications Office of the European Union: Luxembourg, 2016; ISBN 978-92-79-52374-8. [CrossRef]

43. United Nations, DESA Population Division. Population Data. Available online: https://population.un.org/wpp/Download/Standard/Population/ (accessed on 1 August 2019).
44. EXIOBASE Database. Available online: https://www.exiobase.eu/ (accessed on 1 August 2019).
45. KLEMS Database. Available online: http://www.euklems.net/ (accessed on 1 August 2019).

Article

Toward a Low-Carbon Transport Sector in Mexico

Jorge M. Islas-Samperio *, Fabio Manzini and Genice K. Grande-Acosta

Instituto de Energías Renovables, Universidad Nacional Autónoma de México, Priv. Xochicalco s/n, Col. Centro, Temixco 62580, Morelos, Mexico; fmp@ier.unam.mx (F.M.); gkga@ier.unam.mx (G.K.G.-A.)
* Correspondence: jis@ier.unam.mx; Tel.: +52-555-622-9791

Received: 23 July 2019; Accepted: 18 December 2019; Published: 23 December 2019

Abstract: Considering that the world transport sector is the second largest contributor of global greenhouse gas (GHG) emissions due to energy use and the least decarbonized sector, it is highly recommended that all countries implement ambitious public policies to decarbonize this sector. In Mexico the transport sector generates the largest share of greenhouse gas emissions, in 2014 it contributed with 31.3% of net emissions. Two original scenarios for the Mexican transport sector, a no-policy baseline scenario (BLS) and a low carbon scenario (LCS) were constructed. In the LCS were applied 21 GHG mitigation measures, which far exceeds the proposals for reducing transport sector GHG emissions that Mexico submitted in its National Determined Contributions (NDC). As a result, the proposed LCS describes a sector transformation path characterized by structural changes in freight and passenger mobility, new motor technologies for mobility, introduction of biofuels, price signals, transportation practices and regulations, as well as urban planning strategies, which altogether achieve an accumulated reduction of 3166 $MtCO_{2e}$ in a 25 year period, producing a global net benefit of 240,772 MUSD and a GHG emissions' reduction of 56% in 2035 in relation to the BLS.

Keywords: road transport; low carbon scenario; GHG mitigation measures; cost-benefit; mitigation cost; financing; climate change

1. Introduction

On a global scale, the world transport sector generates the second largest share of GHG emissions. In 2014, this sector contributed 21% of global GHG emissions generated by energy use [1]. The rapid growth of the global transport sector's GHG emissions has been driven by road transport, which increased by 71% between 1990 and 2016, and represented 75% of the sector's global emissions in 2016 [2]. If this trend continues, this sector will generate 10,317 million tons of CO_{2e} ($MtCO_{2e}$) by the year 2040, which would represent a growth of 38% compared to transport GHG emissions in 2014 [3].

To know which substantive measures are being implemented currently for mitigating GHG emissions from transport sector that can help us to construct the LCS scenario, a survey of literature from national and international cases was carried out, specially selecting China, Germany, India and USA. Our search process and criteria to define the selection of these countries were the following: first, they are among the eight top world transport sector GHG emitters (see Table 1); second, their road transport sector represented more than 60% of their transport sector GHG emissions according to data available for 2010, third they have prospective studies with great ambition to reduce GHG reductions above 40% in their transport sector by year 2035 in relation to a baseline scenario or a reference year and finally, they present at least two future scenarios and five transport sector mitigation measures. Table 1 shows the four countries and the studies that were selected after applying this search process and mentioned criteria.

Table 1. Countries with the highest proportion of global greenhouse gas (GHG) emissions in the world transport sector and the high ambition GHG reduction prospective country studies in this sector.

Country (World Ranking GHG From Transport Sector) *	GHG Transport Sector Emissions, 2014 * (MtCO$_2$)	GHG Road Transport/ GHG Total Transport Emissions, 2010 **	Scenarios	Measures	2035 GHG Reduction from Baseline or Base Year (MtCO$_{2e}$)	2035 GHG Reduction from Baseline or Base Year	Prospective Country Studies
USA (1)	1,729	79%	2	7	1,595	56%	[4]
CHINA (2)	781	64%	7	11	335	44%	[5]
INDIA (3)	232	83%	2	5	346	51%	[6]
GERMANY (8)	155	78%	3	7	148	88%	[7]

* [8]; ** [9]. Source: Based on [4–9].

Consequently, this review considered the following four country studies with ambitious mitigation scenarios in the transport sector: USA [4], China [5], India [6] and Germany [7]. The proposed mitigation measures of these four studies are diverse and have high potential for reducing GHG emissions. They include traffic optimization; bus rapid transit; light urban train; railways and waterways; subways; increase of public and non-motorized transport, such as the bicycle; vehicle sales' restriction; mode shift; efficient vehicles and trucks; plug-in electric vehicles; plug-in hybrid electric vehicles; battery electric vehicles; fuel cell electric vehicles; electric trains; hybrid buses; diesel hybrid cars; hydrogen hybrid cars; and use of ethanol and biodiesel to substitute fossil fuels.

In relation to GHG emissions from fuel combustion, in 2014 Mexico was ranked eleventh in the world, generating 1.6% of the global volume, equivalent to 422 MtCO$_{2e}$, of which 78% came from two sectors: transport and energy (electricity and fossil fuels) [10]. Transport sector contributed with 31.6% followed by electricity generation with 23.1%.

In year 2013, due to concerns about the growth of criteria air pollutants generated by automobiles in Mexican big cities, which have negative effects on public health, an Official Mexican Standard (NOM in Spanish) was published NOM-163-SEMARNAT-ENER-SCFI-2013 [11], with the purpose of obliging automotive manufacturers to increase the fuel efficiency of light vehicles, thereby reducing the unit consumption of energy and, consequently, the emissions of SOx, NOx, particles, and CO$_{2e}$. Public concern about the problem of the growing emissions of the Mexican transport sector has also led to various studies that include GHG emissions' reduction in the transport sector, [12–15]. These studies consider mitigation measures, such as hybrid and electric cars; biofuels such as ethanol and biodiesel; and management measures and practices for the efficient use of energy, such as the implementation of efficiency standards, increased use of rail for freight transport, and a public bus rapid transport system. Together, these measures from these studies [12–15] have the potential to reduce CO$_2$ emissions by the year 2030 by between 16% to 38%, which is equivalent to 48 and 131 MtCO$_{2e}$ of avoided emissions.

In year 2015, Mexico signed the Paris agreement, a non-mandatory commitment to comply with its national determined contributions (NDC) [16] where Mexico obliges itself to reduce its national GHG emissions by 22% in its unconditioned goal, and 36% GHG reduction in its conditioned goal (that implies greater foreign investment and technology transfer), in relation to an official baseline GHG emissions, which in year 2030 would have emissions of 762 MtCO$_{2e}$; in addition to achieving a maximum emission peak in 2026. In the Mexican NDCs, the transport sector is compelled to achieve 18% reduction in GHG emissions by 2030 as an unconditioned goal, according to its official baseline GHG emissions, which represents a reduction of 48 MtCO$_{2e}$.

According to the literature review on Mexican studies, the National Institute of Ecology and Climate Change (INECC by its acronym in Spanish), carried out a study in 2018 [15] analyzing 8 measures that will reduce emissions by 48 MtCO$_{2e}$ by 2030, to comply with the Mexican NDC commitments; nevertheless, this constitutes the least ambitious scenario among the mentioned authors. Other, more specific studies, cover only one type of measure at a national level, such as focusing on CO$_2$ mitigation through mixing liquid biofuels as additives to fossil fuels in different proportions, either ethanol with gasoline or biodiesel with diesel [17–20]. For studies at cities level with some mitigation measures see [21–23].

This paper describes the construction of a very ambitious but feasible scenario to establish a low carbon transport sector in Mexico through the integration of 21 GHG mitigation measures. To shape the range of measures, both international and domestic experiences that might contribute to increasing mitigation ambition in the proposed new NDC to be reviewed in 2020 were considered. The main objective of this article was developed through the following structure: in the introduction section an overview of world transport sector was displayed, then the literature review findings and the followed steps to determine the selected transport sector prospective studies were described along with the current situation of Mexican transport sector. Then, the general methodology is described: starting from explaining how the original model was implemented using Excel spreadsheets and LEAP software, as well as a clarification of the scope of this article. Subsequently, once defined the reference year, the baseline scenario (BLS) and the alternative low carbon scenario (LCS) were constructed including input data, assumptions for both scenarios and a description of the 21 mitigation measures to the LCS were added. Next, the evolution of the vehicle fleet and the energy consumption calculation model as well as the cost benefit analysis model are described. Finally, results and conclusions are presented.

2. Mexican Transport Sector Current Situation

The energy consumption of the Mexican transport sector increased at an average annual growth rate of 4.4% between 1965–2014, consuming 2246 PJ in 2014, that represented a share of 45.9% of the country´s final energy consumption. Road transport energy consumption was 91.3% of the whole sector, air transport contributed 6% of total consumption, maritime contributed 1.3%, rail contributed 1.2%, and electric transport contributed 0.2%. In terms of fuel type, gasoline was the most consumed fuel, contributing 65% of the total transport energy, followed by diesel, at 27%, kerosene, at 6%, liquefied petroleum gas (LPG), at 1.76%, electricity, at 0.18%, and dry gas and fuel oil together at 0.06% [24]. The number of vehicles increased at an average annual growth rate (AAGR) of 5.9% between 1995–2014, totaling 38 million by the end of that period. Private vehicles accounted for 67.2% of the total, followed by cargo trucks, at 25.9%, motorcycles, at 6.0%, and passenger buses, at 0.9% according to [25]. Carbon dioxide (CO_2) emissions due to road transport increased at an AAGR of 2.4% between 1990–2014, reaching 153.5 $MtCO_{2e}$ by the end of the period, with 71.7% of the total due to gasoline burning, 25.5% to diesel, and 2.5% to LPG. Therefore, gasoline road transport is responsible for 67% of the GHG emissions of the transport sector, with passenger vehicles contributing 38% and cargo trucks contributing 29% [10].

3. General Methodology

The following general methodology is used:

- The year 2010 is established as the reference year, since for that year it was possible to combine all the information needed to represent the Mexican transport sector demand, especially because for this year, the road transport fleet by vehicle type and its survival curves data were available [26], which facilitated the BLS development.
- A BLS is constructed for the period 2011–2035, in which the amount of the most consumed energy carriers, gasoline and diesel, is estimated through an evolution model of the road transport vehicle fleet. Meanwhile, for other transport fuels or energy carriers used, such as kerosene, fuel oil, electricity, natural gas, and liquefied petroleum gas (LPG), their future consumption in this sector is estimated assuming the official consumption prospective of each of these fuels [27,28]. Subsequently, the annual CO_2 emissions are calculated for each of the fuels mentioned in the BLS.
- A low carbon scenario (LCS) is constructed with the 21 mitigation measures, as will be described in the section covering the LCS construction, where the main assumptions are stated.

- Finally, a cost-benefit analysis is carried out to determine the economic viability of the LCS scenario in relation to the BLS scenario. Table 2 shows the fuels levelized prices assumed to be representative of fuels price evolution between 2011–2035 which were incorporated into our calculations in Excel spreadsheets as constants. Last column shows fuel's information sources. All monetary values in this study are expressed in constant year 2007 US dollars (USD).

Table 2. Levelized price of fuels used in Transport sector.

Fuel	USD/GJ	Own Data Based on:
Gasoline	19.0	[29]
Diesel	16.5	[30,31]
Fuel oil	11.2	[30–32]
Natural Gas (NG)	5.1	[33]
Liquid Petroleum Gas (LPG)	20.9	[29,31,34]
Electricity	30.6	[28]

Source: Based on [28–34].

Each phase of the methodology was simulated first in individual Excel spreadsheets, and later, energy results were exported to the Long-Range Energy Alternatives Planning System (LEAP) software [35] to obtain the aggregated energy and GHG emissions in the BLS and LCS scenarios.

3.1. Methodology Development in Excel Spreadsheets

    a)   Baseline scenario (BLS)

- The vehicle stock was classified according to the following vehicle types: Light Duty Vehicles (LDV): subcompact, compact, luxury and sport cars, SUVs and light trucks all gasoline vehicles, and Heavy Freight Vehicles (HDV): freight transport and passenger buses that use diesel (see Table 3).
- For the reference year 2010, the existing vehicle stock was structured in terms of age, type of vehicle and fuel type according to [36] to obtain the vehicle stock in the reference year in the format required by LEAP.
- New vehicle sales specified by type of vehicle and fuel were obtained from [36], then, they were projected with an AAGR of 6% equivalent to the global rate of historical annual sales between 1995 and 2009, according to this same author.
- It was considered that in the reference year 771,135 used vehicles were imported according to [26], an AAGR of 5% after the reference year was adopted according to [36]. It was also assumed from [26], that from total annual imported used vehicles per year, 48% were light trucks, 31% compact cars and SUVs the remaining 21%. Finally, an age composition of vehicles per year of entry was estimated (see Supplementary Material).
- Once this is done, the corresponding vehicle survival curves shown in Figure 1 is used to calculate the stock of vehicles in each year.
- To obtain the annual energy consumption of vehicle stock (existing, new and imported used vehicles) by vehicle type and fuel, was estimated considering the fuel economy and traveled kilometers values showed in Tables 4 and 5, according to model in Section 6.1.

    b)   Low carbon scenario (LCS)

- The LCS was constructed from BLS described previously. For each of the first 17 measures described in Section 5, input data and assumptions were integrated impacting the fuel economy and traveled kilometers, and also to obtain the energy consumed for each of the mitigation measures in the analyzed period. In this simulation the number of vehicles is mostly not affected by modal changes, except in few specified cases.

- For each of these 17 mitigation measures the incremental costs investment, maintenance and fuel (positive or negative) and cost-benefit were calculated according to the assumptions, the economic input data and the calculation model in Section 6.2, also, results were obtained in terms of stock vehicle by type and consumption by energy type.
- In relation to the last four mitigation measures described in Section 5.1, according to [37] the consumptions by biofuel were calculated based on input data and assumptions made to replace gasoline and diesel with biofuels. Likewise, using the economic assumptions and input data from [37], the results of incremental investment, maintenance and fuel costs as well as the cost-benefit were estimated, applying the model of Section 6.2.

3.2. Methodology Development in LEAP Software

The LEAP software [35] is a bottom-up model that allows the energy supply and demand to be tied in a friendly way in trend and alternative scenarios. LEAP is also an accounting framework that allows to be fed with exogenous technical and economic data, contains also an environmental database from the IPCC. These features allow LEAP to carry out the analysis of the trend and alternative scenarios in terms of energy, economic and greenhouse gas emissions from different levels of aggregation. LEAP also allows the analysis of different mitigation measures to be represented in individual scenarios (one for each measure) and later, aggregated to represent an alternative global scenario. So, taking advantage of these features, in this article the LEAP software was used as follows:

- The transport sector was divided into five subsectors, following the National Energy Balance [38], namely: (1) road transport, (2) air, (3) rail, (4) maritime, and (5) electric public transport (subway, trams and trolleybuses).
- The transport sector total energy consumption in 2010 the reference year, was reproduced in LEAP from the energy consumption data of each one of the previously mentioned sub-sectors obtained from [38], so that the aggregate energy consumption of subsectors coincides with the official total energy consumption of the transport sector for that year.
- The BLS was constructed in LEAP representing the future fuel consumption between 2011 and 2035 of each subsector, and was estimated as follows: (1) For road transport subsector, the prospective fuel consumption was estimated in Excel spreadsheets as explained in the previous section and whose results were exported to LEAP. It should be noted that the LEAP software has a module called Transport Analysis which has a similar calculation model as the model in Section 6.1, but it was not used because it was necessary to model the imported used vehicles that are legally introduced to the country and these imported used vehicles currently are not considered in the LEAP transport model, that's why our model was developed in Excel spreadsheets. (2) In the case of the air, rail, maritime and electric public transport subsectors, the future demand for energy from the BLS was projected using future energy consumption rates obtained from the official prospective [27,28], the annual growth rates obtained were introduced into LEAP exogenously.
- Once the energy consumption is obtained in the reference year and the BLS, the GHG emissions in the reference year and the BLS were calculated using emission factors obtained from the technology and environmental database (TED) of LEAP (see Table 6).
- The LCS was also constructed in LEAP applying its characteristic of allowing scenarios aggregation, in this case, one scenario for each mitigation measure was developed; the results of the new energy demands in each mitigation measure calculated, as we mentioned in the previous section, were imported for each subsector specifying the energies involved in the addressed mitigation measure. Finally, from the sum of the 21 scenarios, one for each mitigating measure, the LCS was created combining the results of these 21 scenarios which gives rise to a new total energy demand where the demand for each energy type involved was specified.
- Once calculated the future energy demand, the GHG emissions from the LCS are calculated using the aforementioned TED database.

3.3. Assumptions about the Interaction, Additivity and Linearity of the Transport Systems

Given this methodology, this article does not consider the interactivity or the effects of additionality that real transport systems have, either in the BLS or in the LCS, except for the interactions that are described precisely and clearly in the description of the 21 mitigation measures. Thus, in this article that deals with Mexican transport systems, it is considered that both in the BLS and in particular in the LCS, it is assumed that different transport systems have a dynamic of a linear nature, independent of other transport systems evolution. For this reason, there are no effects of interaction or additionality between Mexican transport systems in most of the 21 mitigation measures analyzed, in this way the demand for energy estimated in each of the transport systems as well as GHG emissions are not influenced by the interaction between transport systems, except for those required in the description of mitigation measures in Section 5, nor by the additive effects that may exist in these systems.

Consequently, this article considers for most of the 21 mitigation measures only direct interventions within the same transport system, except in certain measures where it is explicitly established that there are interactions with other transport systems. These considerations about the low interactivity, the null additionality and the linearity in the dynamics of transport systems could lead to conservative results in the reduction of GHG and in the economic calculations presented in this article, both for each mitigation measure as above all for the evaluation of the LCS.

4. Reference Year Establishment and Construction of Baseline Scenario (BLS)

Table 3 shows the structure of Mexican road transport vehicle fleet in the reference year 2010. From this year´s structure, the energy demand was estimated using equations 1, 2, and 3 toward the year 2035 for each type of vehicle. To do so, first it was necessary to estimate the evolution of the age of the vehicle fleet, the annual national sales and imported used vehicles by type of vehicle, the survival factors, the vehicle fuel economy and the annual average distance travelled by each type of vehicle.

Table 3. Mexican vehicle fleet structure and fuel type in year 2010 (millions).

Vehicle Type	Millions
Compacts (gasoline)	5.7
Subcompacts (gasoline)	4.0
Luxury & Sport (gasoline)	2.1
Sport Utility Vehicles (SUV) (gasoline)	2.2
Light Trucks (gasoline)	7.0
Heavy Duty Vehicles-Freight (HDV-F) (diesel)	0.5
Heavy Duty Vehicles-Passengers (HDF-P) (diesel)	0.3

Source: Data from [26].

The AAGR used to estimate the growth of sales of new LDVs in Mexico was obtained from the historical sales' analysis for the 1995–2009 period, coming to an average annual rate of approximately 6.0%. Concerning the introduction of imported used vehicles to Mexico, an AAGR of 4.3% from 2010 to 2035 was assumed for both light and heavy vehicles.

The survival curves (f) from [26] for the different vehicle types is shown in Figure 1. The f values for vehicles are relatively optimistic because they reflect the conditions and times in which the vehicle stock is used in Mexico and whose main feature is a slow renewal of the vehicle stock, especially the HDV freight, because users try to delay the purchase of new trucks due to their relatively high costs. This situation is worse for the imported used vehicles from the United States to Mexico. Thus, in 2009, the vehicle stock of vehicles that have been sold in Mexico had an average age of 12.98 years while that of imported used vehicles was 18.01 years old, which resulted in a vehicle stock being in circulation that had an average age of 16.34 years according to [39].

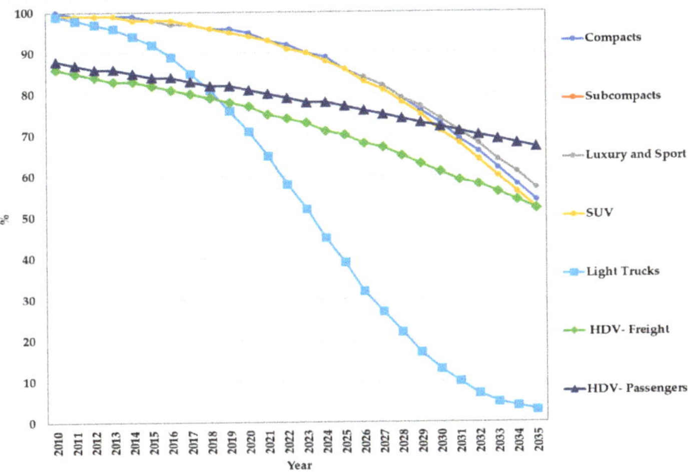

Figure 1. Survival curves by vehicle type. Source: [26].

Regarding fuel economy by vehicle type, the values in Table 4 from [36] were used in the energy consumption calculations in the BLS. These values represent the average performance of the future vehicle fleet comprising: existing, new and imported used vehicles. These values are relatively low because this work is considering a high growth percentage of imported used vehicles (5% per year) and comparable to that of new vehicles (6% per year) as well as the high survival factors that characterize the existing Mexican vehicle fleet. Finally, these values allow that the total volume calculation of gasoline and diesel are in accordance with the official prospective of these fuels [27].

Table 4. Fuel economy evolution by vehicle type.

Vehicle Type	2010	2035	AAGR
	(km/lt)		(%)
Compacts	7.0	9.0	1.0
Subcompacts	9.6	15.1	1.9
Luxury and Sport	6.7	8.7	1.0
SUV	5.5	7.1	1.0
Light Trucks	5.5	5.9	0.3
HDV-F	3.3	7.7	3.4
HDV-P	3.3	7.4	3.3

Source: [36].

Finally, Table 5 shows the evolution of annual kilometers travelled by vehicle type for the period 2010–2035.

Table 5. Evolution of annual kilometers travelled by vehicle type.

Vehicle Type	2010 *	2035	AAGR
	(km/year)		(%)
Compacts	11,290	14,964	1.10%
Subcompacts	11,052	15,626	1.40%
Luxury and Sport	11,964	15,631	1.10%
SUV	12,378	15,804	1.00%
Light Trucks	12,218	15,229	0.90%
HFV-F	65,557	49,977	−1.10%
HDV-P	55,438	41,096	−1.20%

* Reference year. Source: [36].

The data in this table come from [36] which considers the introduction of a degradation factor in the use intensity that depends on vehicle age (see Supplementary Material), which is associated with the high survival factors that characterize the Mexican vehicle fleet. Altogether these assumptions result in a conservative growth in the annual kilometers travelled by LDVs and even a decrease in this parameter by HDVs.

Regarding other fuels that are used for transportation in Mexico, such as Natural Gas, LP gas, fuel oil, kerosene, and electricity, their consumption in this sector in the BLS was estimated following the official consumption outlook of each of these fuels in the transport sector [27,28]. Finally, CO_2 emissions were estimated considering the emission factors shown in Table 6.

Table 6. Emission factors by fuel type.

Fuel Type	(tCO_2/TJ)
Gasoline	68.6
Diesel	73.3
Liquid Petroleum Gas (LPG)	62.7
Natural Gas (NG)	55.8
Fuel oil	72.5
Kerosene	72.5

Source: [40].

5. Construction of a Low Carbon Scenario (LCS)

The following paragraphs will give information on descriptions, assumptions, and costs of the 21 mitigation measures proposed. These measures have been classified into three important groups: those that favor increasing energy efficiency, those that use biofuels, and those that introduce new technologies using electric motors.

5.1. Energy Efficiency

5.1.1. Fuel Economy Standard for Light Duty Vehicles (LDV)

Implementation of fuel economy standards for new LDVs that foster the improvement of LDV energy efficiency by increasing fuel economy by 5% yearly departing from year 3 to 7, generating a 35% increase in relation to the reference year where the LDVs have a fuel efficiency of 12 km/lt. An additional average investment cost of 918 USD is assumed per improved vehicle throughout the first five years [41]. This additional cost was calculated, according to this author, by estimating the global cost for manufacturers to comply with this standard which resulted in a value of 4526 MUSD, which was divided by the number of vehicles sold during those first 5 years of this measure summing up 4.9 million LDV according to [36]. In subsequent years, an average annual performance growth of 2.3% is assumed as an inertial improvement due to the manufacturer's new infrastructure installed to comply with this standard and which was amortized as mentioned in the first 5 years of LDV sales, so that new LDVs from the sixth year on, no longer have an extra unit cost. Finally, the maintenance (M) costs are considered to be similar to those of existing LDVs in the BLS.

5.1.2. Fuel Economy Standard for Heavy Duty Vehicles-Freight (HDV-F)

This measure encourages the implementation of a performance standard that increases the fuel economy of HDV-F by 20% by the fifth year of operation of the regulations, assuming an annual increase of 4% in the fuel economy departing from year 4 to 8. An additional average investment cost of 378 USD per improved vehicle is assumed throughout the mentioned five years, emulating the USA draft fuel economy standard for heavy duty vehicles class 2B-8 according to [42]. This additional cost was calculated, according to [36] by estimating the global cost for manufacturers to comply with this standard which resulted in a value of 160 MUSD, which was divided by the number of vehicles sold during those 5 years of this measure summing up 423,155. An average annual performance increase of 2.5% after the eighth year is assumed as an inertial improvement due to the manufacturer´s new infrastructure installed to comply with this standard and which was amortized in the 5 years mentioned of HDV-F sales, so that new HDV-F from the eighth year on, no longer have an extra unit cost. Finally, the maintenance costs are like those of used vehicles in the BLS.

5.1.3. Gasoline Price without Subventions

In this scenario, with the purpose of reaching a gasoline price without subventions, we assume according to [36] an annual gasoline price increase of 5% over inflation during the first eleven years between 2011 and 2022, thereafter, gasoline price remains constant until the last year of the analyzed period. Under this assumption, a 17% reduction in intensity is achieved; this reduction is calculated using a price–gasoline elasticity of −0.31 according to [43].

5.1.4. Verification and Circulation Restriction in the Main 20 Metropolitan Areas and Five Border Metropolitan Areas

The aim of this measure is the implementation of inspection and maintenance programs, with traffic restriction for highly-polluting vehicles in the 20 Mexican cities with the largest vehicle fleets and also in the five metropolitan zones on the United States Border. This measure consists in establishing a periodic verification of engine performance, this implies a decrease in the use of fuel due to a more efficient combustion and fewer vehicles in circulation, therefore, a decrease in traffic congestion and an increase in the vehicles average speed is expected. It is assumed that this measure applies to 29% of existing national vehicles, which includes 42.2% of imported used vehicles. It is also assumed that 16% of existing national vehicles, including vehicles older than 8 years and 100% of imported used vehicles introduced to the metropolitan areas involved in the program, are verified and necessarily rest one day a week [36]. Finally, it is assumed a verification annual cost per vehicle of 49 USD for verification and 58 USD for maintenance [13].

5.1.5. Border Environmental Customs for Vehicles

This measure aims to establish a vehicle inspection program at the border to prevent the importation of those vehicles that do not comply with the national emissions' regulation. It is assumed that, in the border states, 16% of imported used vehicles do not meet the established standards. This percentage is estimated from a study conducted by [44] which reports that 84% of the inspected vehicles (56% of the total imported vehicles) were in good physical and mechanical condition. Based on this it is inferred that the remaining 16% did not have this condition and consequently, they are the ones that would not comply with the vehicle verification proposed by this measure. Therefore, the proposed environmental customs should prevent the annual importation of that 16% of used vehicles. Finally, it is considered an annual verification cost of 98 USD per imported used vehicle [13].

5.1.6. Optimization of Public Transport Routes in Urban Areas

It is assumed according to [38] that the optimization of public transport routes would result, on the one hand, in a reduction of the urban bus fleet by 20% in large cities (greater than 1 million inhabitants [45]); this process begins in the second year of the period of analysis and the goal is reached gradually in the year 2030. And, on the other hand, in a reduction of public transport by 44% in medium cities (between 500 thousand and 1 million inhabitants [45]) by 2030. This reduction was taken from [46] where it is proposed that this value is feasible for this medium-sized city in Mexico.

5.1.7. Introduction of Hybrid Buses

This article states that the hybrid buses measure is more important than the electric public transport systems (subway and trams -see the following measure 5.1.8), although the electric option has a greater mitigation potential, especially when the GHG emission factor of the power grid is high, and for this reason it is considered as the most important and aggressive low-carbon scenario to achieve in public transport systems. However, in this article, it is considered that hybrid buses have a much greater development than the electric option because the hybrid option avoids the high investments of the development of the rail infrastructure and the acquisition of trains, so from the economic and financial point of view it has a greater viability for a country like Mexico. Hybrid buses are presented in our article as a transition option that has an important development while lowering the costs of electric trains and developing the rail infrastructure to use them.

The introduction of hybrid buses, with the goal of constituting 15% of the total vehicle fleet of passenger buses in medium and large cities by 2035 has been considered. It is assumed that 30% of new buses by the year 2035 will be hybrid [36] and that their fuel economy will be 30% higher than that of conventional engines [47]. It is assumed that this measure applies only to 46% of the national bus fleet for urban use (HDV-P) and that the hybrid bus fleet increases from 41 vehicles in 2012 to 7694 in the year 2035, which implies an average annual growth rate of 25.6%. This measure implies an average differential cost of 282,491 USD for each hybrid bus relative to conventional buses. Finally, M cost for hybrid buses represents 10.7% of the incremental investment cost according to [36].

5.1.8. Electric Public Transport Systems

This measure promotes electric public transport systems (subway and trams) to reduce emissions and noise. According to [36], subway trains are promoted in the largest and most polluted cities in Mexico: Guadalajara, Monterrey, and the Metropolitan Area of Mexico City, and in the rest of the country, extension of the tram network is promoted. It is considered that from the second year, a total of 2.6 km of subway per year would be built in the three mentioned cities. This data is estimated observing the increase in the length of the metro in Mexico City between 1969–2012, during which it extended 226 km over a period of 43 years, according to [48], that is 5.2 km per year. It is assumed that half the extension per year is considered for the mentioned three cities. In the case of trams, 89 wagons were considered in 2012 reaching 960 in 2035 [36]. According to this author, it is assumed that 95% of transport users demand comes from public transport users and 5% from private car users. Also, for subway trains and trams investment costs of 0.371 and 1.297 USD per kilometer travelled per year, respectively, were considered. Finally, a subway and tram systems are assumed to expand to an average of 0.36 and 4.2 million kilometers travelled per year, respectively.

5.1.9. Public Bicycle Systems

The assumption of this measure is to change 0.6% per year of trips made by the sources shown in Table 7 to bicycle trips. It will start in the second year and reach progressively up to 3% in the sixth year and maintain 3% of these annual total trips to bicycle trips for the rest of the period [36]. According to this author, in this article we considered a bicycle average speed of 18 km/h, an average distance of 6 km per trip, and an average 7.1 trips per bicycle, an annual maintenance cost of 42 USD and a cost of

1250 USD for each new bicycle. Finally, it is considered that there is enough infrastructure to incorporate these new bicycle trips, where the new bicycles cost includes improvements of the infrastructure.

Table 7. Bicycle passenger sources by transport mode.

Bicycle Passenger Sources	
Original Transport Mode	**Trips Share**
Private car (sedan)	20%
Bus	26%
Taxi	15%
Non-motorized transport	19%
Not from other transport (new trips)	20%

Source: [36].

5.1.10. Bus Rapid Transit Systems (BRT)

This measure promotes Bus Rapid Transit (BRT) systems fueled by diesel. According to [36] it is considered that 15% of BRT system passengers stopped using their private vehicle and the other 85% change from using conventional public transport to BRT which is more efficient. In addition, these passengers benefit from shorter transfer times. It is also considered, according to the same author, an occupation of 100 passengers per bus used in the BRT, a performance of 1.45 km/L, and a constant traveled kilometers per vehicle per person in all transport modes. Also, a gradual increase in BRT corridors, from 130 km in the reference year to 650 km in the last year of the analysis period, an average cost of 3.9 MUSD per km of corridor, were considered according to [36]. Finally, it is also assumed that the M costs are similar to those of conventional buses in the baseline case.

5.1.11. Clean Freight Transport Program

It is assumed that at the federal level, all new freight vehicles are obliged through a standard to incorporate three technologies to improve performance of new HDV-F vehicles based on [49]: i) improvement in aerodynamics to achieve a 6.9% increase in efficiency; ii) improvement in inflating tire technology, with which an increase in efficiency of 2% is achieved; and iii) installation of an auxiliary power unit for air conditioning when the vehicle is not moving (energy independent of the truck engine), which reduces energy consumption by 75% for this use. For this last technology according to the information from [49], it is considered that the air conditioning unit consumes an average of 5,443 liters of diesel per truck per year, due to the vehicle's engine is turned on idle mode to power it. The implementation of these measures would start from the ninth year of the analysis period. According to the information from [49], we assumed a unitary investment costs of 624 USD due to the new aerodynamic design, 87 USD costs for tire inflating technology, this value is estimated as an average between 58 and 117 USD for class 7 and class 8 trailers, and 4,286 USD cost of an air conditioning unit integrated with an independent auxiliary power unit for class 8 trailers by year 2020.

5.1.12. Urban Development Oriented to Sustainable Transportation (DOST)

According to the study [50] related to urban development in three cities in Mexico, and specifically Mérida, it is assumed that urban development policies may achieve an average improvement of 30% in each of the following four urban indicators applied to new urban developments: population density, mixed land uses index, jobs per house, and employment-workers balance. According to this author, it is assumed that new urban policies affect only new houses and the new population travel patterns, this assumption is conservative as it does not consider interactions with population living outside the new urban developments. The new population growth is estimated considering historical annual growth rates of 56 Mexican urban areas, whose average is 1.5%, according to [51], and a population of 63.8 million people in 2010 [52]. With these data and considering the prospective 2011–2015 of new housing loans [53], projecting the data for 2015 towards the rest of the analysis period, and a constant occupancy of 3.95 persons per housing throughout the period; throughout these assumptions it is possible to obtain the number of new national homes and the number of new homes in the mentioned 56 urban areas, the latter in terms of the percentage of new dwellers in the period. In order to reach a 30% increase in the mixed land use index, it is assumed a 4.3% increase in the area built per person, when compared to the base case, which represents an extra cost of 291.2 USD/m^2 [54]. This is the only additional investment cost considered in this measure.

It is also according to [36] assumed that new houses share, designed with DOST criteria, increases from 1% in the year 1, to 35% in the year 3, to 60% in the year 5, to 90% in year 7 and to 100% from year 9 and forward. In order to quantify the effects of the 4 aforementioned indicators in the travel patterns, both in the average distance and frequency of daily trips, the elasticities in Table 8 were used, originally estimated for the city of Merida by [50]. Finally, the suggested urban development policies have the effect on the trip pattern departing from 2.59 daily trips per person and an average time per trip of 52.73 minutes [55]. Modal changes are not assumed.

Table 8. Effects of new urban development on mobility pattern.

Urban Indicators	Daily Trips Per Person	Average Distance Per Trip
Population density	−0.199	−0.101
Mixed land use index	0.337	−0.177
Jobs per house	-	0.106
Employment-workers balance	−0.336	-

Source: [50].

5.1.13. Freight Rail Systems

This measure assumes promotion of this transport system and its growth in terms of transported freight and reduction in the use of other less efficient road transport modes. It is assumed that the railway network grows by 450 km/year according to the national infrastructure program 2007–2012 [56], therefore grows at an AAGR of 2.57% with an average load of 79.9 tons per wagon, this value was calculated according to the relationship of each type of wagon existing in 2010 (gondolas, hoppers, vans, autoracks, tanks, platforms, piggy back and others) [57] and the estimated average capacity for each type of wagon, based on information from [58,59]. It is also assumed that the fuel performance annual improvement showed between 1999–2009 years, through a linear interpolation departing from 115 ton-km/L [57] in 2010 and increasing at an AAGR of 1.8%. Finally, a cost of investment (including infrastructure and equipment) of 1278 MUSD/year [56] is assumed. From the first to the sixth year, investment costs, including infrastructure and equipment, are assumed at 1,278 MUSD/year and the cost of diesel consumption is 0.61 USD/L.

5.1.14. Freight Transport Companies Integration

In this measure, two actions are undertaken, first, transport companies are integrated to improve travel logistics. In Mexico 36.8% of total freight trips are empty according to [36]. And second, training ecodriving programs for drivers are assumed to save fuel on travel. It is assumed, that there will be a reduction of 15% in diesel consumption in 10 years due to driver training, this value is the average of improvements by training of 9% to 23% indicated by [60], it is also assumed, based on [61] that by the end of year 5 there will be a reduction of 50% of empty trips, starting from the second year of the analyzed period; in this article we consider that after year five the percentage of empty trips will remain constant for the rest of the period. A total of 84 annual ecodriving workshops for 50 drivers each one, are proposed to be held in each one of the 33 Mexican states. A cost of 8333 USD for each course is assumed according to [36].

5.2. Liquid Biofuels

See Table 9 for technical assumptions and assumed costs for these measures.

Table 9. Main technical assumptions and cost structure for biofuels.

	Sugarcane Ethanol	Sorghum Grain Ethanol	Units	*Jatropha curcas* Biodiesel	Palm Oil Biodiesel	Units
Technical Assumptions						
Surface Required	2.9	2.6	Mha	3.2	1.85	Mha
Crop Yield	64	2.6	t/ha	0.7	1.5	t/ha/year
Load Factor	0.85	0.80	-	0.82	0.82	-
Plant Capacity (One Plant)	170	190	ML/plant-year	93,750	140,171	t seed or fruit/year
Annual Production	145	152	ML/plant	37.5	41.6	ML
Industrial Performance	83.2	396.0	L/t sugarcane or sorghum grain	-	-	-
Number of Plants	139	28	-	29	117	-
Transesterification Efficiency	-	-	-	97%	97%	-
Cost Structure (one plant)						
Investment	$ 56.9	$ 70.4	MUSD	$ 25.9	$ 25.9	MUSD
O & M	$ 7.91	$ 10.0	MUSD/year	$ 0.38	$ 0.38	MUSD/year
Raw Material	$ 28.2	$ 224.0	USD/ton	$ 397.0	$ 313.4	USD/ton

Source: [37].

5.2.1. Sugarcane Ethanol

This option considers ethanol produced from sugarcane as a partial substitute for gasoline by mixing up to 10% (in volume basis) of the gasoline, which is an ethanol widely used for blending around the world [62]. In this measure we adopted the technical assumptions used in [37], published in its supporting information appendix (see Table 9), where a surface of 2.9 million hectares (Mha) is required to establish sugarcane crops to introduce sugarcane ethanol. The ethanol production is planned to begin in year 4 with a processing plant that produces 170 million liters (ML)/year and this capacity grows to a total of 139 plants with the same capacity by year 25. Finally, the cost structure given by [37] it is also assumed for this measure.

5.2.2. Sorghum Grain Ethanol

This measure considers the partial substitution of gasoline with ethanol produced from grain sorghum, through gasoline-ethanol blends. In this measure we adopted the technical assumptions used in [37] (see Table 9) for this measure, where it is observed that a surface of 2.6 Mha is required to grow sorghum grain crops to introduce grain sorghum ethanol. The ethanol production is planned to begin in year 6 with a processing plant that produces 190 ML/year and this capacity grows to a total of 28 plants with the same capacity by year 25. Finally, the cost structure given by [37] it is also assumed for this measure.

5.2.3. Jatropha curcas Biodiesel

This measure considers partial substitution of diesel with biodiesel produced from *Jatropha curcas* oil, through diesel-biodiesel blends. In this study we adopted the technical assumptions used in [37] (see Table 9), where a surface required of 3.2 Mha can be used to establish *Jatropha curcas* crops dedicated to biodiesel production. The introduction of *Jatropha* biodiesel is planned to begin in year 8 with a processing plant that produces 93,750 t seed/year and this capacity grows to a total of 29 plants with the same capacity by year 25. Finally, the cost structure given by [37] it is also assumed for this measure.

5.2.4. Palm Oil Biodiesel

This measure considers partial substitution of diesel with biodiesel from oil palm, through diesel-biodiesel blends. In this study we adopted the technical assumptions used in [37] (see Table 9), where a surface required of 1.85 Mha can be used to establish oil palm crops dedicated to biodiesel production. The introduction of oil palm biodiesel is planned to begin in year 8 with a processing plant that produces 140,171 t fruit/year and this capacity grows to a total of 117 plants with the same capacity by year 25. Finally, the cost structure given by [37] it is also assumed for this measure.

5.3. New Electric Motor Technologies

5.3.1. Hybrid Electric Vehicles

This measure promotes the use of hybrid electric vehicles (HEV), which increases the efficiency of fuel use by up to 50% according to [63]. It is assumed that HEV in Mexico will reach 35% of the new vehicle sales by 2035. This percentage is higher than the one estimated for USA [64], where a 24% market share for year 2030 is proposed. Comparing our value with a 40% share in a strong hybrid scenario proposed for USA [65] by 2035, our value is 5% lower but still is an ambitious goal. It is also assumed that the additional investment cost with respect to conventional internal combustion vehicles will be 6250 USD. This value was estimated as follows: according to [66] that proposes an incremental value is 7945 USD which contrasts with the value presented by [67] that proposes an incremental value of 5050 USD, both in year 2010, in our article we propose to use a value close to the average between these two references. Finally, we consider that the incremental value is reduced to 3125 USD by year 2035, which is very close to the estimated by [68], in its electric and hybrid vehicles section, which establishes a value of 3231 USD in 2030.

5.3.2. Plug-in Hybrid Electric Vehicles

This measure assumes a penetration of plug-in hybrid electric vehicles (PHEVs) according to a logistic function that saturates at 40% of the number of LDVs sold in the last year of the period. From a study made for USA by EPRI [64], we found that this market share is near 50% by year 2030, therefore, assuming a 40% PHEV share of new vehicle sales by year 2035 seems ambitious but appropriate for the case of Mexico. It is assumed that the additional investment cost related to conventional internal combustion vehicles is 14,100 USD at the beginning of the period, which is close to the value of 15,000 USD estimated for that same year by [69], and by year 2030 they expect a cost reduction to 7500 USD, this value is not far from to the assumed in this study of 7050 USD by year 2035. Finally, maintenance costs are assumed similar to those of a conventional vehicle.

5.3.3. Battery Electric Vehicles (BEV)

This measure assumes a penetration of battery-powered electric vehicles (BEVs) according to a logistic function which saturates at 10% of the number of new LDVs sold in 2035, this value is between the 7% in 2030 estimated by [70] and 16% by 2035 estimated by [71]. It is also assumed that the additional investment cost related to conventional internal combustion vehicles is 20,000 USD in year 1 as published in [72] and that it decreases by half by 2035 as estimated by [70]. Finally, we assume that the maintenance costs are like those of a conventional vehicle.

6. Calculation Models

6.1. Model Developed to Calculate the Evolution of the Vehicle Fleet and its Energy Consumption

The evolution of the vehicle fleet and the energy consumption by fuel type in the BLS is estimated from the following equations:

$$Stock_{tfn} = \sum Stock_{tfn-1} + V_{tfn} * S_{tfn} + I_{tfn} * S_{tzfn} \tag{1}$$

where: $Stock_{tfn-1}$: Remaining vehicle fleet of type t, using fuel f in year $n-1$; V_{tfn}: Sales of vehicle type t, using fuel f in year n; I_{tfn}: Number of imported used vehicles type t, using fuel f in year n.

S_{tfn}: Survival percentage of vehicles type t, using fuel f in year n; S_{itzfn}: Survival percentage of imported used vehicles type t and model year z, using fuel f in year n.

$$AllStock_n = \sum Stock_{tfn} \tag{2}$$

$AllStock_n$ = Total number of vehicles in year n.

$$E_t = \sum Stock_{tfn} * K_{tfn} * L_{tfn} \tag{3}$$

where: E_t: Energy consumption in year n by vehicle type t; $Stock_{tfn}$: Vehicle fleet of type t, using fuel f in year n; K_{tfn}: Average fuel economy per kilometer travelled for vehicle type t, using fuel f by year n; L_{tfn}: Average annual distance travelled in kilometers by vehicle type t, using fuel f by year n.

6.2. Cost-Benefit Assessment and Mitigation Costs

The following calculation model was used to obtain the cost-benefit and mitigation cost based on [73]:

$$CB_{LCS-BLS} = IC_{LCS-BLS} + OMC_{LCS-BLS} + EC_{LCS-BLS} \tag{4}$$

where: $IC_{LCS-BLS}$: Overall incremental investment costs for all alternative measures in the LCS in present value; $OMC_{LCS-BLS}$: Overall incremental costs of operation and maintenance for all alternative measures in the LCS scenario in present value; $EC_{LCS-BLS}$: Overall avoided costs of energy for all alternative measures in the LCS in present value.

With:

$$IC_{LCS-BLS} = \sum_{i=1}^{M} \sum_{n=1}^{P} \frac{IC_{LCS-BLS_{ni}}}{(1+r)^n} \tag{5}$$

where: $IC_{LCS-BLS_{ni}}$: Annual incremental investment costs in relation to the implementation of the mitigation measure i for any year n in the period P; n: Year, $n = 1, \ldots , P$; i: Mitigating measure, $i = 1, \ldots , M$; r: Discount rate (10%); P: Analyzed period (25 years); M: Number of mitigation measures in LCS (21 measures).

$$OMC_{LCS-BLS} = \sum_{i=1}^{M} \sum_{n=1}^{P} \frac{OMC_{LCS-BLS_{ni}}}{(1+r)^n} \tag{6}$$

where: $OMC_{LCS-BLS_{ni}}$: Incremental costs of operation and maintenance for mitigating measure i in year n:

$$EC_{LCS-BLS} = \sum_{i=1}^{M} \sum_{n=1}^{P} \frac{EC_{LCS-BLS_{ni}}}{(1+r)^n} \tag{7}$$

where: $EC_{LCS-BLS_{ni}}$: Annual cumulative incremental costs of operation and maintenance (O&M) for the mitigating measure i accumulated in the year n in the period P:

$$MC_{LCS-BLS_i} = \frac{TC_{LCS-BLS_i}}{GHG_{LCS-BLS_i}} \tag{8}$$

where: $TC_{LCS-BLS_i}$: Total incremental costs for the mitigating measure i in the LCS in present value; $GHG_{LCS-BLS_i}$: Total avoided GHG emissions for mitigating measure i in the BLS.

With:

$$TC_{LCS-BLS_i} = \sum_{n=1}^{P} \left(\frac{IC_{LCS-BLS_{ni}}}{1+r^n} + \frac{OMC_{LCS-BLS_{ni}}}{1+r^n} + \frac{EC_{LCS-BLS_{ni}}}{1+r^n} \right) \tag{9}$$

$$GHG_{LCS-BLS_i} = \sum_{f} \left(ES_{LCS-BLS_{fi}} * EF_f \right) \tag{10}$$

where: $ES_{LCS-BLSfi}$: Total avoided energy from energy carriers f (gasoline, diesel, fuel oil, NG, LPG, kerosene and electricity), use of which generates emissions in the implementation of measure i during the considered period; EF_f: Carbon Dioxide Emission Factor from fuel f.

7. Results

7.1. Baseline Scenario (BLS)

Table 10 shows the evolution in terms of AAGR of the stock by vehicle category in the BLS in the period 2010–2035 and its comparison with the corresponding historical AAGR in the period 1990–2010.

Table 10. Evolution of the Mexican vehicle fleet structure in the BLS and its comparison with the historical evolution.

Vehicle Category	Vehicle Type	BLS AAGR 2010–2035 (%)	Historical AAGR 1990–2010 (%) *
Motor vehicles	Compacts, subcompacts, L&D and SUV	5.1	6.0
Passenger buses	Heavy Duty Vehicles-Passengers	8.6	6.2
Light Trucks and Heavy Duty Vehicles-Freight	Light Trucks and Heavy Duty Vehicles-Freight	4.4	5.7
Total	-	5.0	5.9

* Source: [25].

Figure 2 shows results of the Mexican vehicle fleet projection in the BLS, which triples by year 2035, reaching 67.3 million vehicles (MVEH): Light trucks go from 6.8 MVEH in 2010 to 18.3 MVEH, compact cars go from 5.5 to 17.4 MVEH, SUVs go from 1.9 to 11.9 MVEH, subcompacts go from 4.1 to 9.9 MVEH, and luxury cars go from 1.9 to 9.8 MVEH. Passenger buses increase from 0.2 to 2.6 MVEH, and heavy trucks go from 0.3 to 3.6 MVEH.

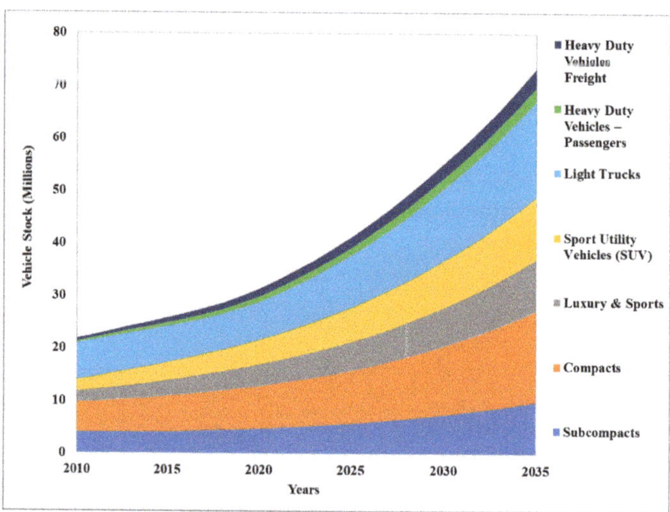

Figure 2. Evolution of the Mexican vehicle fleet in the BLS.

Energy consumption (see Figure 3) continues with the predominance of gasoline and diesel, which represent 65% and 27%, respectively, of total energy consumption in this sector, and the energy requirements total 5879 PJ in 2035, representing a 253% increase when compared to 2010.

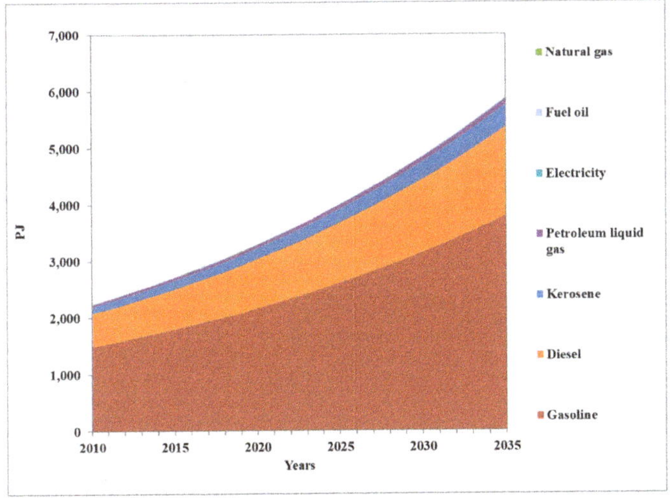

Figure 3. Energy consumption by energy carrier in the BLS.

GHG emissions (see Figure 4) will reach a total 415.1 $MtCO_{2e}$ by 2035, an increase of 259% compared to 2010, of which 63% and 28% come from gasoline and diesel, respectively.

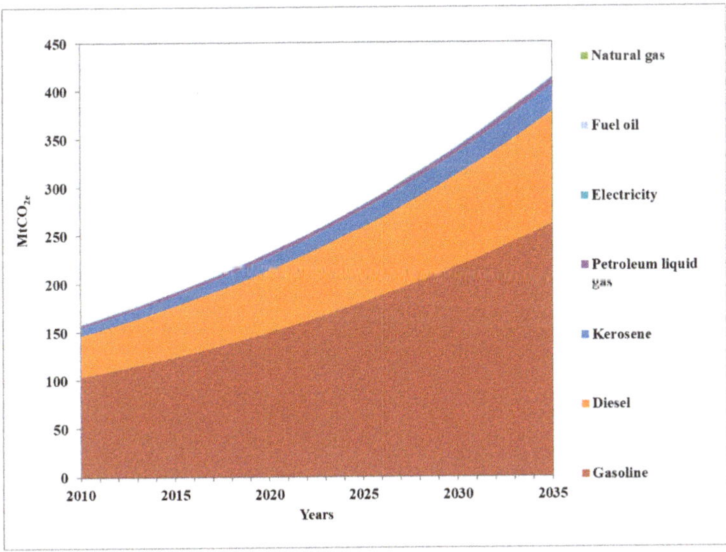

Figure 4. GHG emissions by energy carrier in the BLS.

7.2. Low Carbon Scenario (LCS)

Figure 5 shows the vehicle stock evolution in the LCS, as can be seen the total volume reaches almost 70 million vehicles, 3.7 million vehicles less than BLS.

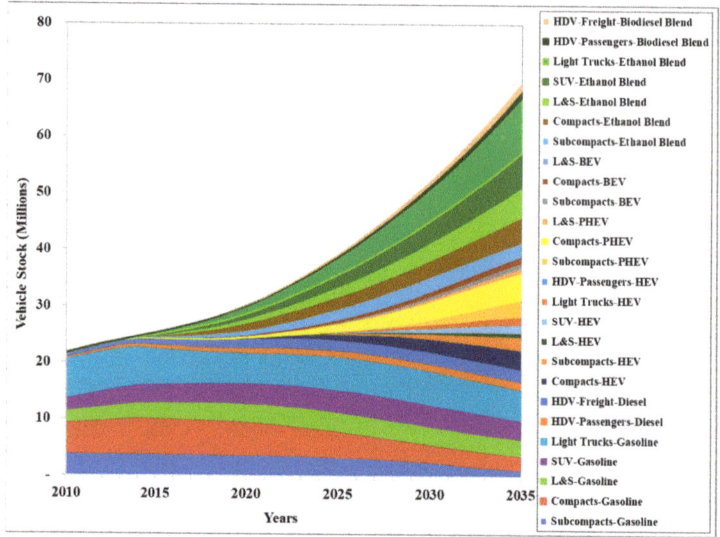

Figure 5. Evolution of the Mexican vehicle fleet in the low carbon scenario.

In addition, there is a significant technological change in the LCS, towards the year 2035, 29% (20.1 million vehicles) are based on new technologies, 44% (30.7 million vehicles) use mixtures with biofuels and the rest 27% (18.9 million vehicles) remain with conventional technologies.

Figure 6 shows energy consumption results of the LCS, where gasoline and diesel contribute 51.4% and 16.4% of total energy consumption, respectively, followed to a lesser extent by kerosene, at 11.1%; sugarcane ethanol, at 9.8%; electric power, at 3.3%; LPG, at 3.1%; sorghum ethanol, at 1.2%; and jatropha biodiesel, at 1.0%. Fuel oil and NG contribute 0.18% and 0.04%, respectively. By year 2035, transport sector energy requirements total 3,468 PJ in the LCS, which represents a reduction of 41% compared to the BLS.

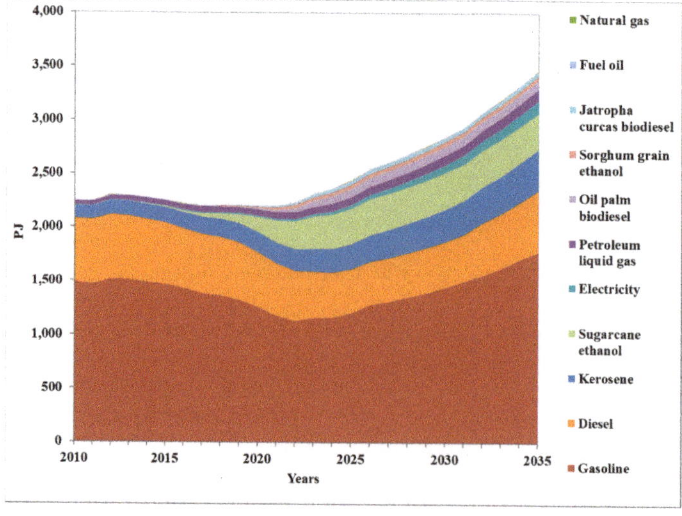

Figure 6. Energy consumption by energy carrier in the LCS.

Figure 7 shows results of the GHG reductions in the LCS scenario. In terms of energy efficiency measures, gasoline price without subventions contributes most, at 571 $MtCO_{2e}$, followed by the integration of transport companies, at 415 $MtCO_{2e}$; Customs vehicular environmental in the border, at 259 $MtCO_{2e}$; rail systems for cargo, at 236 $MtCO_{2e}$; federal clean transportation program, at 226 $MtCO_{2e}$; sustainable oriented transport development, at 210 $MtCO_{2e}$; performance standard for light vehicles, at 120 $MtCO_{2e}$; performance standard for cargo vehicles, at 116 $MtCO_{2e}$; optimization of public transportation routes, at 98 $MtCO_{2e}$; verification and restriction of traffic in the main 20 metropolitan zones and 5 border metropolitan zones, at 17 $MtCO_{2e}$; public bicycle system, at 9 $MtCO_{2e}$; rapid transport systems, at 5 $MtCO_{2e}$; introduction of hybrid buses, at 5 $MtCO_{2e}$ and, finally, public transport electric systems, at 3 $MtCO_{2e}$.

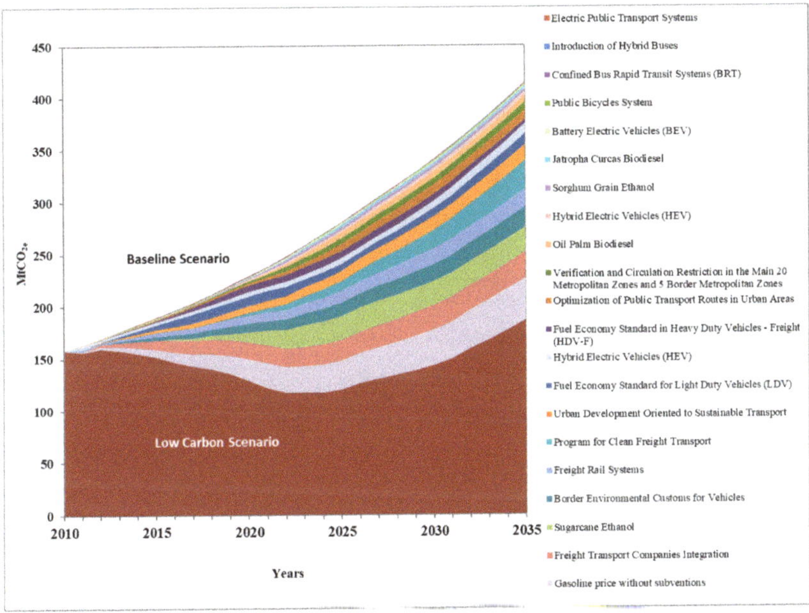

Figure 7. GHG emissions' reduction by measure in the LCS.

Regarding biofuels' use, the one that reduces GHG most is sugarcane ethanol, at 355 $MtCO_{2e}$, followed by palm oil biodiesel, at 82 $MtCO_{2e}$; ethanol from sorghum grain, at 43 $MtCO_{2e}$; and biodiesel from *Jatropha curcas*, at 35 $MtCO_{2e}$.

Regarding the new electric motor technologies, the HEV reduces GHG by 127 $MtCO_{2e}$, followed by the PHEV, at 45 $MtCO_{2e}$; and, finally, the BEV, at 23 $MtCO_{2e}$.

To summarize, the cumulative total of the emissions avoided in the analysis period amounted to 3165.9 $MtCO_{2e}$ in the LCS, which represents a total mitigation potential of 46.3% when compared to the emissions from the BLS. At 2035 levels, 229 $MtCO_{2e}$ are mitigated, which corresponds to a 59.3% GHG emissions' reduction relative to a BLS.

The resulting 10 best GHG mitigation measures in the Mexican transport sector, representing a total of 85% of avoided emissions, are: the gasoline price without subventions, at 18.0%; freight transport companies integration, at 13.1%; sugarcane ethanol, at 11.2%; border environmental customs for vehicles, at 8.2%; freight rail systems, at 7.5%; program for clean freight transport, at 7.1%; urban development oriented to sustainable transport, at 6.6%; fuel economy standard for LDV, at 5.4%; HEV, at 4.0%; and, finally, fuel economy standard in freight HDV, at 3.8%.

Considering the classification from the study by [6], in relation to an ambitious transport sector low carbon scenario in India, and adding our results according to this classification, Table 11 was developed in order to compare our results for Mexico with that of this author.

Table 11. Comparison of mitigation results of similar transport sector´s Low Carbon Scenarios in 2035 from India [6] and Mexico (this work) in absolute and relative values relative to a baseline scenario.

Measures Category	INDIA		MEXICO	
	[6]		This Work	
	Absolute GHG Reductions		Absolute GHG Reductions	
	$MtCO_2$	%	$MtCO_2$	%
Sustainable Mobility *	50	7%	26	6%
Freight Logistics **	18	3%	69	17%
Fuel Economy ***	120	18%	83	20%
Biofuels ****	18	3%	35	8%
Electric Vehicles *****	140	20%	15	4%
Total GHG Reductions from a Baseline Scenario	346	51%	229	56%

* Sustainable mobility: Public bicycles system, confined bus rapid transit systems (BRT), urban development planning oriented to sustainable transportation, optimization of public transport routes in urban areas; ** Freight logistics: Program for clean freight transport, freight rail systems, freight transport companies' integration; *** Fuel economy: Fuel economy standard for light duty vehicles (LDV), gasoline price without subventions, verification and circulation restriction in the main 20 metropolitan areas and five border metropolitan areas, fuel economy standard in heavy duty vehicles-freight (HDV), border environmental customs for vehicles; **** Biofuels use: Sugarcane ethanol, sorghum grain ethanol, *Jatropha curcas* biodiesel, oil palm biodiesel; ***** Electric mobility: Hybrid electric vehicles (HEV), plug-in hybrid electric vehicles (PHEV), battery electric vehicles (BEV), introduction of hybrid buses, electric transport systems.

According to the percentage data shown in the mentioned table, in the mitigation measures related to sustainable mobility, our results are similar with respect to the mentioned study. The same can be said about the measures that concern the fuel economy, however, the set of mitigation measures of our work is more ambitious regarding freight logistic and biofuels measures of the low-carbon scenario carried out for the India's transport sector, achieving in the first case a higher mitigation by a factor of 5 and in the second case, higher by a factor of 3, in percentage terms. Finally, it is observed that our measures concerning electric vehicles have less ambition to mitigate in the Mexican transport sector than that established for the study of India mentioned, being smaller by a factor of almost 6, in percentage terms.

In global terms, the two low-carbon scenarios of the two countries show important similarities in terms of ambition to mitigate. When comparing CO_2 reductions in year 2035 in percentage terms, we can observe that the mitigation potential identified for Mexico represents a 56% of CO_2 reduction when compared to the BLS, while for the case of India, this CO_2 reduction is 51% compared to a baseline scenario.

Table 12 shows the results of the incremental costs of investment, maintenance, avoided fuel, avoided subsidies, co-benefits, cost-benefit, mitigation cost, and avoided emissions from the 21 mitigation measures considered. According to this table, only 11 measures have a net investment cost; these are Introduction of hybrid buses, electric public transport systems, confined BRT systems, urban development oriented to sustainable transport, in addition to all measures corresponding to the use of liquid biofuels and those of new electric mobility technologies.

Table 12. Avoided GHG costs and emissions from mitigation measures in the transport sector in Mexico during 2010–2035 period.

Mitigation Measure	I	M	S	CPS	AF	C-B	MC	AE
	MUSD						USD/tCO$_{2e}$	MtCO$_{2e}$
Energy Efficiency	35,897	6905	−171,057	0	−136,631	−264,888	−108	2457
Fuel Economy Standard for LDV	3085	0	0	0	−16,727	−13,642	−79	172
Gasoline Price Without Subventions	0	0	−171,057	0	−38,175	−209,232	−366	571
Verification and Circulation Restriction in the Main 20 Metropolitan Zones and 5 Border Metropolitan Zones	0	4775	0	0	−5498	−723	−7	98
Fuel Economy Standard in HDV-F	90	0	0	0	−6101	−6011	−50	120
Border Environmental Customs for Vehicles	6522	2092	0	0	−16,787	−8174	−32	259
Optimization of Public Transport Routes in Urban Areas	0	0	0	0	−5835	−5835	−50	116
Introduction of Hybrid Buses	356	38	0	0	−152	242	48	5
Electric Public Transport Systems	656	0	0	0	−209	446	149	3
Public Bicycles System	544	0	0	0	−1375	−832	−48	17
Confined BRT Systems	1275	0	0	0	−893	382	43	9
Program for Clean Freight Transport	2730	0	0	0	−8482	−5751	−25	226
Urban Development Oriented to Sustainable Transport	2231	0	0	0	−1150	1081	5	210
Freight Rail Systems	18,244	0	0	0	−14,018	4226	18	236
Freight Transport Companies Integration	164	0	0	0	−21,229	−21,065	−51	415
Liquid Biofuels	**3048**	**35,912**	**0**	**−3662**	**−28,487**	**6814**	**13**	**514**
Sugarcane Ethanol	2371	24,984	0	−1930	−20,928	4498	13	355
Sorghum Grain Ethanol	369	4576	0	−1500	−2497	949	22	43
Jatropha curcas Biodiesel	232	2340	0	−202	−1519	852	24	35
Oil Palm Biodiesel	76	4012	0	−30	−3543	515	6	82
New Electric Mobility Technologies	**25,559**	**0**	**0**	**0**	**−8258**	**17,302**	**89**	**194**
Hybrid Electric Vehicles (HEV)	6297	0	0	0	−4047	2250	18	127
Plug-in Hybrid Electric Vehicles (PHEV)	13,218	0	0	0	−1922	11,296	251	45
Battery Electric Vehicles (BEV)	6044	0	0	0	−2289	3756	163	22
Totals	**64,504**	**42,817**	**−171,057**	**−3662**	**−173,376**	**−240,772**	**−76.05**	**3166**

Note: I: Investment cost, M: Maintenance cost, S: Subsidies, CPS: Co-products and Sales incomes, AF: Avoided Fuel cost, C-B. Cost Benefit, MC: Mitigation Cost and AE: Accumulated Avoided Emissions.

The rest of the measures considered have no net investment cost, and they have benefits; these are fuel economy standard for LDVs, gasoline price without subventions, verification and circulation restriction in the main 20 metropolitan zones and five border metropolitan zones, fuel economy standard in HDV-F, border environmental customs for vehicles, optimization of public transport routes in urban areas, public bicycles system, program for clean freight transport, and freight transport companies' integration. To summarize, the 21 measures studied represent an investment cost of 13,135 MUSD, a maintenance cost of 6867 MUSD and an avoided fuel cost of −120,209 MUSD. From the economic analysis results a cost-benefit of −271,265 MUSD and mitigation costs from −366 to −7 USD/tCO$_{2e}$. Finally, the cumulative mitigation potential is 1994 MtCO$_{2e}$.

Figure 8 shows the marginal cost curve of the measures analyzed and, in turn, indicates a route to follow for the implementation of mitigation measures based mainly on those that have no cost but have great potential for reducing emissions, which could initiate a transition toward a low carbon transport sector.

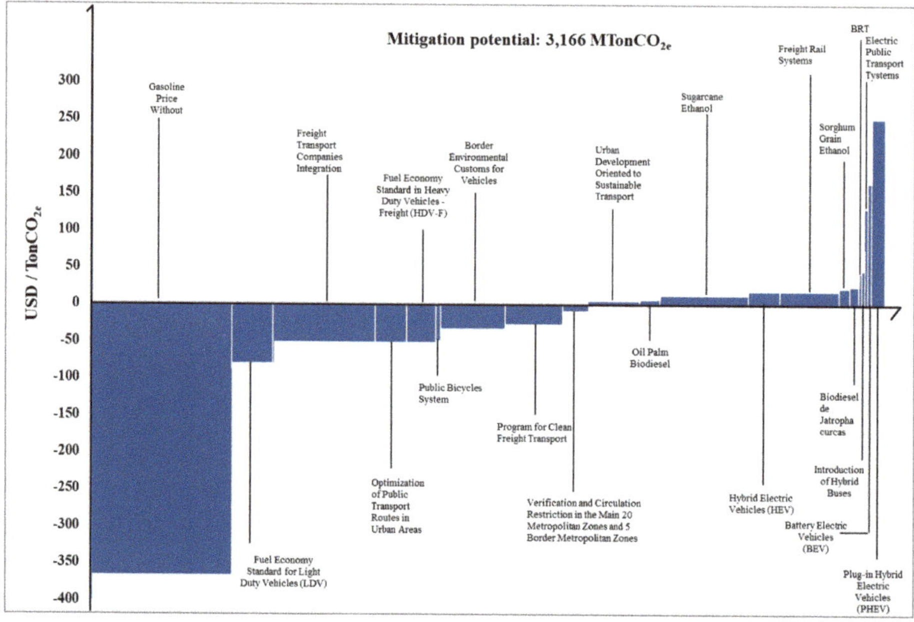

Figure 8. Marginal abatement cost curve for transport sector in Mexico.

8. Conclusions

In this article, a low carbon scenario (LCS) is proposed for the Mexican transport sector through the integration of 21 greenhouse gas (GHG) mitigation measures. As a result, we have an LCS that describes a transport sector transformation path characterized by structural changes in passenger and freight mobility; use of new mobility technologies, with electric motors; biofuels' introduction; price signals; and changes in transport practices, emission regulations, and urban planning.

The economic and environmental analysis of joint and parallel implementation of the selected mitigation measures shows, in year 2035, accumulated benefits for $-240{,}772$ MUSD, an average mitigation cost of -76.0 USD / tCO_{2e}, and an accumulated value of GHG emissions' reduction of 3166 $MtCO_{2e}$ (equivalent to a reduction of accumulated GHG emissions of 46.3% when compared to BLS, an average annual reduction of 126.7 $MtCO_2$, and a 59.3% reduction of GHG emissions relative to the BLS in the year 2035). We believe that the GHG reductions' portfolio of mitigation measures analyzed in this article will help Mexico, and other countries in the world, to establish more robust, more ambitious, and faster energy transitions to limit GHG emissions in this key sector to restrain global climate change. However, this article shows that the great challenge is to raise the significant investment required to achieve this energy transition in a very capital-intensive sector such as the transport sector, as shown by the Mexican case, where an accumulated investment of 64,326 MUSD is needed to establish a low carbon transport sector that will contribute to the actions to restrain climate change.

These results should be considered carefully since the interaction between the 21 mitigation measures is limited and restricted to one or two modes of transport at best, and the possible additivity effects of these measures were not studied and considered in this article. The representations of the measures were consequently assumed of linear nature, so our results turn out to be conservative and do not necessarily represent what could happen in the real-world transport systems. Taking into account the additive effects of the 21 mitigation measures analyzed, which would reflect the greater interactivity between transport systems and their non-linear nature, would result in an improvement in the overall results (cumulative emissions, emissions to the year 2035, cost-benefit, mitigation costs and investments) that we have presented in this article.

Supplementary Materials: The following are available online at http://www.mdpi.com/1996-1073/13/1/84/s1, Figure S1: Degradation factor in the use intensity per vehicle age, Table S1: Age composition of imported used vehicles per year of entry.

Author Contributions: J.M.I.-S. and F.M. conceived the presented study. Jorge Islas developed energy, economic and environmental assessment methodology, G.K.G.-A. performed the computations in LEAP, Jorge Islas supervised the study, F.M. wrote the original manuscript, J.M.I.-S. and F.M. performed the formal analysis of the mitigation measures. All authors discussed the results and contributed to the final manuscript.

Funding: Authors thank SENER CONACyT Energy Sustainability Fund projects 117808 and 246911.

Acknowledgments: Authors thank María de Jesús Pérez Orozco for the technical support in scenario simulation using LEAP model. Also, we thank to the editor Johnson Wang and the three anonymous reviewers for their valuable commentaries.

Conflicts of Interest: The authors, Jorge M. Islas-Samperio, Fabio Manzini and Genice K. Grande-Acosta declare no conflict of interest, also we declare that the funders had no role in the design of the study; in the collection, analyses, or interpretation of data; in the writing of the manuscript, an in the decision to publish the results.

References

1. World Research Institute (WRI). CAIT Climate Data Explorer. 2015. Available online: http://cait.wri.org (accessed on 11 August 2017).

2. International Energy Agency (IEA). World Energy Outlook 2016. 2016. Available online: https://www.iea.org/publications/freepublications/publication/WorldEnergyOutlook2016ExecutiveSummaryEnglish.pdf (accessed on 15 June 2018).

3. International Energy Agency (IEA). Statistics Key CO Emissions Trends. Excerpt from CO_2 Emissions From Fuel Combustion. 2016. Available online: http://www.iea.org/publications/freepublications/ (accessed on 18 November 2018).

4. Melaina, M.; Heath, G.; Sandor, D.; Steward, D.; Vimmerstedt, L.; Warner, E.; Webster, K.W. Alternative Fuel Infrastructure Expansion: Costs, Resources, Production Capacity, and Retail Availability for Low-Carbon Scenarios. Transportation Energy Futures Series; U.S. Department of Energy by National Renewable Energy Laboratory: Golden, CO, USA, 2013. Available online: https://www.nrel.gov/docs/fy13osti/55640.pdf (accessed on 15 October 2019).

5. Hao, H.; Geng, Y.; Li, W.; Guo, B. Energy Consumption and GHG Emissions from China's Freight Transport Sector: Scenarios through 2050. *Energy Policy* **2015**, *85*, 94–101. [CrossRef]

6. Dhar, S.; Shukla, P.R. Low Carbon Scenarios for Transport in India: Co-Benefits Analysis. *Energy Policy* **2015**, *81*, 186–198. [CrossRef]

7. Schmid, E.; Knopf, B. Ambitious Mitigation Scenarios for Germany: A Participatory Approach. *Energy Policy* **2012**, *51*, 662–672. [CrossRef]

8. World Resources Institute (WRI). *Climate Watch*; World Resources Institute: Washington, DC, USA, 2018; Available online: https://www.climatewatchdata.org/ghg-emissions (accessed on 18 October 2019).

9. International Transport Forum (ITF); Organisation for Economic Co-Operation and Development (OECD). Transport Greenhouse Gas Emissions: Country Data 2010. 2011. Available online: https://www.itf-oecd.org/sites/default/files/docs/10ghgcountry.pdf (accessed on 19 October 2019).

10. Instituto Nacional de Ecología y Cambio Climático (INECC). Inventario Nacional de Gases y Compuestos de Efecto Invernadero. *Tabla del INEGyCEI 1990–2015*. 2018. Available online: http://www.inecc.gob.mx/descargas/cclimatico/INEGEI_2014_EMISIONES_QUEMA_COMBUSTIBLES_FOSILES_1.pdf (accessed on 26 March 2018).

11. Diario Oficial de la Federación (DOF). NOM-163-SEMARNAT-ENER-SCFI-2013, Emisiones de Bióxido de Carbono (CO2) Provenientes del Escape y su Equivalencia en Términos de Rendimiento de Combustible, Aplicable a Vehículos Automotores Nuevos de peso Bruto Vehicular de Hasta 3857 kg. 2013. Available online: http://dof.gob.mx/nota_detalle.php?codigo=5303391&fecha=21/06/2013 (accessed on 19 June 2019).

12. Secretaría de Medio Ambiente y Recursos Naturales (SEMARNAT). *El Cambio Climático en México y El Potencial de Reducción de Emisiones Por Sectores*; Secretaría de Medio Ambiente y Recursos Naturales (SEMARNAT): Mexico City, Mexico, 2008.

13. Johnson, T.; Alatorre, C.; Romo, Z.; Liu, F. *Mexico: Estudio Sobre La Disminución de Emisiones de Carbono*; The International Bank: Washington, DC, USA, 2009; p. 185.

14. Solís Ávila, J.C.; Sheinbaum Pardo, C. Consumo de Energía y Emisiones de CO_2 del Autransporte En México y Escenarios de Mitigación. *Rev. Int. Contam. Ambient.* **2016**, *32*, 7–23.

15. Instituto Nacional de Ecología y Cambio Climático (INECC). Costos de las Contribuciones Nacionalmente Determiandas de México. Medidas Sectoriales no Condicionadas. 2018. Available online: https://www.gob.mx/cms/uploads/attachment/file/330857/Costos_de_las_contribuciones_nacionalmente_determinadas_de_M_xico__dobles_p_ginas_.pdf (accessed on 19 June 2019).

16. Gobierno de la República. Intended Nationally Determined Contribution. Gobierno de la República, México. Available online: https://www.gob.mx/cms/uploads/attachment/file/162973/2015_indc_ing.pdf (accessed on 20 October 2019).

17. García, C.; Manzini, F.; Islas, J. Air Emissions Scenarios from Ethanol as a Gasoline Oxygenate in Mexico City Metropolitan Area. *Renew. Sustain. Energy Rev.* **2010**, *14*, 3032–3040. [CrossRef]

18. Lozada, I.; Islas, J.; Grande, G. Environmental and Economic Feasibility of Palm Oil Biodiesel in the Mexican Transportation Sector. *Renew. Sustain. Energy Rev.* **2010**, *14*, 486–492. [CrossRef]

19. García, C.; Manzini, F. Environmental and Economic Feasibility of Sugarcane Ethanol for the Mexican Transport Sector. *Sol. Energy* **2012**, *86*, 1063–1069. [CrossRef]

20. García, C.; Riegelhaupt, E.; Masera, O. Escenarios de Bioenergía En México: Potencial de Sustitución de Combustibles Fosiles y Mitigación de GEI. *Rev. Mex. Física* **2013**, *59*, 93–103.

21. Manzini, F. Inserting Renewable Fuels and Technologies for Transport in Mexico City Metropolitan Area. *Int. J. Hydrog. Energy* **2006**, *31*, 327–335. [CrossRef]

22. Argelia, M.; Margarito, Q.; Moisés, G. Análisis de Las Estrategias de Mitigación y Adaptación Del Sector Transporte En La Ciudad de Mexicali. *Estud. Front.* **2013**, *14*, 79–105.

23. Jazcilevich, A.D.; Reynoso, A.G.; Grutter, M.; Delgado, J.; Ayala, U.D.; Lastra, M.S.; Zuk, M.; Oropeza, R.G.; Lents, J.; Davis, N. An Evaluation of the Hybrid Car Technology for the Mexico Mega City. *J. Power Sour.* **2011**, *196*, 5704–5718. [CrossRef]

24. Secretaría de Energía (SENER). Balance Nacional de Energía 2014. 2015. Available online: https://www.gob.mx/cms/uploads/attachment/file/44353/Balance_Nacional_de_Energ_a_2014.pdf (accessed on 15 August 2018).

25. Instituto Nacional de Estadística y Geografía (INEGI). Vehículos de Motor Registrados en Circulación. 2018. Available online: https://www.inegi.org.mx/sistemas/olap/Proyectos/bd/continuas/transporte/vehiculos.asp?s=est# (accessed on 6 June 2018).

26. Instituto Mexicano del Petróleo (IMP). *Información de Autotransporte para la Prospectiva Energética*; Dirección de Estudios Económicos; Instituto Mexicano del Petróleo (IMP): Mexico City, Mexico, 2011.

27. Secretaría de Energía (SENER). *Prospectiva de Petrolíferos 2010–2025*; Secretaría de Energía (SENER): Mexico City, Mexico, 2011.

28. Secretaría de Energía (SENER). *Prospectiva del Sector Eléctrico 2010–2025, México*; Secretaría de Energía (SENER): Mexico City, Mexico, 2011.

29. Secretaría de Energía (SENER). Sistema de Información Energética; Precios de los Petrolíferos. 2011. Available online: http://sie.energia.gob.mx (accessed on 24 January 2011).

30. US Energy Information Administration (US EIA). Petroleum and Other Liquids. 2011. Available online: http://tonto.eia.doe.gov/dnav/pet/hist/LeafHandler.ashx?n=pet&s=eer_epd2dxl0_pf4_rgc_dpg&f=a (accessed on 27 January 2011).

31. Comisión Federal de Electricidad (CFE). *Evolución de Precios Entregados y Fletes de Combustibles 2003–2004, México*; Comisión Federal de Electricidad (CFE): Mexico City, Mexico, 2005.

32. Secretaría de Energía (SENER). *Pronóstico de Precios de los Petrolíferos*; Secretaría de Energía (SENER): Mexico City, Mexico, 2005.

33. Comisión Federal de Electricidad (CFE). *Costos y Parámetros de Referencia para la Formulación de Proyectos de Inversión*; Generación; Escenario de Precios Medio; Subdirección de Programación, Comisión Federal de Electricidad: Mexico City, México, 2012.

34. US Energy Information Administration (US EIA). World Energy Outlook 2010. 2010. Available online: https://www.iea.org/publications/freepublications/publication/weo2010.pdf (accessed on 15 December 2010).

35. Heaps, C. *Long-Range Energy Alternatives Planning (LEAP) System*; Version 2008.0.0.33 2008; Environment Institute: Stockholm, Sweden, 2008.

36. Centro de Transporte Sustentable; Instituto de Energías Renovables; Universidad Nacional Autónoma de México (CTS-IER-UNAM). *Sector Transporte: Identificación de Acciones y Medidas de Mitigación de GEI. Evaluación Económica y Ambiental de Escenarios al 2030 de la Inserción de Fuentes Alternas de Eficiencia Energética en el Sistema Energético Mexicano en Base a Potencial de Reducción de GEI, Reporte Final para CONACYT-SENER Sustentabilidad—Energética*; Proyecto No. 117808; Centro de Transporte Sustentable: Mexico City, Mexico; Instituto de Energías Renovables: Temixco, Mexico; Universidad Nacional Autónoma de México: Mexico City, Mexico, 2013.

37. García, C.A.; Riegelhaupt, E.; Ghilardi, A.; Skutsch, M.; Islas, J.; Manzini, F.; Masera, O. Sustainable bioenergy options for Mexico: GHG mitigation and costs. *Renew. Sustain. Energy Rev.* **2015**, *43*, 545–552. [CrossRef]

38. Secretaría de Energía (SENER). *Balance Nacional de Energía 2010*; Secretaría de Energía (SENER): Mexico City, Mexico, 2011.

39. Centro de Estudios Sociales y de Opinión Pública (CESOP). Los Vehículos Usados de Procedencia Extranjera en México. Cámara de Diputados, LX Legislatura, México. 2012. Available online: https://www.google.com/url?sa=t&rct=j&q=&esrc=s&source=web&cd=1&ved=2ahUKEwj9z5-Wt9vlAhVHnp4KHQ6nCoEQFjAAegQIARAC&url=http%3A%2F%2Fwww3.diputados.gob.mx%2Fcamara%2Fcontent%2Fdownload%2F293625%2F956425%2Ffile%2FVehiculos-usados-extranjeros-docto142.pdf&usg=AOvVaw0nimoK2k_w27RMfxejLk6a (accessed on 10 October 2019).

40. Intergovernmental Panel on Climate Change (IPCC). Guidelines for National Greenhouse Gas Inventories. Chapter 2: Stationary Combustion. 2006. Available online: https://www.ipcc-nggip.iges.or.jp/public/2006gl/pdf/2_Volume2/V2_2_Ch2_Stationary_Combustion.pdf (accessed on 12 June 2018).

41. Centro de Transporte Sustentable de México (CTS). *Modelo de Evaluación Económica para Norma de Eficiencia en Vehículos Ligeros*; EMBARQ-CTS México: Mexico City, Mexico, 2011.

42. Federal Register. Rules and Regulations. On Line via the Government Publishing Office; 2011. Available online: https://www.epa.gov/regulations-emissions-vehicles-and-engines/regulations-greenhouse-gas-emissions-passenger-cars-and (accessed on 3 October 2019).

43. Haro, R.A.; Ibarrola, J.L. Cálculo de la elasticidad precio de la demanda de gasolina en la zona fronteriza norte de México. *Gac. Econ.* **2000**, *6*, 237–262.

44. Secretaría del Medio Ambiente y Recursos Naturales (SEMARNAT). Importación Definitiva de Vehículos Usados. Consecuencias e Impactos Ambientales. 2008. Available online: http://207.248.177.30/mir/uploadtests/24183.177.59.1.Importaci%C3%B3n_de_veh%C3%ADculos_usados_Versi%C3%B3n_Noviembre_2008.pdf (accessed on 31 May 2013).

45. Secretaría de Desarrollo Social, Consejo Nacional de Población (SEDESOL-CONAPO). Catálogo. Sistema Urbano Nacional 2012. 2012. Available online: https://www.gob.mx/cms/uploads/attachment/file/112772/Catalogo_Sistema_Urbano_Nacional_2012.pdf (accessed on 15 March 2015).

46. Gobierno de Querétaro. *Plan Integral de Transporte Colectivo de la Zona Metropolitana de Querétaro*; Reporte Técnico; Gobierno de Querétaro: Querétaro, México, 2005.

47. Macías, J.; Martínez, H.; Unal, A. *Bus Technology Analysis*; CTS-EMBARQ: Mexico City, Mexico, 2007.

48. Gobierno de la Ciudad de México. Operación. Cronología del Metro. 2019. Available online: https://www.metro.cdmx.gob.mx/cronologia-del-metro (accessed on 26 September 2019).

49. Environmental Protection Agency (EPA). Draft Regulatory Impact Analysis. Proposed Rulemaking to Establish Greenhouse Gas Emissions Standards and Fuel Efficiency Standards for Medium- and Heavy-Duty Engines and Vehicles. 2010. Available online: https://tinyurl.com/y3mvr9re (accessed on 30 September 2019).

50. Centro de Transporte Sustentable de México (CTS). *Hacia Ciudades Competitivas Bajas en Carbono*; Centro de Transporte Sustentable: Mexico City, Mexico, 2011.

51. Secretaría de Desarrollo Social; Consejo Nacional de Población; Instituto Nacional de Estadística; Geografía e Informática (SEDESOL-CONAPO-INEGI). Delimitación de las Zonas Metropolitanas de México 2005. 2007. Available online: http://www.conapo.gob.mx/work/models/CONAPO/zonas_metropolitanas/completoZM2005.pdf (accessed on 20 October 2019).

52. Secretaría de Desarrollo Social; Consejo Nacional de Población; Instituto Nacional de Estadística; Geografía e Informática (SEDESOL, CONAPO, INEGI). Delimitación de las Zonas Metropolitanas de México 2010. 2012. Available online: https://www.gob.mx/cms/uploads/attachment/file/112786/1_DZM_2010_PAG_1-34.pdf (accessed on 15 October 2019).

53. Instituto del Fondo Nacional de la Vivienda para los Trabajadores (INFONAVIT). Infonavit. Resultados y Perspectivas. Programas y Acciones de Financiamiento al Sector de la Vivienda. 2011. Available online: http://www.shcp.gob.mx/ApartadosHaciendaParaTodos/banca_desarrollo/vivienda/5_infonavit_11012011_perspectivas_sector_vivienda.pdf (accessed on 23 October 2013).

54. Instituto del Fondo Nacional de la Vivienda para los Trabajadores (INFONAVIT). Plan Financiero 2016–2020. 2015. Available online: https://portal.infonavit.org.mx/ (accessed on 23 October 2016).

55. Instituto Nacional de Geografia y Estadistica (INEGI). Encuesta Origen-Destino de la Zona Metropolitana del Valle de Mexico 2007. 2007. Available online: https://es.slideshare.net/borisdahl/encuesta-origen-destino-zmvm-2007 (accessed on 12 March 2014).

56. Secretaría de Comunicaciones y Transporte (SCT). Programa Nacional de Infraestructura 2007–2012. 2007. Available online: http://www.sct.gob.mx/fileadmin/ProgramaNacional/pni.pdf (accessed on 12 November 2018).

57. Secretaría de Comunicaciones y Transportes (SCT). Anuario Estadístico Ferroviario 2010. 2011. Available online: http://www.sct.gob.mx/fileadmin/DireccionesGrales/DGTFM/Anuarios_DGTFM/Anuarios_pdf/Anuario_2010.pdf (accessed on 20 August 2012).

58. Ferromex. Equipos que Utilizamos para Mover tu Carga. Ferromex, 2015. Available online: https://www.ferromex.com.mx/ferromex-lo-mueve/flota.jsp (accessed on 15 May 2015).

59. Secretaría de Comunicaciones y Transportes (SCT). Capítulo 4. Modos de Transporte. 2000. Available online: https://www.puertoensenada.com.mx/upl/sec/Capitulo_04_Modos_de_Transporte.pdf (accessed on 31 August 2015).

60. Secretaría de Medio Ambiente y Recursos Naturales (SEMARNAT). Guía del Taller Transportista Eficiente (Como Ahorrar Diesel en el Autotransporte). 2010. Available online: https://www.gob.mx/cms/uploads/attachment/file/210345/Transportista_eficiente.pdf (accessed on 25 November 2015).

61. Secretaría de Comunicaciones y Transportes; Instituto Mexicano del Transporte (SCT-IMT). Base de Datos del Estudio Estadístico de Campo del Autotransporte Nacional. 2011. Available online: https://www.imt.mx/micrositios/seguridad-y-operacion-del-transporte/estadisticas/consulta_del-eecan.html (accessed on 26 November 2015).

62. Chen, L.; Stone, R.; Richardson, D. A study of mixture preparation and PM emissions using a direct injection engine fuelled with stoichiometric gasoline/ethanol blends. *Fuel* **2012**, *96*, 120–130. [CrossRef]

63. US DOE. Vehicle Technologies Office. 2014. Available online: https://www.energy.gov/eere/vehicles/vehicle-technologies-office (accessed on 30 November 2014).

64. Electric Power Research Institute (EPRI). Environmental Assessment of Plug-In Hybrid Electric Vehicles. Volume 1: Nationwide Greenhouse Gas Emissions. 2007. Available online: https://www.energy.gov/sites/prod/files/oeprod/DocumentsandMedia/EPRI-NRDC_PHEV_GHG_report.pdf (accessed on 10 January 2013).

65. Bandivadekar, A.P. Evaluating the Impact of Advanced Vehicle and Fuel Technologies in US Light-Duty Vehicle Fleet. Ph.D. Thesis, Massachusetts Institute of Technology, Cambridge, MA, USA, 2008.

66. Electric Power Research Institute (EPRI). Comparing the Benefits and Impacts of Hybrid Electric Vehicle Options. 2001. Available online: http://www.ourenergypolicy.org/wp-content/uploads/2011/11/2001_07_EPRI_ComparingHybridElectricVehicleOptions.pdf (accessed on 15 January 2013).

67. Carlsson, F.; Johansson-Stenman, O. Costs and Benefits of Electric Vehicles. A 2010 Perspective. *J. Transp. Econ. Policy* **2003**, *37*, 1–28.

68. McKinsey Global Institute. Resource Revolution: Meeting the World's Energy, Materials, Food, and Water Needs. 2011. Available online: https://www.mckinsey.com/~{}/media/McKinsey/Business%20Functions/Sustainability/Our%20Insights/Resource%20revolution/MGI_Resource_revolution_full_report.ashx (accessed on 20 January 2013).

69. International Energy Agency (IEA). *World Energy Outlook 2007*; International Energy Agency (IEA): Paris, France, 2007.

70. International Energy Agency (IEA). *World Energy Outlook 2009*; International Energy Agency (IEA): Paris, France, 2009.

71. International Energy Agency (IEA). *World Energy Outlook 2010*; International Energy Agency (IEA): Paris, France, 2010.
72. US National Research Council (NRC). *Transitions to Alternative Transportation Technologies—Plug-in Hybrid Electric Vehicles*; US National Research Council (NRC): Washington, DC, USA, 2010.
73. Grande-Acosta, G.; Islas-Samperio, J. Towards a Low-Carbon Electric Power System in Mexico. *Energy Sustain. Dev.* **2017**, *37*, 99–109. [CrossRef]

Article

Exploring the "Energy-Saving Personality Traits" in the Office and Household Situation: An Empirical Study

Qian-Cheng Wang [1,†], Yi-Xuan Wang [2,†], Izzy Yi Jian [2,3,†], Hsi-Hsien Wei [2,*], Xuan Liu [2,*] and Yao-Tian Ma [4]

[1] Department of Land Economy, University of Cambridge, Cambridge CB3 9EP, UK; qw250@cam.ac.uk
[2] Department of Building and Real Estate, The Hong Kong Polytechnic University, Hong Kong 518000, China; wendyyixuan.wang@connect.polyu.hk (Y.-X.W.); yi.jian@connect.polyu.hk (I.Y.J.)
[3] Research Institute of Sustainable Urban Development, The Hong Kong Polytechnic University, Hong Kong 518000, China
[4] Department of Architecture, University of Cambridge, Cambridge CB2 1PX, UK; my361@cam.ac.uk
* Correspondence: hsi-hsien.wei@polyu.edu.hk (H.-H.W.); irene.l.liu@connect.polyu.hk (X.L.)
† These authors contributed equally to this study.

Received: 11 May 2020; Accepted: 6 July 2020; Published: 9 July 2020

Abstract: Behavior-driven energy conservation has been a promising strategy for reducing building energy consumption as well as carbon emissions. With the intention of revealing the impacts of an individual's personality basis on energy conservation behavioral attitudes and intentions in households and offices, the present study proposes and conducts an experiment in Xi'an, China with two groups for the investigation of such attitudes towards household energy-saving behavior (HESB) and office energy-saving behavior (OESB), respectively. The research adopts structural equation modeling for experiment data analysis. The analysis results suggest that the two personality traits, Agreeableness and Neuroticism, are significantly related to both HESB and OESB attitudes. Especially, agreeable people tend to present stronger energy-saving attitudes, while individuals with higher Neuroticism are less likely to do so. The results indicate that the impacts of these two traits on energy-saving attitude are found to be less influenced by different environment settings. Further, the results find that Extraversion positively influences energy-saving attitude in the office environment, while Openness only significantly works in the household environment. It is hoped that the findings of the present study can provide informative references to energy-saving intervention design as well as further studies on the spillover of pro-environmental behaviors.

Keywords: energy-saving; attitude; Big Five; personality traits; office; household; pro-environment

---

1. Introduction

Energy security plays a critical role in social, economic and environmental development [1]. An adequate energy supply can not only reduce the risk of political extortion, but also control the cost of industrial development. Besides, energy consumption is closely related to greenhouse gas (GHG) emissions as well as a series of environmental pollution. In addition to developing novel energy-efficient techniques [2], in the last few decades, governments have paid much attention to efficient energy consumption and resource conversation. More and more governments implement policies to encourage business owners and households to use equipment with higher energy efficiency. The European Union, for example, has set up several directives and projects to reduce energy consumption by 20% [3]. Through these energy policies, these countries or regions have not only reduced their dependence on energy imports, but also effectively promoted environmental sustainability.

Pro-environmental behaviors (PEBs) refer to "individual behaviors to enhance environmental sustainability" [4]. Promoting energy conservation and other PEBs seems a promising strategy for sustainable development: compared with other means [5], behavior-driven approaches generally work with less initial investments and quick returns [6]. Behavior-driven energy-saving solutions have been well considered in several green building schemes [7]. PEBs can be predicted by some psychological and demographic items [8–10]. The theory of personality trait, for example, explains the highly stable individual differences in PEB intentions [11,12]. Personality traits refer to factors reflecting individual characteristic patterns of thoughts as well as feelings and further reflecting individual behavioral patterns [13]. Personality traits might influence one's attitude towards PEBs [14–16], and then further contribute to sustainable intentions and behaviors directly or indirectly [17,18]. Some studies support that people with some specific personality characteristics (so-called "green personalities") tend to present stronger attitudes towards PEBs and have more environmental protection potential [12]. This is also in line with some studies on PEB spillover: people performing one PEB have higher likelihoods of performing other PEBs [19,20]. Based on the personality trait theories, research works have further made use of personality characteristics as well as interventions in PEB promotion [21,22].

However, there is growing evidence that people may present different intentions towards different PEBs and in different environments [9,23]. Wells et al. [9], for example, found the spillover effect of different resource conservation behaviors in different situations less significant. Especially, people present a remarkable difference between their PEB intention inside and outside their homes. Tudor et al. [24] also reported the connections and differences between PEBs at work and at home, which supported the above statement. Besides, people with similar personality characteristics might present individual differences in different PEBs. For example, Tang and Lam [25] noted that agreeable people who consider more about others' feelings are more likely to pay for green hotels consuming less energy and resources. However, Kamal and Barpanda [26] found that not agreeable but extroverted students tend to save energy at school. Shen et al. [27] found the contribution of both of the above two personality traits to household energy-saving behavior (HESB) insignificant. Thorough understanding of the individual difference in energy-saving attitudes and intentions at different situations is important for effective energy policy-making as well as energy-saving intervention development. Thus, it is necessary to reveal the contribution of "energy-saving personality traits" to the mentioned individual differences.

This study aims to reveal the personality basis of individual differences in attitudes as well as intentions towards energy conservation inside and outside the family environment. There are only a few pro-environmental behavior studies focusing on the developing areas. To bridge this gap, this study, however, conducts the field experiment with 800 participants in Xi'an, a typical city in the northwest region, the most undeveloped area in China. Participants are divided into two groups focusing on HESB and office energy-saving behavior (OESB), respectively. HESB is a typical respective of in-home pro-environmental behavior (IHPEB), which refers to "individual actions for environmental protection at households or private spaces". OESB represents out-of-home pro-environmental behavior (OHPEB), which refers to "individual actions for pro-environmental purpose in public spaces". The study first reviews the recent literature on the relationship between personality traits and PEBs in different situations and introduces the design and conduct process of the experiment in detail. This research then analyzes the experiment data with the structural equation modeling (SEM) technique. Based on the analysis results, the paper discusses the potential causes of the results, the policy implications of the findings and the limitations of this study. This study investigates the influence of personality traits on energy-saving attitude in office and household environments. The findings of this study would provide a personality explanation of the individual differences in energy conservation attitudes in different situations. The findings would be helpful in energy-saving intervention and policy-making.

2. Literature Review

Broadly speaking, personality traits refer to the highly stable, individual characteristic set of behaviors, cognitions and emotional patterns that evolve from biological and environmental factors [28].

Personality traits can influence attitude directly [29]. Several studies have provided sufficient evidence supporting the significant contribution of personality traits to PEBs directly [30] or mediated by attitude [12,31].

The Big Five personality trait theory (also called Five-Factor Model of Personality, FFM) is one of the most commonly used models in the psychological field [32]. The FFM comprises five personality traits: Extraversion (E), Agreeableness (A), Conscientiousness (C), Neuroticism (N) and Openness (O). Extraversion refers to the individual tendency to be outgoing, energetic and assertive to the outer world [33]. Thus, extraverted individuals often present a positive attitude to social connection engagement and show the ability to garner energy from socializing. Agreeableness is a trait manifesting itself in individual behavioral characteristics that are perceived as kind, sympathetic and cooperative [34]. Agreeableness individuals are caring for the well-beings of others and present strong humility and trustfulness. Besides, Conscientiousness is defined as the tendency of being organized and obeying obligations and goals [34]. Therefore, Conscientiousness individuals are more likely to accomplish assigned tasks and respect for the disciplines that encourage organizational achievements [35]. Neuroticism refers to the individual tendency to experience negative emotions [34]. Individuals with higher scores on Neuroticism tend to present lower emotional stability. Openness is the tendency to embrace knowledge, interest in generating novel configurations within practices and appreciation for variety thinking and experiences.

The literature so far critically employed the FFM in the field of PEBs, such as energy-saving [27,36,37], recycling [38], paying for green hotels [25] and sustainable tourism [31,39]. Additionally, there are several studies connecting the Big Five personality traits with PEB-related psychological factors. For instance, some previous studies linked Big Five personality traits to attitude towards PEBs [14,15,40].

Considerable efforts have also been made to explore the impacts of personality traits on the IHPEBs. Several studies indicate that Agreeableness, Conscientiousness as well as Openness are potentially related to attitude and intention towards household PEBs. Markowitz et al. [17], for example, found significant relationships between the three personality traits (i.e., Agreeableness, Conscientiousness and Openness) and the HESB intention. Busic-Sontic and Brick [41] noted that individuals with higher Openness tend to accept green household installations in the UK, which is also supported by He and Veronesi's [42] study on household renewable energy technology adoption in mainland China. Shen et al. [27] argued that Conscientiousness presents the most consistency in its correlation with HESB, and Agreeableness and Openness also present positive contributions. Swami et al. [43] and Zhang et al. [44] believe that Conscientiousness is an important predictor of household waste management behavior. However, empirical findings are inconsistent. For example, both Brick and Lewis [12] and White and Hyde [45] found the relationship between Conscientiousness and household pro-environmental behavior less significant. Further, Swami et al. [43] suggested that the link between Agreeableness and in-home waste management is insignificant. The roles of Extraversion and Neuroticism, the other two personality traits, in the in-home PEB process seem unclear. Limited evidence suggested that Neuroticism is weakly but positively related to household sustainable installation behavior while the contribution of Extraversion presents as negative [41]. Individuals with high Extraversion scores are more likely to turn off lights when nobody is at home. The same law applies to individuals with lower Neuroticism [27].

Further, several personality studies focus on the OHPEBs. There is growing evidence showing that Agreeableness and Extraversion potentially contribute to out-of-home PEBs. For instance, Sun et al. [15] and Luchs [46] indicate that both Agreeableness and Extraversion positively affect the attitude towards green buying. This is in unison with the conclusion of Tang and Lam [25] who provide empirical evidence confirming the positive relationship when examining people's willingness to pay for green hotels. Yazdanpanah and Hadji Hosseinlou [47] found that extraverted people are more likely to choose public transport means. They, together with agreeable individuals, tend to present higher acceptability of sustainable transport policies, where higher levels of trust in the government would be a potential explanation [48]. Kvasova [39] indicates that both Agreeableness and Extraversion positively contribute to pro-environmental tourist behaviors in Cyprus. However, there also exist some different statements. For example, Passafaro et al.'s study [31] on sustainable tourism found that neither Extraversion nor Agreeableness directly contributes to pro-environmental attitudes or OHPEBs.

Yet, there exists little consensus about the relationships between the other three traits (i.e., Openness, Neuroticism and Conscientiousness) and the OHPEBs. Sun et al. [15] found that Openness and Conscientiousness also positively contribute to green consumption intention. Yazdanpanah and Hadji Hosseinlou [47] noted that Neuroticism presents a negative relationship with public transport choosing intention while the contributions of Agreeableness, Openness and Conscientiousness seem insignificant. Kvasova [39] noted that Neuroticism and Conscientiousness are positively associated with green tourist behavior while the link between Openness and the OHPEB is less significant. Chiang et al. [49] found that emotional stability positively contributes to OHPEBs. These findings highlight the complex attribute of personality traits. Further investigations on the roles of the above-mentioned three traits are necessary.

Previous research has well explored the role of the Big Five personality traits in pro-environmental attitudes and behaviors. Many studies have thoroughly analyzed the impacts of personality traits on various typical environmental-friendly attitudes or behaviors such as garbage recycling, saving resources and purchasing green products. The results of these studies indicate that personality traits have different influences on different environmental-friendly behaviors. For example, in three personality studies on energy conservation behavior, Shen et al. [27], Tiefenbeck et al. [50] and Markowitz et al. [17] presented different findings with different backgrounds. Further, participants in two studies on recycling, by Poškus and Žukauskienė [38] and Swami et al. [43], also present different attitudes with different environments (i.e., school and family). These findings show that the role of personality traits on the same pro-environment behavior in different environments might be different, thus requiring further explorations. However, there is only a limited number of studies focusing on the energy-saving attitudes in different environments and their personality explanation. Based on the above literature review, this study aims to reveal the personality basis of individual differences in energy-saving attitude in family and office environments. The researchers propose a theoretical framework connecting the Big Five personality traits with PEB intentions via attitudes. Further, the study puts several hypotheses forward on HESB and OESB, respectively. These hypotheses are shown in Table 1.

Table 1. The hypotheses put forward in the study.

	Hypothesis
H_1	Agreeableness contributes to HESB by positively affecting attitude towards HESB
H_2	Conscientiousness contributes to HESB by positively affecting attitude towards HESB
H_3	Openness contributes to HESB by positively affecting attitude towards HESB
H_4	Extraversion contribute to HESB by negatively affecting attitude towards HESB
H_5	Neuroticism contribute to HESB by negatively affecting attitude towards HESB
H_6	Agreeableness contributes to HESB by positively affecting attitude towards OESB
H_7	Conscientiousness contributes to HESB by positively affecting attitude towards OESB
H_8	Openness contributes to HESB by positively affecting attitude towards OESB
H_9	Extraversion contribute to HESB by negatively affecting attitude towards OESB
H_{10}	Neuroticism contribute to HESB by negatively affecting attitude towards OESB

The hypotheses and relationships between variables in the research framework are presented graphically in Figure 1. Figure 1a shows the research model for HESB, while Figure 1b shows the research model for OESB.

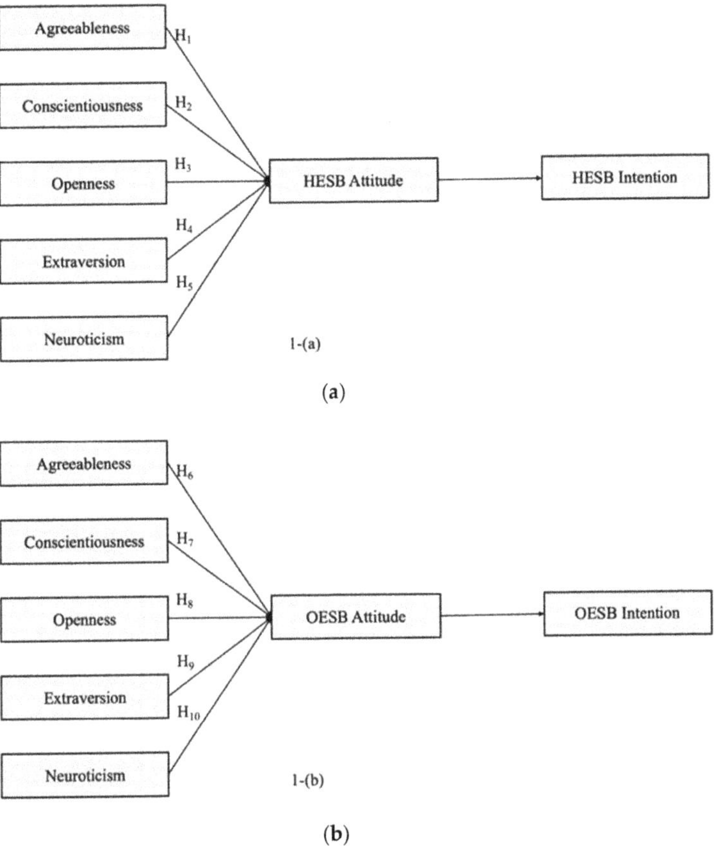

Figure 1. The research model. (**a**) The research model for HESB; (**b**) The research model for OESB.

3. Methodology

3.1. Study Overview

This research conducted a field experiment in Xi'an, one of the most densely populated cities located in Shaanxi Province, northwest China, and selected two typical residential communities in Yan-Ta district (urban area in Xi'an) as the study areas. Notably, these two communities are located in the same district and the residents in the selected study areas share similar socioeconomic characteristics. Especially, the participants have similar cultural and economic backgrounds as well as a built surrounding environment. Before substantial data gathering, the researchers invited five experts to review as well as revise the questionnaire in terms of its structure, wording and comprehensibility to make sure the questionnaire was understandable for the general public. These experts included two researchers in psychology, two experienced community workers in the study areas, as well as one property manager. The revised questionnaire was pilot-tested by 30 individuals to give revision suggestions to the final version of the questionnaire.

In this study, the community workers played the role as the gatekeeper. The study used the two-stage sampling method. Residents in the selected communities were encouraged to participate in this research by the staff in the neighborhood committee via a Wechat group to maximize the sample size. The researchers called for participants in the study areas through the local residential committee and there were 853 households signed-up. The researchers selected 800 households from the signed-up group. The selection was based on two standards: (1) the participant had lived in the investigation sites for at least six months; and (2) the participant had a steady job and had worked at the current company for at least six months so that they are familiar enough with the environments.

The participants in the experiment were purposely divided into two groups (i.e., Group A and Group B) according to their sociodemographic backgrounds to maximize the representativeness of the diverse background and to reduce the possibility of sampling bias. The respondents in different groups were required to complete the different questionnaires with different themes. The first version of the questionnaire concerns HESB (so called HESB-Q) and was prepared for Group A. The second version of the questionnaire concerns OESB (so called OESB-Q) and was prepared for Group B.

The responses from the participants were collected during January to March 2020. Due to the coronavirus (i.e., COVID-19) outbreak in China, the researchers adopted online tools to avoid face-to-face interaction from late January. Especially, the researchers employed Wechat, one of the most popular instant message Apps for consumer smart devices (i.e., smart phones and tablets) in China, to distribute the questionnaire and to collect the responses. The high usage rate makes Wechat (7.0.12, Tencent, Shenzhen, China) a suitable tool to approach potential participants in Chinese cities [8,43,51]. Sun et al. [51] provide sufficient evidence showing the feasibility of questionnaire distribution with Wechat. In this study, potential respondents were sent the online questionnaire which is able to be opened by clicking the link or scanning a QR-code on a prepared survey card to reach the online questionnaire. The structural equation modeling (SEM) approach is then adopted to delineate the causal mechanisms and numerous relationships between personality traits and different types of energy-saving behaviors in this study.

3.2. Questionnaire Design

Based on an extensive review of the relevant literature [6,52,53], the questionnaire survey was chosen as the main data instrument to solicit the opinions on the subject matter. In order to encourage participation, the questionnaire was completely anonymous and voluntary, respondents were assured of the confidentiality of their data [54]. No reminder letter was sent to the respondents during the data collection process. Vouchers were given to participants who completed the questionnaire and provided valid feedback successfully. Both of the questionnaires for Group A and Group B encapsulated three sections.

The first sections in the two versions are the same, where the researchers employed the Chinese version of the Ten-Item Personality Inventory (i.e., TIPI-C) to evaluate the personality characteristics of the participants. The TIPI is a short inventory for the Five-Factor Personality assessment with only 10 items. The TIPI inventory was developed by Gosling, Rentfrow and Swann [34] and has been widely employed in psychological studies on PEBs [17,55]. Lavelle-Hill et al. [56], for example, employed the TIPI in a study on plastic bag consumption. In addition, Komatsu and Nishio [37] also adopted the Japanese version of the TIPI in a study on HESB. There are several studies on the validity and reliability of the Chinese form of TIPI [57,58]. All items in the first section used a 5-point Likert rating scale (i.e., 1 = Strongly Disagree, 2 = Disagree, 3 = Neutral, 4 = Agree, 5 = Strongly Agree).

In the second section, the two questionnaires focus on the attitudes towards household and office energy-saving behavior, respectively. Each of them includes four items using the 5-point Likert rating scale. The items employed in the questionnaires for the two groups are presented in Table 2. Besides, the researchers also provided a clear definition of energy-saving behaviors as well as some typical HESBs and OESBs (see Table 3) to the participants. The HESB-Q categorizes sixteen typical HESBs into four family scenes: kitchen, living room, bedroom and bathroom. The OESB-Q categorizes nine typical HESBs into two office scenes: office desk and pantry. The questionnaires then tested the energy-saving behavioral intentions of the participants in different scenes. Different from the first section, the items as well as the checklist in this section are original.

The final section was designed to collect the socio-demographic characteristics of the respondents. The items in the third section include questions about gender, age, education level and income level.

Table 2. The items in Section 2 of the Uniform Questionnaire

Version	Group	Item	Code
HESB-Q	A	HESBs are valuable for environmental protection;	ATT-1A
		HESBs are important for environmental protection;	ATT-2A
		HESBs are wise actions;	ATT-3A
OESB-Q	B	OESBs are valuable for environmental protection;	ATT-1B
		OESBs are important for environmental protection;	ATT-2B
		OESBs are wise actions;	ATT-3B

Table 3. The household energy-saving behavior (HESB) and office energy-saving behavior (OESB) checklist.

Version	Scene	Item
HESB-Q	Kitchen	Keep the fridge door closed after taking food. Reduce the flame when boiling starts. Cool down hot food before storing in the fridge. Allow some space all around the fridge and keep the fridge far from the heater. Not overfill the fridge and allow some gaps between fridge and food.
	Living room	Turn off the television when not in use. Close windows and doors when using heating or cooling system. Set the heating system below 20 °C in winter and set the air-condition around 25 °C in summer. Set computer on energy-saving mode.
	Bedroom	Use natural light instead of artificial light in the daytime. Use task lighting before plan to sleep. Turn off light when sleeping. Turn off the air-conditioning system when leaving the room.
	Bathroom	Take a shower rather than a bath. Control the showering time. Turn on the water heater only when necessary.
OESB-Q	Office Desk	Set computer on energy-saving mode when leaving for a short time and switch off the computer when leaving for a long time. Close windows and doors when using heating or cooling system. Set the heating system below 20 °C in winter and set the air-condition around 25 °C in summer. Use task lighting for activities requiring a small amount of focus light. Turn off air-conditioning and light when leaving the office.
	Pantry	Switch on the water heater only when necessary and turn it off when not use. Cool down hot food before storing in the fridge. Turn off light when leaving. Heat enough water without too much unused.

3.3. Respondents Profile

The study distributed 800 questionnaires in the two groups. A few participants declined to participate or filled in the questionnaire incompletely. In total, the researchers collected 753 responses, 683 (i.e., 90.70%) of them were valid. There were 335 valid responses from Group A, where 168 were male (i.e., 50.51%) and 167 were female (i.e. 49.85%). Group B has a valid sample of 348: the numbers of male and female were 177 (i.e., 50.86%) and 171 (i.e., 49.14%), respectively. In Group A, 219 (i.e., 65.37%) respondents held a bachelor or higher-level degree. There were 235 (i.e., 48.56%) bachelor or higher-level degree holders in Group B. The annual income level of most participants ranged from CNY 50,000 to 150,000 per year. The respondent profile and the statistical data in Xi'an and mainland China are compared in Table 4 in detail. The sociodemographic information of Xi'an city was found from the 2010 population census of the People's Republic of China [59] as well as the statistical communique of Xi'an on national economic and social development in 2019 [60]. The national sociodemographic information comes from the 2010 population census of the People's Republic of China [59] and the People's Republic of China on national economic and social development in 2019 [61]. All the participants in this study had stable jobs. Thus, the percentages of adolescents and senior people are less than the national average level. Besides, the average income level of the participants was significantly higher than the national level for the same reason.

Table 4. Socio-demographic characteristics of respondents.

Item	Range	Group A		Group B		Statistic Data (Xi'an)	Statistic Data (China)
		Frequency	Percentage	Frequency	Percentage	Percentage	Percentage
Age	<18	5	1.49%	8	2.30%	25.32%	24.10%
	18–25	70	20.90%	67	19.25%	10.62%	9.56%
	26–30	79	23.58%	83	23.85%	7.87%	7.58%
	31–40	92	27.46%	88	25.29%	16.12%	16.14%
	41–50	51	15.22%	58	16.67%	15.45%	17.28%
	51–60	33	9.85%	35	10.06%	12.26%	12.01%
	>60	5	1.49%	9	2.59%	12.37%	13.32%
Gender	Male	168	50.15%	177	50.86%	51.26%	51.19%
	Female	167	49.85%	171	49.14%	48.74%	48.81%
Education Level	Secondary School or Below	27	8.06%	24	6.90%	N/A	N/A
	High School or equivalent	89	26.57%	89	25.57%	20.66%	15.02%
	Bachelor's degree or equivalent	165	49.25%	169	48.56%		
	Master's degree or equivalent	52	15.52%	62	17.82%	22.00%	9.53%
	Doctor's degree or above	2	0.60%	4	1.15%		
Income Level (CNY Per Year)	<30,000	48	14.33%	54	15.52%		
	30,000–50,000	34	10.15%	38	10.92%		
	50,000–100,000	93	27.76%	83	20.03%		
	100,000–150,000	71	21.19%	72	20.69%	The average disposable income of urban residents is 41,850 while for rural residents is 14,588.	The average disposable income of urban residents is 42,359 while for rural residents is 16,021.
	150,000–250,000	41	12.24%	48	13.79%		
	250,000–300,000	15	4.48%	15	4.31%		
	>300,000	12	3.58%	12	3.45%		
	N/A	21	6.27%	26	7.47%		

3.4. Data Analysis

This study employed structural equation modelling (SEM) for data analysis. SEM consists of two types, namely covariance-based SEM and partial least squares SEM (PLS-SEM) [62]. As a useful statistical tool for testing the formulated hypotheses, PLS-SEM was selected in this study to quantify the impacts of different constructs [63]. PLS-SEM has been widely employed in behavioral sciences-related research with the ability to handle non-normal data and avoid many restrictive data assumptions. For example, Liu et al. [8] adopted PLS-SEM to investigate the psychological factors influencing HESBs. Another example is that Tan [64] employed PLS-SEM to predict sustainable real estate purchasing intention with personal values and attitudes. Nomura et al. [65] also revealed the psychological driving force behind household recycling behavior. This study employs the software Smart-PLS 3 as a tool for data analysis.

The analysis process involves two steps: (1) assess the measurement model and (2) evaluate the structural model [8,66]. For the measurement modeling, composite reliability (CR), convergent validity (CV) and discriminant validity (DV) are the common criteria to indicate the model's validity and reliability. Normally, a satisfactory value for CR varies between 0.7 to 0.9 [66]. CV can be assessed with the values of average variance extracted (AVE) and the measurement items loadings.

AVE refers to a measure of the amount of variance which is captured by a latent construct in relation to the amount of variance due to measurement error [67]. The acceptable AVE value of each element should exceed 0.5 [63], while the measurement loading of a specific item should be larger than 0.4 [8,68]. DV assessment is used to confirm that each latent variable is not correlated with other latent variables [68]. The heterotrait–monotrait (HTMT) ratio is one of the common methods for DV measurement, and its index should not exceed 0.9 [8,69]. Using the bootstrap re-sampling technique, the path coefficients (*t*-value) and levels of significance (*p* value), construct reliability and validity (CR and CV) and discriminant validity (DV) were generated in the structural model [66,70]. The analysis results are discussed in the succeeding sections.

4. Results

In this section, the study first shows the measurement modeling results to ensure all measurement items and constructs are statistically reliable and valid for modeling, which includes two steps: (1) the CV and CR assessment, and (2) the DV measurement. The study analyzes the factor loading of each item as well as AVE, CR, CV and heterotrait–monotrait (HTMT) ratio. Then, the study conducts the structural modeling and analysis.

4.1. Group A: HESB

Tables 5 and 6 illustrate the results of the measurement modeling. Table 5 pays attention to the factor loadings as well as to two criteria in the CV and CR assessment (i.e., AVE and CR). The data show that the measurement loadings of all items are greater than or equal to 0.614 (i.e., >0.4). Further, the AVEs and CRs of the constructs well met the statistical requirements. Besides, Table 6 focuses on the DV assessment results by illustrating the HTMT ratio. The result analysis suggests that the measurement in this study fulfills the discriminant validity requirements.

Subsequent to the measurement modeling process, this study then conducts the PLS structural modelling analysis and illustrates the analysis results in Table 7. In addition to the statistical significance of the hypothesized relationships, the table also presents the sample mean and standard deviation (STDEV) of each relationship.

Table 5. The convergent validity (CV) and composite reliability (CR) assessment results of Group A.

Constructs	Item	FL	AVE	Composite Reliability
Attitude (ATT)	ATT-1A	0.644	0.586	0.810
	ATT-2A	0.771		
	ATT-3A	0.862		
Behavioral Intention (BI)	HESBI-1	0.720	0.546	0.767
	HESBI-2	0.739		
	HESBI-3	0.669		
	HESBI-4	0.614		
Extraversion (E)	E-1A	0.867	0.785	0.865
	E-2A	0.915		
Openness (O)	O-1A	0.884	0.656	0.788
	O-2A	0.728		
Agreeableness (A)	A-1A	0.949	0.598	0.721
	A-2A	0.712		
Conscientiousness (C)	C-1A	0.887	0.685	0.832
	C-2A	0.751		
Neuroticism (N)	N-1A	0.844	0.705	0.821
	N-2A	0.824		

Note: FL refers to factor loading; AVE refers to average variance extracted; HESBI refers to household energy-saving behavioral intention.

Table 6. Discriminant validity (DV) assessment results of Group A.

	ATT	HESBI	A	C	E	N	O
ATT							
HESBI	0.745						
A	0.539	0.762					
C	0.400	0.197	0.456				
E	0.204	0.075	0.604	0.619			
N	0.405	0.558	0.711	0.548	0.730		
O	0.524	0.430	0.558	0.316	0.527	0.463	

Note: ATT refers to attitude towards HESB; HESBI refers to household energy-saving behavioral intention; A refers to Agreeableness; C refers to Conscientiousness; E refers to Extraversion; O refers to Openness; N refers to Neuroticism.

Table 7. Structural modeling analysis results of Group A.

Hypothesis	β	Sample Mean	STDEV	*t*-Value	*p*-Value
H_1: A→HESB ATT	0.190	0.195	0.063	3.027	0.003 **
H_2: C→HESB ATT	0.045	0.050	0.070	0.647	0.518
H_3: O→HESB ATT	0.236	0.238	0.062	3.840	<0.001 ***
H_4: E→HESB ATT	−0.025	−0.018	0.060	0.421	0.674
H_5: N→HESB ATT	−0.154	−0.152	0.073	2.107	0.035 *
HESB ATT→HESBI	0.367	0.383	0.061	6.067	<0.001 ***

*: $p < 0.05$; **: $p < 0.01$; ***: $p < 0.001$.

The results suggest that HESB attitude is an effective predictor of HESB intention, for the correlation between HESB attitude and intention is significant (i.e., $\beta = 0.367$, $p < 0.001$). There are only two personality traits positively correlated with HESB attitude, including Agreeableness and Openness. It seems that Openness presents the strongest contribution (i.e., $\beta = 0.236$, $p < 0.001$) to HESB attitude among the five personality traits. Besides, Agreeableness has a positive effect (i.e., $\beta = 0.190$, $p = 0.003$) on HESB attitude as well. On the contrary, Neuroticism shows a significant but negative relationship (i.e., $\beta = -0.154$, $p = 0.035$) with HESB attitude. The analysis results indicate that the effect of Conscientiousness on HESB attitude is small and less significant (i.e., $\beta = 0.045$, $p = 0.518$). The negative influence of Extraversion on HESB attitude seems small and insignificant as well.

4.2. Group B: OESB

Tables 8 and 9 present the measurement modeling results of Group B. Similar to Table 5 in Section 4.1, Table 8 shows the factor loading of each item and the CV and CR assessment results. The factor loadings of the employed items vary between 0.708 and 0.949 and are significantly greater than 0.4. Further, the employed constructs show satisfactory AVE (i.e., no less than 0.5) as well as CR (i.e., between 0.7 and 0.9) levels. Table 9 presents the HTMT ratio of the constructs, which supports that the measurement meets the DV requirements.

Then, this study conducts the structural modeling analysis and illustrates the statistical significance of the hypothesized relationships in Table 10.

Similar to the HESB intention, the results indicate that OESB intention can be well predicted by OESB attitude. The relationship between OESB attitude and OESB intention is positive and significant (i.e., $\beta = 0.587$, $p < 0.001$). There are two personality traits positively contributing to OESB attitude as well: Agreeableness and Extraversion. Agreeableness presents a relatively strong and positive effect (i.e., $\beta = 0.181$, $p = 0.011$) on OESB attitude, while Extraversion presents a positive and significant relationship (i.e., $\beta = 0.139$, $p = 0.043$) with OESB attitude. By contrast, Neuroticism plays a negative role in OESB attitude. The relationship between Neuroticism and OESB attitude is negative but significant (i.e., $\beta = -0.133$, $p = 0.042$). The contribution of Conscientiousness and Openness seems positive but less significant.

Table 8. The CV and CR assessment results of Group B.

Constructs	Item	FL	AVE	Composite Reliability
Attitude (ATT)	ATT-1B	0.888	0.724	0.887
	ATT-2B	0.907		
	ATT-3B	0.748		
Behavioral Intention (BI)	OESBI-1	0.913	0.751	0.857
	OESBI-2	0.818		
Extraversion (E)	E-1B	0.712	0.693	0.816
	E-2B	0.938		
Openness (O)	O-1B	0.780	0.517	0.731
	O-2B	0.966		
Agreeableness (A)	A-1B	0.949	0.513	0.735
	A-2B	0.754		
Conscientiousness (C)	C-1B	0.896	0.674	0.804
	C-2B	0.738		
Neuroticism (N)	N-1B	0.708	0.620	0.748
	N-2B	0.920		

Note: FL refers to factor loading; AVE refers to average variance extracted; OESBI refers to office energy-saving behavioral intention.

Table 9. DV assessment results of Group B.

	ATT	OESBI	A	C	E	N	O
ATT							
OESBI	0.778						
A	0.766	0.781					
C	0.266	0.107	0.461				
E	0.276	0.102	0.682	0.389			
N	0.274	0.261	0.701	0.551	0.426		
O	0.270	0.207	0.556	0.270	0.763	0.625	

Note: ATT refers to attitude towards HESB; HESBI refers to household energy-saving behavioral intention; A refers to Agreeableness; C refers to Conscientiousness; E refers to Extraversion; O refers to Openness; N refers to Neuroticism.

Table 10. Structural modeling analysis results of Group B.

Hypothesis	β	Sample Mean	STDEV	*t*-Value	*p*-Value
H_6: A→OESB ATT	0.181	0.179	0.071	2.538	0.011 [a]
H_7: C→OESB ATT	0.034	0.043	0.061	0.555	0.579
H_8: O→OESB ATT	0.013	0.032	0.055	0.245	0.806
H_9: E→OESB ATT	0.138	0.139	0.068	2.022	0.043 *
H_{10}: N→OESB ATT	−0.133	−0.139	0.065	2.038	0.042 *
OESB ATT→OESBI	0.587	0.587	0.050	11.681	<0.001 ***

*: $p < 0.05$; ***: $p < 0.001$.

The analysis results suggest that two personality traits, Agreeableness and Neuroticism, play critical roles in general energy conservation behavior: those two traits are significantly related to both HESB and OESB attitude. However, Agreeableness plays a positive role in the energy-saving behavioral process, while the role Neuroticism plays is negative. Among all the traits, Openness represents the strongest predictor of HESB attitude. However, the effect of Openness on OESB attitude is small and less significant. Besides, Extraversion presents a significant correlation with OESB attitude, while the influence of Extraversion on HESB attitude seems less significant. The effect of Conscientiousness on both HESB and OESB attitude is insignificant.

5. Discussion

5.1. The Impacts of Personality Traits

The results of the current study suggest that the influence of Agreeableness and Neuroticism on energy conservation behavioral attitudes and intentions less depends on the external environment. These two "energy-saving personality traits" can work as predictors reflecting overall energy-saving attitudes. However, the relationships between some personality traits, such as Openness and Extraversion, and energy-saving personality only become significant with some specific external environment. Those personality traits may predict the energy-saving potentials of individuals in some specific situations. The study finds that Agreeableness and Neuroticism are significantly correlated with both HESB and OESB attitudes. These findings indicate that the impacts of Agreeableness and Neuroticism on energy-saving attitude might not depend on the surrounding environment. The positive correlation between Agreeableness and energy-saving attitude is in line with previous studies on the personality basis of PEBs (such as [15,25,27]). Agreeable people are more likely to present stronger empathic concern [71] and tend to consider collective interests and others' feelings. Individuals with higher Agreeableness levels might extend their empathic concerns to energy-saving. This is also supported by Berenguer [72] and Luch et al. [46], whose findings suggested that empathic concerns work positively in the environmental protection behavioral process. Surprisingly, Neuroticism presents significant but negative relationships with both HESB and OESB attitudes. This finding suggests that neurotic people present a weaker energy conservation attitude in both the office and the family environment. Individuals with higher Neuroticism show higher tendency to present negative emotional characteristics, such as sensitivity, nervousness and insecurity [73]. Thus, emotional stability might play the role as the predictor of energy conservation intention. This is also supported by Chiang et al. [49] and Van Egeren [74], who believe that Neuroticism positively contributes to the external locus of control (ELC) while ELC is negatively related to PEB attitudes. People with stronger emotional uneasiness are more likely to be influenced by ELC.

The study results demonstrate a significant correlation between Extraversion and OESB attitude, which is consistent with some previous evidence that supports the positive correlation between Extraversion and OHPEB attitudes and intentions (such as [25] and [48]). However, the study also found that the effect of Extraversion on IESB attitude is small and less significant. Extroverts generally present stronger sociability [75]. Individuals with higher Extraversion tend to engage in specific behaviors to integrate into the collective. Previous studies also found that Extraversion is one of the predictors of country-level environmental engagement [30]. Further, Kamal and Barpanda [26] note that Extraversion bridges social aspiration and environmental concern. Compared with the office, the family environment emphasizes privacy. At home, people receive fewer expectations and evaluations from others, and do not need to perform special actions for social engagement. This might partially explain the reason why people with higher Extraversion present higher energy-saving potential than others at the office but not at home.

Openness only presents a significant correlation with HESB attitude. Openness reflects the degree of intellectual curiosity, creativity and a preference for novelty and variety of an individual. Therefore, people with a higher level of openness-to-experience are more natural to accept environmental protection concepts. Besides, people with higher openness are more likely to be interested in novel environmental-friendly technologies [41]. At home, more open individuals have more opportunities to practice different environmental protection solutions or install novel environmental protection technologies. This might explain the significant correlation between Openness and household energy-efficient technology installation attitudes in two studies in the UK [41] and China [42]. However, office facilities are relatively monotonous with complex rules. It could be difficult for people to apply novel energy-saving ideas to the office environment. This may limit the willingness of more open employees to save energy at the office.

The results reveal that personality traits obviously influence the energy-saving process in both household and office environments. The contributions of some specific personality traits to the pro-environmental attitude are not dependent on the situation and environment. Those personality traits can well predict the overall environmental protection intentions as well as the potentials of individuals. However, some personality characteristics are only able to predict the pro-environmental behavior in some specific situations. In addition, the researchers found out and thus prospect that it would be valuable to customize environmental protection schemes in the future: the same strategies are not suitable to apply to each individual or in each situation. It would be better to develop specific informatic interventions according to the personality characteristics of targets as well as the surrounding environment. Although the evaluation of personality and other demographic characteristics of each individual is a challenge to decision-makers, the effectiveness and efficiency of those schemes would be highly strengthened by achieving the target for the individuals with the most substantial energy-saving potential and the targets who are able to save more with assistance. Besides, the findings in this study would provide references for behavior simulation in the built environment in further studies.

5.2. Limitations and Further Studies

This research has some limitations as well. First, the findings in this study cannot be generalized to the population in other settings. The field experiment in this study was conducted in Xi'an, China. Due to the different social and economic backgrounds, the results should be empirically validated with caution before applying in different contexts. Second, despite the efforts made by the researcher to maximize the potential respondents, the sample size for the study is relatively small. The sample sizes for the two groups were 335 and 348, respectively. Although the sample size meets the requirement of statistical data analysis, further studies would benefit from a larger sample size. Moreover, the selected study area is a typical urban area in a well-developed Chinese city, therefore the percentages of bachelor's or higher-level degree holders and high-income families in this study exceed the average level in China since the rural population constitutes approximately 45% in China. Follow-up studies are therefore recommended to take into account energy-saving studies in other regions and pay attention to rural areas, which are conducive to the comprehensiveness of the research. In addition, the authors recommend that future research would benefit from employing different personality inventories to avoid excessive two-variables-based latent constructs.

6. Conclusions

This study presents a field experiment in Xi'an, China to explore the personality basis of the individual difference between energy-saving attitudes in office and household environments. The SEM analysis results indicate that Agreeableness and Neuroticism are significantly correlated with energy-saving attitudes in both office and household environments. The finding suggests that the influence of these two personality traits on energy-saving attitude might not depend on the surrounding environment. By contrast, the external environment plays a more critical role in the relationship between energy-saving attitudes and the other two personality traits, Openness and Extraversion. Openness only positively contributes to energy-saving attitudes and intentions in the household environment, while Extraversion only presents a significant relationship to ones' energy-saving attitudes in the office. It is hoped that the findings would provide references for the energy-saving schemes as well as simulations in the future. The effectiveness of further energy-saving interventions would be much improved by targeting individuals with more energy-saving potentials. Besides, based on the findings in this study, further studies would explore the personality basis of the spillover effect of PEBs.

Author Contributions: Conceptualization, Q.-C.W. and Y.-X.W.; methodology, Q.-C.W.; software, Y.-X.W.; validation, Y.-X.W. and Y.-T.M.; formal analysis, Y.-X.W. and I.Y.J.; investigation, Y.-X.W. and Y.-T.M.; resources, Q.-C.W., I.Y.J. and X.L; data curation, I.Y.J.; writing—original draft preparation, Q.-C.W., Y.-X.W. and I.Y.J.; writing—review and editing, Q.-C.W., H.-H.W. and X.L.; visualization, I.Y.J. and Y.-T.M.; supervision, H.-H.W.

and X.L.; project administration, H.-H.W. and X.L. All authors have read and agreed to the published version of the manuscript.

Funding: The research project is partially supported by the China Scholarship Council (CSC).

Acknowledgments: The authors would like to thank the CSC for financial support. Besides, the authors would like to thank Meng Ni from the Hong Kong Polytechnic University who guided the project. Further, the authors would like to further acknowledge Jia-Yu Chen, Rui-Zhen Hu, Hao-Zhe Li, Yi-Nuo Su, Josephine Yim, Ya-Ni Zhang, Fang Lee and Justin Foo from the University of Cambridge as well as Frankie Leung from Harvard University. The project would not have completed without their warm encourage and invaluable suggestions.

Conflicts of Interest: The authors declare no conflicts of interests.

References

1. Winzer, C. Conceptualizing energy security. *Energy Policy* **2012**, *46*, 36–48. [CrossRef]
2. Yu, C.R.; Guo, H.S.; Wang, Q.C.; Chang, R.D. Revealing the Impacts of Passive Cooling Techniques on Building Energy Performance: A Residential Case in Hong Kong. *Appl. Sci.* **2020**, *10*, 4188. [CrossRef]
3. Augutis, J.; Krikstolaitis, R.; Martisauskas, L.; Peciulyte, S. Energy security level assessment technology. *Appl. Energy* **2012**, *97*, 143–149. [CrossRef]
4. Ones, D.S.; Wiernik, B.M.; Dilchert, S.; Klein, R. Pro-environmental behavior. In *International Encyclopedia of the Social & Behavioral Sciences*, 2nd ed.; Elsevier Inc.: Amsterdam, The Netherlands, 2015; pp. 82–88.
5. Chang, S.; Wang, Q.; Hu, H.; Ding, Z.; Guo, H. An NNwC MPPT-based energy supply solution for sensor nodes in buildings and its feasibility study. *Energies* **2019**, *12*, 101. [CrossRef]
6. Shen, M.; Lu, Y.; Kua, H.W.; Cui, Q. Eco-feedback delivering methods and psychological attributes shaping household energy consumption: Evidence from intervention program in Hangzhou, China. *J. Clean. Prod.* **2020**, *265*, 121755. [CrossRef]
7. Chang, R.; Wang, Q.; Ding, Z. How is the Energy Performance of Buildings Assessed in Australia?— A Comparison between four Evaluation Systems. *Int. J. Struct. Civil Eng. Res.* **2019**, *8*, 133–137. [CrossRef]
8. Liu, X.; Wang, Q.; Wei, H.-H.; Chi, H.-L.; Ma, Y.; Jian, I.Y. Psychological and Demographic Factors Affecting Household Energy-Saving Intentions: A TPB-Based Study in Northwest China. *Sustainability* **2020**, *12*, 836. [CrossRef]
9. Wells, V.K.; Taheri, B.; Gregory-Smith, D.; Manika, D. The role of generativity and attitudes on employees home and workplace water and energy saving behaviours. *Tour. Manag.* **2016**, *56*, 63–74. [CrossRef]
10. Yazdanpanah, M.; Forouzani, M.; Abdeshahi, A.; Jafari, A. Investigating the effect of moral norm and self-identity on the intention toward water conservation among Iranian young adults. *Water Policy* **2016**, *18*, 73–90. [CrossRef]
11. Busic-Sontic, A.; Czap, N.V.; Fuerst, F. The role of personality traits in green decision-making. *J. Econ. Psychol.* **2017**, *62*, 313–328. [CrossRef]
12. Brick, C.; Lewis, G.J. Unearthing the "green" personality: Core traits predict environmentally friendly behavior. *Environ. Behav.* **2016**, *48*, 635–658. [CrossRef]
13. Bak, W. Personality predictors of anger. The role of FFM traits, shyness, and self-esteem. *Pol. Psychol. Bull.* **2016**, *47*, 373–382. [CrossRef]
14. Pavalache-Ilie, M.; Cazan, A.-M. Personality correlates of pro-environmental attitudes. *Int. J. Environ. Health Res.* **2018**, *28*, 71–78. [CrossRef] [PubMed]
15. Sun, Y.; Wang, S.; Gao, L.; Li, J. Unearthing the effects of personality traits on consumer's attitude and intention to buy green products. *Nat. Hazards* **2018**, *93*, 299–314. [CrossRef]
16. Johansson, M.V.; Heldt, T.; Johansson, P. The effects of attitudes and personality traits on mode choice. *Transp. Res. Part A Policy Pract.* **2006**, *40*, 507–525. [CrossRef]
17. Markowitz, E.M.; Goldberg, L.R.; Ashton, M.C.; Lee, K. Profiling the "pro-environmental individual": A personality perspective. *J. Personal.* **2012**, *80*, 81–111. [CrossRef]
18. Ribeiro, J.D.A.; Veiga, R.T.; Higuchi, A.K. Personality traits and sustainable consumption. *Rev. Bras. Mark.* **2016**, *15*, 297–313. [CrossRef]
19. Xie, X.; Lu, Y.; Gou, Z. Green building pro-environment behaviors: Are green users also green buyers? *Sustainability* **2017**, *9*, 1703. [CrossRef]

20. Truelove, H.B.; Carrico, A.R.; Weber, E.U.; Raimi, K.T.; Vandenbergh, M.P. Positive and negative spillover of pro-environmental behavior: An integrative review and theoretical framework. *Glob. Environ. Chang.* **2014**, *29*, 127–138. [CrossRef]

21. Wang, Z.; Guo, D.; Wang, X.; Zhang, B.; Wang, B. How does information publicity influence residents' behaviour intentions around e-waste recycling? *Resour. Conserv. Recycl.* **2018**, *133*, 1–9. [CrossRef]

22. Bhushan, N.; Steg, L.; Albers, C. Studying the effects of intervention programmes on household energy saving behaviours using graphical causal models. *Energy Res. Social Sci.* **2018**, *45*, 75–80. [CrossRef]

23. Larson, L.R.; Stedman, R.C.; Cooper, C.B.; Decker, D.J. Understanding the multi-dimensional structure of pro-environmental behavior. *J. Environ. Psychol.* **2015**, *43*, 112–124. [CrossRef]

24. Tudor, T.; Barr, S.; Gilg, A. A tale of two locational settings: Is there a link between pro-environmental behaviour at work and at home? *Local Environ.* **2007**, *12*, 409–421. [CrossRef]

25. Tang, C.M.F.; Lam, D. The role of extraversion and agreeableness traits on Gen Y's attitudes and willingness to pay for green hotels. *Int. J. Contemp. Hosp. Manag.* **2017**, *29*, 607–623. [CrossRef]

26. Kamal, A.; Barpanda, S. Factors influencing the energy consumption behavior pattern among the Indian higher education institution students. In Proceedings of the 2017 International Conference on Technological Advancements in Power and Energy (TAP Energy), Kollam, India, 21–23 December 2017; pp. 1–6.

27. Shen, M.; Lu, Y.; Tan, K.Y. Big Five Personality Traits, Demographics and Energy Conservation Behaviour: A Preliminary Study of Their Associations in Singapore. *Energy Procedia* **2019**, *158*, 3458–3463. [CrossRef]

28. Phares, E.J. *Introduction to Personality*; Scott, Foresman & Co.: Glenview, IL, USA, 1988.

29. Balderjahn, I. Personality variables and environmental attitudes as predictors of ecologically responsible consumption patterns. *J. Bus. Res.* **1988**, *17*, 51–56. [CrossRef]

30. Milfont, T.L.; Sibley, C.G. The big five personality traits and environmental engagement: Associations at the individual and societal level. *J. Environ. Psychol.* **2012**, *32*, 187–195. [CrossRef]

31. Passafaro, P.; Cini, F.; Boi, L.; D'Angelo, M.; Heering, M.S.; Luchetti, L.; Mancini, A.; Martemucci, V.; Pacella, G.; Patrizi, F. The "sustainable tourist": Values, attitudes, and personality traits. *Tour. Hosp. Res.* **2015**, *15*, 225–239. [CrossRef]

32. Shen, M.; Lu, Y.; Wei, K.H.; Cui, Q. Prediction of household electricity consumption and effectiveness of concerted intervention strategies based on occupant behaviour and personality traits. *Renew. Sustain. Energy Rev.* **2020**, *127*, 109839. [CrossRef]

33. Robinson, O.C.; Lopez, F.G.; Ramos, K. Parental antipathy and neglect: Relations with Big Five personality traits, cross-context trait variability and authenticity. *Personal. Individ. Differ.* **2014**, *56*, 180–185. [CrossRef]

34. Gosling, S.D.; Rentfrow, P.J.; Swann, W.B., Jr. A very brief measure of the Big-Five personality domains. *J. Res. Personal.* **2003**, *19*, 139–152. [CrossRef]

35. Lay, C.; Kovacs, A.; Danto, D. The relation of trait procrastination to the big-five factor conscientiousness: An assessment with primary-junior school children based on self-report scales. *Personal. Individ. Differ.* **2020**, *25*, 187–193.

36. Bergmann, N.; Schacht, S.; Gnewuch, U.; Mädche, A. Understanding the Influence of Personality Traits on Gamification: The Role of Avatars in Energy Saving Tasks. In Proceedings of the 38th International Conference on Information Systems (ICIS), Seoul, South Korea, 10–13 December 2017.

37. Komatsu, H.; Nishio, K.-I. An experimental study on motivational change for electricity conservation by normative messages. *Appl. Energy* **2015**, *158*, 35–43. [CrossRef]

38. Poškus, M.S.; Žukauskienė, R. Predicting adolescents' recycling behavior among different big five personality types. *J. Environ. Psychol.* **2017**, *54*, 57–64. [CrossRef]

39. Kvasova, O. The Big Five personality traits as antecedents of eco-friendly tourist behavior. *Personal. Individ. Differ.* **2015**, *83*, 111–116. [CrossRef]

40. Pavalache-Ilie, M.; Cazan, A. Measuring ecological attitudes in a Romanian context. *Bull. Transilv. Univ. Bras.* **2016**, *9*, 85–90.

41. Busic-Sontic, A.; Brick, C. Personality trait effects on green household installations. *Collabra Psychol.* **2018**, *4*. [CrossRef]

42. He, P.; Veronesi, M. Personality traits and renewable energy technology adoption: A policy case study from China. *Energy Policy* **2017**, *107*, 472–479. [CrossRef]

43. Swami, V.; Chamorro-Premuzic, T.; Snelgar, R.; Furnham, A. Personality, individual differences, and demographic antecedents of self-reported household waste management behaviours. *J. Environ. Psychol.* **2011**, *31*, 21–26. [CrossRef]

44. Zhang, Y.; Wu, S.; Rasheed, M.I. Conscientiousness and smartphone recycling intention: The moderating effect of risk perception. *Waste Manag.* **2020**, *101*, 116–125. [CrossRef]

45. White, K.M.; Hyde, M.K. The role of self-perceptions in the prediction of household recycling behavior in Australia. *Environ. Behav.* **2012**, *44*, 785–799. [CrossRef]

46. Luchs, M.G.; Mooradian, T.A. Sex, personality, and sustainable consumer behaviour: Elucidating the gender effect. *J. Consum. Policy* **2012**, *35*, 127–144. [CrossRef]

47. Yazdanpanah, M.; Hadji Hosseinlou, M. The role of personality traits through habit and intention on determining future preferences of public transport use. *Behav. Sci.* **2017**, *7*, 8. [CrossRef] [PubMed]

48. Kim, J.; Schmöcker, J.-D.; Bergstad, C.J.; Fujii, S.; Gärling, T. The influence of personality on acceptability of sustainable transport policies. *Transportation* **2014**, *41*, 855–872. [CrossRef]

49. Chiang, Y.-T.; Fang, W.-T.; Kaplan, U.; Ng, E. Locus of control: The mediation effect between emotional stability and pro-environmental behavior. *Sustainability* **2019**, *11*, 820. [CrossRef]

50. Tiefenbeck, V.; Degen, K.; Tasic, V.; Goette, L.; Staak, T. On the Effectiveness of Real-time Feedback: The influence of Demographics, Attitudes and Personality Traits. Final report to the Swiss Fedearl Office of Energy, Bern. 2014. *Bits Energy Lab. Zur.* **2014**, *2*, 1–50.

51. Sun, Z.-J.; Zhu, L.; Liang, M.; Xu, T.; Lang, J.-h. The usability of a WeChat-based electronic questionnaire for collecting participant-reported data in female pelvic floor disorders: A comparison with the traditional paper-administered format. *Menopause* **2016**, *23*, 856–862. [CrossRef]

52. Milfont, T.L.; Milojev, P.; Greaves, L.M.; Sibley, C.G. Socio-structural and psychological foundations of climate change beliefs. *N. Zealand J. Psychol.* **2015**, *44*, 17–30.

53. Shen, M.; Cui, Q.; Fu, L. Personality traits and energy conservation. *Energy Policy* **2015**, *85*, 322–334. [CrossRef]

54. Kingsford Owusu, E.; Chan, A.P. Barriers Affecting Effective Application of Anticorruption Measures in Infrastructure Projects: Disparities between Developed and Developing Countries. *J. Manag. Eng.* **2018**, *35*, 04018056. [CrossRef]

55. Menardo, E.; Brondino, M.; Pasini, M. Adaptation and psychometric properties of the Italian version of the Pro-Environmental Behaviours Scale (PEBS). *Environ. Dev. Sustain.* **2019**, *21*, 1–24. [CrossRef]

56. Lavelle-Hill, R.E.; Smith, G.; Bibby, P.; Clarke, D.; Goulding, J. Psychological and Demographic Predictors of Plastic Bag Consumption in Transaction Data. In Proceedings of the 3rd World Conference on Personality, Hanoi, Vietnam, 2–6 April 2019.

57. Carciofo, R.; Yang, J.; Song, N.; Du, F.; Zhang, K. Psychometric evaluation of Chinese-language 44-item and 10-item big five personality inventories, including correlations with chronotype, mindfulness and mind wandering. *PLoS ONE* **2016**, *11*, e0149963. [CrossRef] [PubMed]

58. Peng, K.-H.; Liou, L.-H.; Chang, C.-S.; Lee, D.-S. Predicting personality traits of Chinese users based on Facebook wall posts. In Proceedings of the 2015 24th Wireless and Optical Communication Conference (WOCC), Taipei, Taiwan, 23–24 October 2015; pp. 9–14.

59. China, N.B.o.S.o. *The 2010 Population Census of the People's Republic of China*; Bureau, N.S., Ed.; National Bureau of Statistics of China: Beijing, China, 2010.

60. Xi'an, T.S.B.o. *Statistical Communique of Xi'an on National Economic and Social Development in 2019*; Xi'an Municipal Statistics Bureau: Xi'an, China, 2019.

61. China, N.B.o.S.o. *Statistical Communique of the People's Republic of China on National Economic and Social Development in 2019*; National Bureau of Statistics of China: Beijing, China, 2020.

62. Hair, J.F.; Ringle, C.M.; Sarstedt, M. PLS-SEM: Indeed a silver bullet. *J. Mark. Theory Pract.* **2011**, *19*, 139–152. [CrossRef]

63. Owusu, E.K.; Chan, A.P.; Hosseini, M.R. Impacts of anti-corruption barriers on the efficacy of anti-corruption measures in infrastructure projects: Implications for sustainable development. *J. Clean. Prod.* **2020**, *246*, 119078. [CrossRef]

64. Tan, T.H. Use of structural equation modeling to predict the intention to purchase green and sustainable homes in Malaysia. *Asian Soc. Sci.* **2013**, *9*, 181. [CrossRef]

65. Nomura, H.; Takahashi, Y.; Yabe, M. Psychological driving forces behind households' behaviors toward municipal organic waste separation at source in Vietnam: A structural equation modeling approach. *J. Mater. Cycles Waste Manag.* **2017**, *19*, 1052–1060.

66. Yen, Y.; Wang, Z.; Shi, Y.; Xu, F.; Soeung, B.; Sohail, M.T.; Rubakula, G.; Juma, S.A. The predictors of the behavioral intention to the use of urban green spaces: The perspectives of young residents in Phnom Penh, Cambodia. *Habitat Int.* **2017**, *64*, 98–108. [CrossRef]
67. Afthanorhan, W. A comparison of partial least square structural equation modeling (PLS-SEM) and covariance based structural equation modeling (CB-SEM) for confirmatory factor analysis. *Int. J. Eng. Sci. Innov. Technol.* **2013**, *2*, 198–205.
68. Hair, J.F.; Ringle, C.M.; Sarstedt, M. Partial least squares structural equation modeling: Rigorous applications, better results and higher acceptance. *Long Range Plan.* **2013**, *46*, 1–12. [CrossRef]
69. Henseler, J.; Ringle, C.M.; Sarstedt, M. A new criterion for assessing discriminant validity in variance-based structural equation modeling. *J. Acad. Mark. Sci.* **2015**, *43*, 115–135. [CrossRef]
70. Wang, D.; Brown, G.; Liu, Y.; Mateo-Babiano, I. A comparison of perceived and geographic access to predict urban park use. *Cities* **2015**, *42*, 85–96. [CrossRef]
71. Graziano, W.G.; Habashi, M.M.; Sheese, B.E.; Tobin, R.M. Agreeableness, empathy, and helping: A person× situation perspective. *J. Personal. Soc. Psychol.* **2007**, *93*, 583. [CrossRef] [PubMed]
72. Berenguer, J. The effect of empathy in proenvironmental attitudes and behaviors. *Environ. Behav.* **2007**, *39*, 269–283. [CrossRef]
73. Costa, P.T.; McCrae, R.R. Influence of extraversion and neuroticism on subjective well-being: Happy and unhappy people. *J. Personal. Soc. Psychol.* **1980**, *38*, 668. [CrossRef]
74. Van Egeren, L.F. A cybernetic model of global personality traits. *Personal. Soc. Psychol. Rev.* **2009**, *13*, 92–108. [CrossRef] [PubMed]
75. Saboor, A.; Arfeen, M.I.; Mohti, W. Sociability Impact on Learner's Personality in Classroom and E-learning Environments: A Comparative Analysis to Help HRM Decisions. *J. Indep. Stud. Res. Manag. Soc. Sci. Econ.* **2017**, *15*, 78–90. [CrossRef]

MDPI

St. Alban-Anlage 66

4052 Basel

Switzerland

Tel. +41 61 683 77 34

Fax +41 61 302 89 18

www.mdpi.com

Energies Editorial Office

E-mail: energies@mdpi.com

www.mdpi.com/journal/energies

9 783039 435579